How well do you know your physiology?

Use our online study aid to find out.

Congratulations on purchasing the perfect study tool!

Physiology–An Illustrated Review is a two-part package that will be invaluable for your course and exams. In addition to the wealth of illustrations and information in this book, you also get access to 400 review questions online at **WinkingSkull.com**.

These questions (200 from the book plus 200 more) have answer explanations for right and wrong answers. Test yourself and get immediate feedback, quickly identifying areas of mastery or for further study.

Use the scratch-off code below to get started today!

Thieme's new illustrated *Review Series* will help you master the essentials of the basic sciences — *Pharmacology, Physiology, Anatomy, Biochemistry, Microbiology & Immunology,* and *Pathology*.

To find out more about this essential series, visit **www.thieme.com/IllustratedReview**

Get started today!

Scratch off the panel below to reveal your access code and then log on to www.WinkingSkull.com to get started.

110508

This book cannot be returned if the access code panel is scratched off.

Physiology
An Illustrated Review

Physiology
An Illustrated Review

Roger TannerThies, PhD

Professor Emeritus of Physiology
The University of Oklahoma Health Sciences Center
Oklahoma City, Oklahoma

Thieme
New York • Stuttgart

QT 18.2 T167p 2012
TannerThies, Roger
Physiology :
10031791

Thieme Medical Publishers, Inc.
333 Seventh Ave.
New York, NY 10001

Vice President and Editorial Director, Educational Products: Anne T. Vinnicombe
Developmental Editor: Julie O'Meara
Production Editor: Megan Conway
Senior Vice President, International Sales and Marketing: Cornelia Schulze
Director of Sales: Ross Lumpkin
Chief Financial Officer: Sarah Vanderbilt
President: Brian D. Scanlan
Compositor: Manila Typesetting Company
Printer: Transcontinental Interglobe Printing Inc.

Library of Congress Cataloging-in-Publication Data

TannerThies, Roger.
Physiology : an illustrated review / Roger TannerThies.
       p. ; cm. – (Thieme illustrated review series)
     Includes index.
Summary: "Physiology—An Illustrated Review helps you master important physiologic facts and concepts and teaches you how to apply that knowledge successfully on course exams and in daily practice. This indispensable review book includes 200 spectacular, full-color illustrations depicting cardiologic, cellular, and renal function. It also includes all the regular series features, including 400 print and online review questions and explanatory answers"—Provided by publisher.
     ISBN 978-1-60406-202-1 (pbk.)
 1. Human physiology—Atlases. 2. Human physiology—Examinations, questions, etc. I. Title. II. Series: Thieme illustrated review series
     [DNLM: 1. Physiological Phenomena—Atlases. 2. Physiological Phenomena–Examination Questions. QT 18.2]
     QP40.T36 2012
     612—dc23
                                                                                    2011027808

Important note: Medical knowledge is ever-changing. As new research and clinical experience broaden our knowledge, changes in treatment and drug therapy may be required. The authors and editors of the material herein have consulted sources believed to be reliable in their efforts to provide information that is complete and in accord with the standards accepted at the time of publication. However, in view of the possibility of human error by the authors, editors, or publisher of the work herein or changes in medical knowledge, neither the authors, editors, nor publisher, nor any other party who has been involved in the preparation of this work, warrants that the information contained herein is in every respect accurate or complete, and they are not responsible for any errors or omissions or for the results obtained from use of such information. Readers are encouraged to confirm the information contained herein with other sources. For example, readers are advised to check the product information sheet included in the package of each drug they plan to administer to be certain that the information contained in this publication is accurate and that changes have not been made in the recommended dose or in the contraindications for administration. This recommendation is of particular importance in connection with new or infrequently used drugs.

Some of the product names, patents, and registered designs referred to in this book are in fact registered trademarks or proprietary names even though specific reference to this fact is not always made in the text. Therefore, the appearance of a name without designation as proprietary is not to be construed as a representation by the publisher that it is in the public domain.

Printed in Canada

5 4 3 2 1

ISBN 978-1-60406-202-1

This book is dedicated to my family:
  My wife, Nancy
  My sons, Eric and David
  My stepchildren, Curtis and Christina

Contents

VI Gastrointestinal Physiology

VII Endocrine Physiology

VIII Reproductive Physiology

Preface

Understanding human physiology, the function of the body, from subcellular processes to those of the whole organism, is essential for any health professional. Health depends upon the proper functioning of all body systems. When there is a malfunction, our bodies have many methods of self-correction, but they also benefit from interventions by skilled health professionals. In order to intervene successfully, the health professional has to correctly diagnose the malfunction, provide the appropriate treatment, and monitor and recognize when function returns to normal. This chain of events is all dependent on an understanding of physiology.

Our understanding of physiology has developed over the last century and a half due to the efforts of thousands of medical scientists. The French physiologist, Claude Bernard, summarized his understanding of the functions of the body with a book entitled, "The Basis of Experimental Medicine" 150 years ago. His best known concept is that of the internal environment and its relative constancy. All cells in the body are bathed in a dilute salt solution similar to that of sea water eons ago. Our bodies regulate the composition of this solution to maintain the viability of all our cells. One part of physiology is the study of how the body regulates the composition of this solution. About 100 years ago, Walter Cannon, MD, coined the term "homeostasis" to describe this constancy. Such control is not static, but requires energy to maintain the steady state of the bathing solution. About 50 years ago, Arthur Guyton, MD, the great physiology teacher, and others applied engineering control systems theory to physiological processes. Negative feedback control is one concept they applied to physiological systems. So physiological understanding continues to grow.

There have been many breakthroughs in our understanding of physiology in recent years. We can now describe many of the mechanisms by which the brain and psychological states influence functions of other organs and body systems. Molecular genetics has discovered the genetic defects that lead to many pathophysiological processes, and ways of treating such defects have been devised. Satellite cells that were originally described in skeletal muscle tissue 40 years ago have now been seen in most tissues and provide a means of regeneration of tissues and possibly organs. As physiological understanding continues to expand, medical science will find treatments to maintain longer and better human life.

Physiology—An Illustrated Review covers the facts of physiology and integrates the concepts that you must master for success in the classroom and for the USMLE. It provides a concise study aid for physiology courses. It can also be used as a succinct source of key knowledge for daily clinical practice.

This book is in a streamlined bullet-point format and includes hundreds of full-color illustrations that demonstrate physiologic processes.

Sidebars connect physiologic concepts in the text with normal function (orange), fundamental biochemical, genetic, and embryologic processes (green), and disease and treatment (blue).

For self-testing, the text includes both factual and USMLE-style questions. All of the questions are accompanied by explanatory answers. The 200 questions and answers in the text are supplemented by an additional 200 questions and answers online at WinkingSkull. com via the scratch-off code in the book. The questions provide intensive practice, offer immediate feedback, and will allow you to quickly identify areas for further study.

As you use this book, please send comments or suggestions you have for improvement to IllustratedReview@thieme.com.

Acknowledgements

This book grew out of a previous review book, *Oklahoma Notes Physiology*, which went through four editions between 1987 and 1995. That book was written by all faculty in the Department of Physiology, namely Kirk W. Barron, Robert C. Beesley, Siribhinya Benyajati, Robert W. Blair, Kenneth J. Dormer, Jay P. Farber, Robert D. Foreman, Kennon M. Garrett, Stephen S. Hull, Jr., Philip A. McHale, Y. S. Reddy, Rex D. Stith, and Roger Thies. Some years later the author adapted *Oklahoma Notes Physiology* into outline form and matched the text to 2,000 questions from *Biotest Physiology*. At that time Stephen S. Hull, Jr., made a major contribution by rewriting the chapters on cardiovascular and respiratory physiology. Norman Levine, Carl L. Thompson, and Norman Weisbrodt reviewed individual chapters.

The following faculty and students reviewed various chapters of this book and updated the information: John P. Pooler (Emory University School of Medicine), Henry Edinger (New Jersey Medical School), Ronaldo P. Ferraris (New Jersey Medical School), Paul Greenman (Nova Southeastern University), Norman Levine (Nova Southeastern University), Nicholas Lufti (Nova Southeastern University), Michael Markus (Wright State University), Kenneth D. Mitchell (Tulane University School of Medicine), James Michael Olcese (Florida State University College of Medicine), William H. Percy (Sanford School of Medicine at the University of South Dakota), Dexter Speck (University of Kentucky College of Medicine), Carl Thompson (New York Medical College), Gabi N. Waite (Indiana University School of Medicine), Douglas Wangensteen (University of Minnesota), Anthony Cheng (Feinberg School of Medicine, Northwestern University), Catherine Howard (Tulane University School of Medicine), Chris Lee (Harvard Medical School), Joshua Lennon (Albany Medical College), and Kelly Wright (John H. Stroger, Jr., Hospital of Cook County).

The author appreciates the helpful contributions of the Thieme editorial staff to this book. Cathrin Weinstein, MD, provided initial guidance. Anne Vinnicombe improved the organization and presentation in many ways. The first developmental editor, Rebecca McTavish, contributed to this book by organizing the material, requesting more explanations, and generally giving life to the words. The second developmental editor, Julie O'Meara, made the presentation more focused, systematic, and clinically relevant. Avalon Garcia improved the questions and explanations. Megan Conway guided this book to its final form.

Roger TannerThies
Jefferson City, MO

1 The Cell Membrane

1.1 Structure of the Cell Membrane

Membrane Components

Lipid Bilayer

Cell membranes are made up of amphipathic phospholipids with polar (hydrophilic) heads and apolar (hydrophobic) fatty acid tails (**Fig. 1.1**). Amphipathic molecules tend to arrange themselves to minimize the contact of hydrophobic portions with water. This causes the spontaneous formation of a lipid bilayer. Cholesterol molecules in the lipid bilayer affect membrane fluidity.

Proteins

Integral membrane proteins are embedded in the lipid bilayer and have contact with both the extracellular and intracellular fluid. Most have components that span the bilayer multiple times. Peripheral membrane proteins are associated with either the phospholipid (hydrophilic) heads or other embedded proteins.

Carbohydrates

Oligosaccharide residues combine with lipids and proteins on the outer cell membrane surface to form glycolipids and glycoproteins.
— Glycolipids and glycoproteins both contribute to structural stability of cell membranes.
— Glycoproteins are also important for cell recognition and immune response.

Intercellular Connections

Tight Junctions

Tight junctions are attachments between epithelial cells.
— True "tight" junctions prevent the movement of dissolved molecules and water from one side to the other.
— "Leaky" tight junctions act as a pathway for solutes and water to cross epithelial cell layers.

Gap Junctions

Gap junctions are channels between cells that permit intercellular communication.
— Small molecules (ions, adenosine triphosphate [ATP], cyclic adenosine monophosphate [cAMP], etc.) can pass through gap junctions.
— Gap junctions electrically couple cells, so they act together as a functional syncytium (e.g., in the heart and smooth muscles of the gut).

1.2 Transport across Membranes

Selective transport of substances across cell membranes allows cells to regulate their internal content and to carry out crucial functions such as secretion and absorption, which are controlled by neural and hormonal activity.

Fig. 1.1 ▶ **Structure of the cell membrane.**
The cell membrane consists of a phospholipid bilayer. Each phospholipid molecule has a glycerol head (hydrophilic) and two fatty acid tails (hydrophobic). These hydrophobic tails are arranged so that they face each other in the bilayer. Integral proteins and cholesterol are embedded within the bilayer. Other proteins may lie peripherally. Carbohydrate moieties may bind to lipids and proteins on the extracellular surface of the membrane, forming glycolipids and glycoproteins.

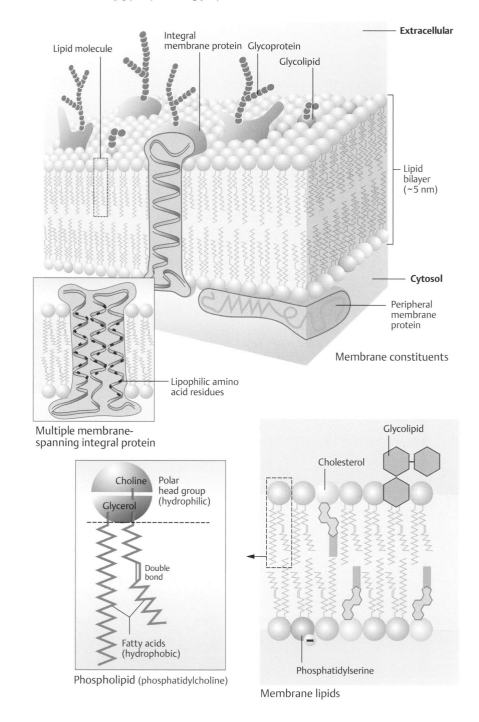

Extracellular

Lipid molecule

Integral membrane protein   Glycoprotein

Glycolipid

Lipid bilayer (~5 nm)

Cytosol

Peripheral membrane protein

Membrane constituents

Lipophilic amino acid residues

Multiple membrane-spanning integral protein

Choline

Glycerol

Polar head group (hydrophilic)

Double bond

Fatty acids (hydrophobic)

Phospholipid (phosphatidylcholine)

Glycolipid

Cholesterol

Phosphatidylserine

Membrane lipids

Free Diffusion

Free diffusion is the migration of molecules from a region of higher concentration to one of lower concentration as a result of random motion (**Fig. 1.2**).
— Free diffusion does not require external energy and is therefore passive.
— *Example:* Oxygen (O_2) and carbon dioxide (CO_2) move across cell membranes down their concentration gradients by diffusion.

Fick's Law

Fick's first law of diffusion states that the net flow of a substance (J) is proportional to the membrane permeability (P), the concentration gradient (ΔC), and the available area for diffusion (A):

$$J = PA(\Delta C)$$

where
J = net flow (mmol/s)
P = membrane permeability (cm/s)
A = area (cm²)
ΔC = concentration difference (mmol/cm³)

Membrane permeability.   *Membrane permeability* is a variable in Fick's law and is increased by
— ↑ lipid solubility of the solute
— ↓ membrane thickness
— ↓ size of the solute
  Lipid-soluble, small, nonionized substances are most permeable (**Fig. 1.3**).

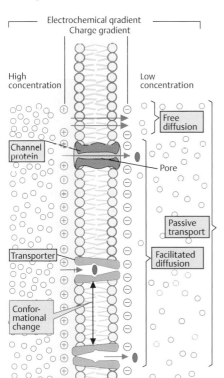

Fig. 1.2 ▶ **Passive transport: free diffusion and uniport.**
Free diffusion occurs when a substance moves across a membrane down its electrochemical gradient. When this process requires a transport (carrier) protein, it is known as uniport or facilitated diffusion. It is a conformational change in the transport protein that permits this membrane transport. Both of these forms of membrane transport are passive because they do not require energy.

Fig. 1.3 ► Permeability of membranes.
Small apolar and polar uncharged molecules can diffuse freely through cell membranes. Large molecules and charged molecules cannot diffuse freely and must be transported to cross a membrane.

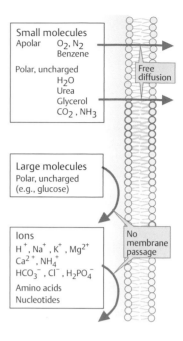

Carrier-mediated Transport

Carriers are integral membrane proteins that transport substances that are hydrophilic or too large to cross the membrane by simple diffusion. They also permit faster transport of lipid-soluble substances than simple diffusion. Carrier proteins possess the following characteristics:
— *Selectivity:* Most carriers exhibit a preference for just one or a small class of solutes.
— *Competition for binding sites:* Structurally related compounds can compete for binding sites and inhibit the binding of the related solute to the carrier protein; for example, glucose and galactose compete with each other for absorption into enterocytes by Na^+-dependent cotransport (SGLT1).
— *Saturation of carrier proteins:* The rate of carrier-mediated transport may show saturation at high solute concentrations, as the number of carrier proteins is finite, and the cycling of carrier proteins is limited.

Uniport

Uniport (formerly called facilitated diffusion) is a carrier-mediated transport mechanism that moves solutes down their electrochemical gradients (**Fig. 1.2**).
— It does not use metabolic energy and is therefore passive.
— *Example:* In the transport of glucose into red blood cells, L-glucose cannot enter red blood cells by simple diffusion. D-glucose enters via a protein glucose transporter that transports other sugars poorly. D-glucose transport saturates when all the transporters are being used.

Primary Active Transport

Primary active transport is a carrier-mediated transport mechanism that moves molecules against their electrochemical gradients (**Fig. 1.4**).
— It requires metabolic energy and uses ATP as the direct energy source.
— *Example:* Na^+–K^+ ATPase (carrier protein) pumps Na^+ out of the cell and K^+ into the cell against their concentration gradients. It maintains a low intracellular $[Na^+]$ and high intracellular $[K^+]$ ratio. Three Na^+ ions are transported out of the cell for every two K^+ ions that are transported into the cell (**Fig. 1.5**).
— *Example:* Ca^{2+}–ATPase transports Ca^{2+} back into Ca^{2+} stores in a muscle cell and out of a muscle cell after the influx of Ca^{2+} triggers muscle contractions.

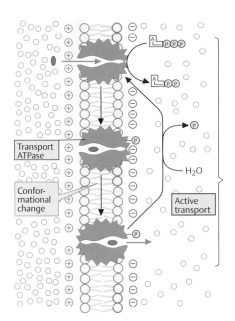

Fig. 1.4 ► **Active transport.**
Active transport occurs when a substance is transported across a membrane against its electrochemical gradient by transport proteins. This process requires energy in the form of adenosine triphosphate (ATP), therefore it is active. The transport protein (an ATPase) binds the substance on one side of the membrane, and ATP-dependent phosphorylation causes a conformational change that releases it on the other side of the membrane.

Secondary Active Transport

Secondary active transport is a carrier-mediated transport mechanism that uses the downhill movement of one substance to move another substance uphill. The flow of one species down its electrochemical gradient powers the actively transported species against its electrochemical gradient (**Fig. 1.6**). The electrochemical gradient for Na^+ is usually maintained by the Na^+–K^+ ATPase pump.

Symporters carry the substrate and cosubstrate in the same direction.
- *Example:* In Na^+–glucose cotransport in the small intestine and kidney, Na^+ transported into cells brings glucose with it.
- *Antiporters* carry the substrate and cosubstrate in opposite directions.
  - *Example:* HCO_3^-–Cl^- countertransport at red blood cell membranes.
  - *Example:* $3Na^+$–Ca^{2+} countertransport at cardiac muscle cell membranes.

Fig. 1.5 ► **Na^+–K^+ ATPase.**
The Na^+–K^+ ATPase (Na^+–K^+ pump) is present in all cell membranes. It consists of two α subunits and two β subunits. The α subunits are phosphorylated, causing a conformational change that allows them to form the ion transport pathway. During one transport cycle, three Na^+ ions are pumped out of the cell, and two K^+ ions are pumped into the cell by the Na^+–K^+ ATPase using one molecule of adenosine triphosphate (ATP). Both Na^+ and K^+ ions are transported against their concentration gradients. (ADP, adenosine diphosphate)

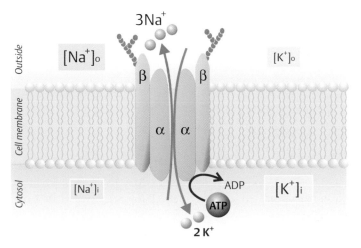

Digoxin

Digoxin is a cardiac glycoside that was once one of the first-line agents used in the treatment of heart failure. Its use is now reserved for cases when symptoms are not fully treated by standard therapies or in cases of severe heart failure while standard therapies are initiated. The therapeutic and toxic effects of digoxin are attributable to inhibition of Na^+–K^+ ATPase (the digitalis receptor) located on the outside of the myocardial cell membrane. When the pump is inhibited, Na^+ accumulates intracellularly. The decreased Na^+ gradient that results from this affects Na^+–Ca^{2+} exchange, and Ca^{2+} accumulates intracellularly. Consequently, more Ca^{2+} (stored in the sarcoplasmic reticulum) is available for release and interaction with the contractile proteins in these cells during the excitation–contraction coupling process. At therapeutic doses of digoxin, there is an increase in contractile force. Toxicity to digoxin also relates to inhibition of Na^+–K^+ ATPase. Inhibition of the Na^+–K^+ pump affects the K^+ gradient; this may lead to a significant reduction of intracellular K^+, predisposing the heart to arrhythmias. Likewise, high levels of Ca^{2+} intracellularly may contribute to serious arrhythmias.

Fig. 1.6 ► **Secondary active transport.**
Secondary active transport occurs when uphill transport of a substance via a carrier protein (e.g., sodium–glucose transport type 2 [SGLT2]) is coupled with the downhill transport of an ion (Na$^+$ in this example) (**1**). In this case, the electrochemical Na$^+$ gradient into the cell (maintained by Na$^+$–K$^+$ ATPase) provides the driving force for the cotransport of glucose into the cell. The SGLT2 is an example of a symporter, as Na$^+$ and glucose are transported in the same direction. Examples **2** and **3** also illustrate symport. Antiport occurs when the compound and driving ions are transported in opposite directions. For example, when an electrochemical Na$^+$ gradient drives H$^+$ in the opposite direction (**4**).

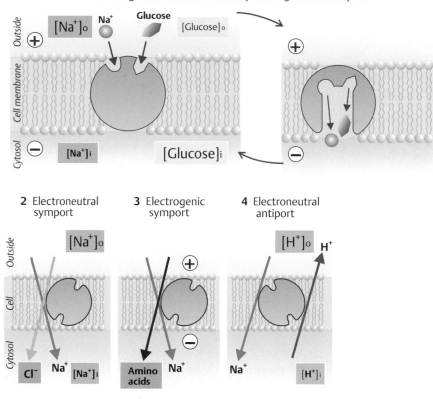

1 Electrochemical Na$^+$ gradient drives secondary active glucose transport

2 Electroneutral symport   **3** Electrogenic symport   **4** Electroneutral antiport

Table 1.1 compares the different membrane transport mechanisms.

Table 1.1 ► Comparison of Membrane Transport Mechanisms				
Transport Mechanism	**Electrochemical Gradient**	**Energy**	**Carrier-mediated**	**Classification**
Simple diffusion	Downhill	None	No	Passive
Facilitated diffusion	Downhill	None	Yes	Passive
Primary active transport	Uphill	ATP hydrolysis	Yes	Active
Secondary active transport	Substrate uphill Cosubstrate downhill	Energy exchange	Yes	Active

Abbreviation: ATP, adenosine triphosphate.

Transport of Water across Membranes

Osmosis

Osmosis is the net diffusion of water across a semipermeable (permeable to water but not solutes) membrane. The osmolarity of a solution is the concentration of osmotically active particles in the solution. The net movement of water across a semipermeable membrane is due to the concentration differences of the nonpenetrating solutes. Water diffuses from a low

Fig. 1.7 ► **Water output and intake from the cell by osmosis.**
In a hypertonic environment, there is a higher concentration of solutes outside the cell than inside, so water moves out of the cell by osmosis. In a hypotonic environment, there is a higher concentration of solutes inside the cell, so extracellular water moves into the cell by osmosis.

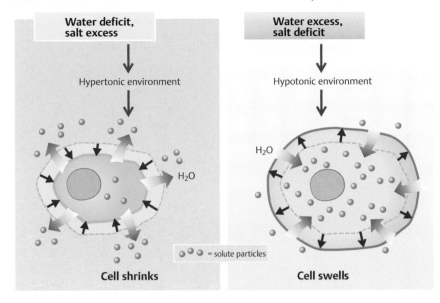

osmolarity solution (high water concentration, low solute concentration) to a high osmolarity solution (low water concentration, high solute concentration) in attempting to achieve equal water concentrations on both sides of the membrane (**Fig. 1.7**).

Osmotic Pressure. The *osmotic pressure* of a solution is the theoretical hydrostatic pressure that would be required to just prevent the osmotic flow of water across a semipermeable membrane. Numerically, it is simply a constant multiplied by the osmolality.

The flow of water through a membrane is expressed by the *van't Hoff equation*:

$$J_v = K_c \times \Delta\pi$$

where
J_v = water flow (mL/min)
K_c = hydraulic conductivity of the membrane (mL/mm/mmHg)
$\Delta\pi$ = osmotic pressure difference (mOsm/g water)

Reflection Coefficient. The *reflection coefficient* is a number between 0 and 1 that describes the ability of a membrane to prevent diffusion of a solute relative to water.
— If the reflection coefficient is 1, the solute is completely impermeable and will not pass through the membrane. Serum albumin has a reflection coefficient that is close to 1. This explains why albumin is retained in the vascular compartment and exerts an osmotic effect.
— If it is 0, the solute will pass through the membrane as easily as water and will not exert any osmotic effect (i.e., it will not cause water to flow).

1.3 Receptors and Signal Transduction

Types of Receptor

Ligand-gated Ion Channels
Ligand-gated ion channels are specialized membrane pores made up of multisubunit proteins. Binding of ligands (e.g., hormones or neurotransmitters) to these receptors opens or

Fig. 1.8 ▶ **Ligand-gated ion channel.**
An example of a ligand-gated ion channel is the nicotinic receptor of the motor end plate. When two acetylcholine (ACh) molecules bind to this receptor simultaneously (at the α subunits), the inner pore opens; Na$^+$ then enters the cell, and K$^+$ leaves the cell. This causes membrane depolarization.

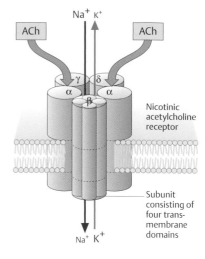

closes the pores, thus changing the permeability of the membrane to Na$^+$, K$^+$, Cl$^-$, or other ions (**Fig. 1.8**).

– *Examples:* Nicotinic receptor for acetylcholine (see **Fig. 2.2**), the glutamate receptor, and the gamma-aminobutyric acid type A (GABA$_A$) receptor

G-Protein Coupled Receptors. G proteins facilitate signal transduction that is initiated by ligand-receptor binding and culminates in a cellular response. The mechanisms of G-protein signal transduction are discussed on page 9.

Voltage-dependent Ion Channels

Voltage-dependent ion channels open or close in response to changes in the membrane potential.

— *Example:* Depolarization opens the activation gate of Na$^+$ channels, allowing Na$^+$ to flow into cells.

Enzyme-linked Membrane Receptors

When a ligand binds to this receptor, it causes an enzyme to become "switched on" intracellularly. This enzyme then catalyzes the formation of other signal proteins that ultimately lead to the drug's effect.

— *Example:* Insulin "switches on" the tyrosine kinase activity of the insulin receptor to affect glucose uptake into cells (**Fig. 1.9**).

■ **Calcium channel blockers**

Calcium channel blockers (e.g., verapamil and nifedipine) inhibit Ca^{2+} entry into cells via voltage-dependent ion channels. In smooth muscle cells, this produces arterial vasodilation, which leads to reduced coronary artery spasm, decreased blood pressure, and reduced cardiac work. In cardiac muscle cells, these agents inhibit cardiac functions, causing decreased heart rate, atrioventricular conduction, and contractility. Nifedipine acts predominantly on smooth muscle cells to produce vasodilation and has almost no effect on cardiac function at therapeutic doses. Verapamil acts on both smooth muscle cells and heart muscle cells.

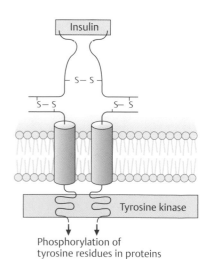

Fig. 1.9 ▶ **Enzyme-linked membrane receptor.**
Insulin binding to the receptor causes the enzyme, tyrosine kinase, to phosphorylate tyrosine residues in proteins. These proteins can then signal other proteins to be formed, thus exerting the physiological effect.

Fig. 1.10 ► **Intracellular receptor.**
Lipophilic substances, such as steroid hormones and thyroid hormones, can diffuse through the cell membrane and interact with receptors in the cytoplasm or nucleus. The hormone-receptor complex then alters gene transcription causing proteins that exert the physiological effect to be made. The hormone-receptor complex interacts with DNA in pairs that may be identical (homodimeric) or nonidentical (heterodimeric).

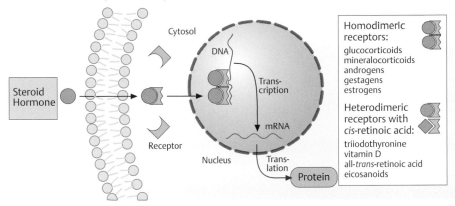

Intracellular Receptors

Lipid-soluble substances diffuse through cell membranes and bind either to receptors in the cellular cytosol or in the nucleus. Gene expression is altered, and protein synthesis is either increased or decreased, which causes the cellular response (**Fig. 1.10**).
— *Examples:* Steroid hormones, calcitriol, and thyroxine

G-Protein Coupled Receptors and Signal Transduction

Heterotrimeric G proteins couple to membrane receptors (e.g., α-adrenergic receptors). When the receptor binds a ligand, this causes the α-subunit of the G protein to split from the β and γ subunits. The now free subunits then interact with other proteins in the membrane that may produce second messengers (**Fig. 1.11**). These second messengers are cAMP, diacylglycerol (DAG), and inositol 1,4,5-triphosphate (IP$_3$).
— G$_s$ proteins activate adenylate cyclase.
— G$_i$ proteins inhibit adenylate cyclase.
— G$_q$ proteins activate phospholipase C, which then activates DAG and IP$_3$.

Fig. 1.11 ► **Signal transduction by G proteins.**
A substance binding to a G-protein coupled receptor alters its conformation and causes the α subunit of the attached G protein to exchange guanosine diphosphate (GDP) for guanosine triphosphate (GTP) (**1**). The G protein then separates from the receptor and dissociates into an α subunit and a βγ subunit. In the case illustrated, the α subunit activates adenylate cyclase, which promotes cyclic adenosine monophosphate (cAMP) production (**2**). The cAMP then acts as a second messenger, activating protein kinase A, which, in turn, activates further proteins (see **Fig. 1.12**). The intrinsic GTPase activity of the α subunit hydrolyzes bound GTP to GDP, thereby terminating the effect of the G protein. (ATP, adenosine triphosphate; PP$_i$, diphosphate)

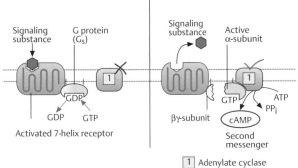

1.   2.

When G proteins are activated, guanosine triphosphate (GTP) replaces guanosine diphosphate (GDP) on the α subunit. Following activation of G proteins, GTP is rapidly degraded to inactive GDP by the activity of the α-subunit GTPase.

Second Messenger Systems

Adenylate Cyclase System. G_s-activating substances bind to a receptor that activates G_s, which, in turn, stimulates adenylate cyclase to convert ATP to cAMP, which then activates protein kinase A. This phosphorylates proteins, resulting in the physiologic response. Following its activation, cAMP is degraded to 5'AMP by phosphodiesterase (**Fig. 1.12**).

G_i-activating substances bind to a receptor that activates G_i, which inhibits adenylate cyclase ($\downarrow$ cAMP). Therefore, protein kinase A is not activated, and proteins are not phosphorylated.

DAG and IP$_3$ System. The amplifier enzyme phospholipase C produces the second messengers IP$_3$ and DAG from a single precursor (**Fig. 1.13**).

Hydrophilic IP$_3$ diffuses from the membrane to organelles containing Ca^{2+} and releases it. The Ca^{2+} released can then cause physiologic effects in the following ways:
— Interaction with the cAMP system
— Activation of protein kinase C (with DAG) leading to the phosphorylation of proteins
— Binding to calmodulin with the resultant complex mediating further effects, for example, production of nitric oxide (**Fig. 1.14**)

Lipophilic DAG has two functions:
— Activation of protein kinase C (this process is Ca^{2+} dependent)
— Formation of arachidonic acid (an eicosanoid precursor) following its degradation by DAG lipase.

Phosphodiesterase inhibitors

Phosphodiesterase inhibitors inhibit the degradation of cAMP and cyclic guanosine monophosphate (cGMP). Drugs that specifically inhibit phosphodiesterase type 5 (e.g., sildenafil citrate [Viagra]) cause prolonged vasodilation of penile arteries and are therefore used to treat erectile dysfunction. Phosphodiesterase inhibitors type 3 (e.g., milrinone and inamrinone) increase cAMP levels in cardiac cells. This causes an increase in intracellular Ca^{2+}, resulting in increased heart rate and force of contraction (positive chronotropic and inotropic effects). They also cause vasodilation of blood vessels. These agents are used as adjuvants in heart failure therapy. Side effects of phosphodiesterase inhibitors include headache and cutaneous flushing.

Fig. 1.12 ► **Cyclic AMP.**
Adenylate cyclase synthesizes cAMP by cleaving diphosphate (PPi) from ATP. Adenylate cyclase is regulated by G proteins (G_s and G_i), which are controlled by substances attaching to G-protein coupled 7-helix receptors. The cAMP activates protein kinase A (PKA), which then phosphorylates proteins, including enzymes, transcription factors, and ion channels.

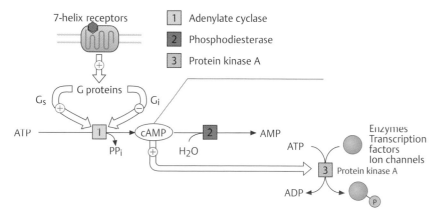

Fig. 1.13 ► Diacylglycerol (DAG) and inositol 1,4,5-triphosphate (IP$_3$).
Binding of a molecule to a G protein activates phospholipase C, which, in turn, activates DAG and IP$_3$. Both DAG and IP$_3$ act to phosphorylate effector proteins. DAG also forms eicosanoids. Ca^{2+} exerts further effects by forming a complex with calmodulin. (ECF, extracellular fluid; PIP$_2$, phosphadidylinositol-4, 5-biphosphonate; cGMP, cyclic guanosine monophosphate)

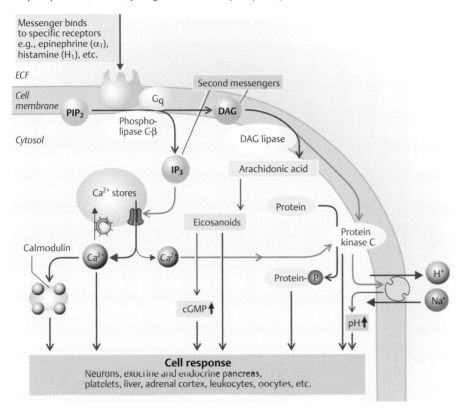

Fig. 1.14 ► Nitric oxide (NO) as a transmitter substance.
Ca^{2+}–calmodulin complex activates nitric oxide synthase, which catalyzes the formation of NO from arginine. NO is then able to diffuse to other cells, where it activates guanylate cyclase, which converts GTP to cyclic guanosine monophosphate (cGMP). The cGMP activates protein kinase G, which, in turn, decreases the intracellular Ca^{2+} concentration by blocking the IP$_3$ receptors on Ca^{2+} stores. This culminates in vasodilation. (ECF, extracellular fluid; NADPH, reduced form of nicotinamide adenine dinucleotide phosphate; PIP$_2$, phosphadidylinositol-4,5-biphosphonate, PLC, phospholipase C)

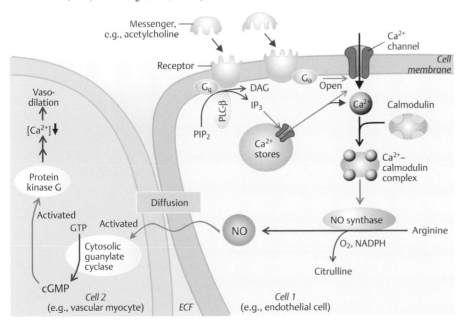

1.4 Membrane Potentials

Equilibrium Potential

Ion movement or flux is controlled by both concentration gradients and electrical gradients. If these gradients are equal but opposite in direction for a particular ion, then its total electrochemical potential is zero, and there is no net current flow. This is *electrochemical equilibrium.*

The *equilibrium potential* is the membrane potential for an ion that would exactly oppose the tendency of an ion to move down its concentration gradient. It is calculated using the *Nernst equation*:

$$E = -61/z \times \log [ion]_{out}/[ion]_{in}$$

where

E = equilibrium potential
z = the charge of the ion

— *Example*: The equilibrium potential for K^+ is calculated as follows:

$$E = -61 \times \log (150 \text{ mM}/4 \text{ mM})$$
$$= -96 \text{ mV}$$

Resting Membrane Potential

The *resting membrane potential* (RMP) is the normal potential difference across a membrane (in millivolts). It is established as a result of the difference in concentration of permeable ions across the membrane (**Fig. 1.15**) as each of these ions tries to reach its equilibrium potential.
— Skeletal muscle cell membranes at rest are 30 times more permeable to K^+ than Na^+, so the RMP of −90 mV approaches the equilibrium potential for K^+ of −96 mV.

Fig. 1.15 ▶ **Resting potential.**
The resting membrane potential results from the uneven distribution of positively and negatively charged ions inside and outside the cell. The Na^+–K^+ ATPase pump establishes concentration differences by pumping three Na^+ ions out of the cell and two K^+ ions into the cell. Some K^+ ions flow back down this concentration gradient and leave the cell via K^+ channels. The protein anions that predominate inside the cell cannot follow them. The result is a slight excess of positively charged ions outside the cell, with a slight excess of anions inside the cell. The force of attraction between positive and negative ions creates the resting membrane potential (RMP).

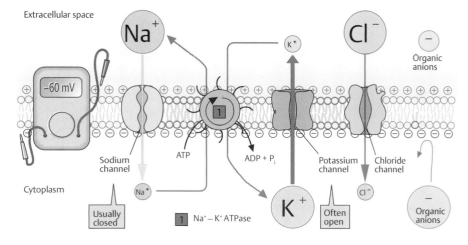

— Neural cells are only 6 times as permeable, so they have RMPs of −70 mV. Cl⁻ distributes passively at the potential determined by primarily K⁺ and secondarily by Na⁺.
— The Na⁺–K⁺ ATPase pump is responsible for generating the concentration differences needed to cross the membrane.
— The RMP is stored energy (a battery) whose brief discharge generates an action potential that can transmit a signal rapidly for a long distance through the body.

1.5 Action Potentials

Table 1.2 defines terms that must be understood when considering action potentials.

Table 1.2 ▶ **Definitions of Terms for Discussing Action Potentials**		
Term	**Definition**	**Comments**
Action potential	Rapid changes in membrane permeability characterized by depolarization of the membrane potential from negative to positive, followed by a return to a negative membrane potential	Action potentials are a means of high-speed communication in excitable nerve and muscle cells. Action potentials are all-or-nothing events. Each one is of a specified magnitude and duration.
Depolarization	Inflow of positively charged Na⁺ ions causes the membrane potential to become less negative.	Partial membrane depolarization is the stimulus for initiation of an action potential.
Hyperpolarization	Outflow of positively charged K⁺ ions causes the membrane potential to become more negative.	Hyperpolarization makes action potential initiation less likely (i.e., more inflow of positively charged ions is needed to reach threshold).
Threshold	The membrane potential at which an action potential is initiated	The threshold is ~15 mV less negative than the resting potential.

Action Potential Events

Depolarization

Depolarization due to a graded potential reduces the RMP approximately −15 mV. Some fast voltage-gated Na⁺ channels are activated and open, allowing some Na⁺ to enter the cell (**Fig. 1.16**).

When the depolarization threshold is reached, the action potential is initiated. Inward current causes the membrane potential difference to move rapidly toward the Na⁺ equilibrium potential of +55 mV. During this rising phase of depolarization, the relative conductance of the membrane to Na⁺ increases to ~50 times the conductance of K⁺. In the overshoot phase of an action potential, the membrane potential passes zero and is reversed (positive).

Repolarization

Na⁺ channels close spontaneously within a millisecond after opening. This process is called inactivation. The conductance of K⁺ also increases. The decrease of Na⁺ conductance relative to K⁺ causes the membrane to repolarize back toward the RMP, which facilitates the repolarization.

The electrochemical gradient for Na⁺ is usually maintained by the Na⁺–K⁺ ATPase pump.

Hyperpolarization

The membrane hyperpolarizes briefly at the end of the action potential as it passes the RMP. This is due to the greater than normal conductance of K⁺ during repolarization. The RMP is reestablished as Na⁺ and K⁺ conductances return to their resting states.

Fig. 1.16 ▶ Action potential and ion conductivity.

Following the binding of a neurotransmitter to an inotropic receptor on the postsynaptic membrane, the following steps occur that culminate in an action potential: (**1**) Voltage-gated Na⁺ channels open, and due to their high equilibrium potential, Na⁺ ions flow into the cell, causing depolarization. (**2**) The Na⁺ channels immediately close again, so the influx of positive charges is very brief. (**3**) Voltage-dependent K⁺ channels open, and K⁺ flows out of the cell. This results in repolarization of the membrane. (**4**) This briefly leads to the potential falling below resting membrane potential (RMP), and the membrane is hyperpolarized. The K⁺ channels then close, and if there are gap junctions, the neuron is ready for restimulation. The Na⁺–K⁺-ATPase (labeled in the membrane) operates continuously to maintain the concentration gradient for Na⁺ and K⁺.

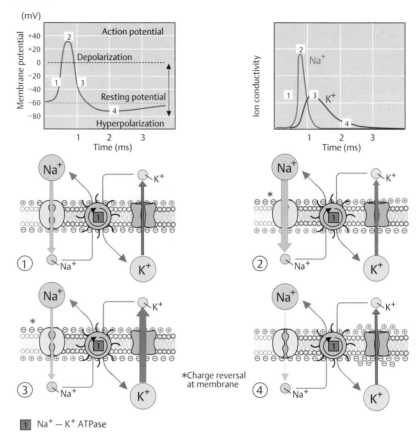

Absolute Refractory Period

The *absolute refractory period* is the period when the membrane cannot be stimulated to produce a second action potential regardless of the stimulus strength. It is the duration of the action potential and is due to the inactivation of the majority of membrane Na⁺ channels.

Relative Refractory Period

The *relative refractory period* occurs after the absolute refractory period. During this time, a significantly larger depolarization stimulus (inward flow) can initiate an additional action potential. This is due to a prolonged increase in K⁺ conductance during this time, which opposes depolarization as the membrane is farther from threshold.

Accommodation

Accommodation is a slight increase in threshold in response to prolonged subthreshold stimulation.

- It is caused by a continuous activation and inactivation of some Na⁺ channels and K⁺ channels, so fewer Na⁺ channels are available for activation. It may occur in smooth muscle cells of the gut.

Propagation of Action Potentials

When an action potential depolarizes one portion of a membrane, local currents (Na^+ influx) depolarize adjacent areas of the membrane, bringing them to threshold. Thus, the action potentials are propagated.

Conduction Velocity

The conduction velocity of action potentials moving along nerve axons is increased by
— ↑ *axon radius:* Thicker axons have lower intracellular resistance to local current flow, so conduction velocity is faster.
— *Myelination:* Conduction velocity is greatly increased in myelinated neurons due to the insulating effects of the myelin and because myelin allows for saltatory conduction. In saltatory conduction, Na^+ currents can only flow at the nodes of Ranvier, where the myelin sheath is absent, so the action potential "jumps" from node to node (**Fig. 1.17**). This significantly increases conduction velocity and requires less energy.

Fig. 1.17 ► **Continuous (1a, 1b) and saltatory (2) propagation of action potentials.**
(**1a**) Na^+ enters the nerve fiber, causing depolarization of the adjacent membrane. (**1b**) Na^+ then enters the adjacent membrane. The region of depolarization moves. (**2**) Myelin insulation forces current to flow to the next node of Ranvier, causing depolarization there. Na^+ then enters at the next node of Ranvier, ~1 mm along the nerve fiber.

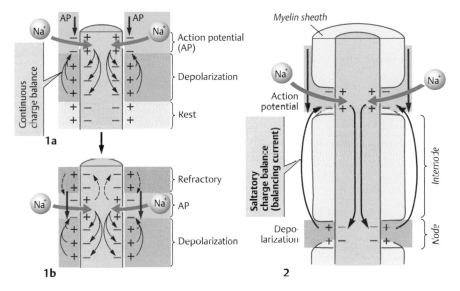

2 Neurotransmission

Propagated action potentials carry information through axons over long distances, but they do not transfer electrical impulses directly to other neurons, glands, or muscle cells. Communication between most nerve cells is accomplished via neurotransmitter molecules released at synapses.

2.1 Neurotransmitters

Acetylcholine

Synthesis. Acetylcholine is synthesized from acetyl coenzyme A (CoA) and choline by the enzyme choline acetyltransferase in the presynaptic terminal. The uptake of choline is the rate-limiting step.

Degradation. Breakdown is rapid via acetylcholinesterase to produce acetate and choline. Acetylcholinesterase is located on neuronal membranes, muscle cell membranes, and red blood cells. Pseudocholinesterases (nonspecific) and butyrylcholinesterases, which are more widely distributed, can also hydrolyze acetylcholine.

Release. Acetylcholine is the neurotransmitter released from the following neurons (see also page 35 and **Fig. 4.1**):
— Pre- and postganglionic parasympathetic neurons
— Preganglionic sympathetic neurons
— Postganglionic sympathetic neurons that innervate sweat glands
— Motoneurons at the neuromuscular junction

Norepinephrine

Synthesis. Norepinephrine is synthesized from the precursor amino acid tyrosine by hydroxylation to dihydroxyphenylalanine (dopa) in the postganglionic neuron. Dopa is decarboxylated to dopamine, which is oxidized to norepinephrine and packaged in vesicles.

Degradation. Termination of action is primarily by reuptake (60–90%) into nerve terminals. Secondary degradation is by monoamine oxidase (MAO) and catechol O-methyltransferase (COMT).

Release. Norepinephrine is the main neurotransmitter released from postganglionic sympathetic neurons. It is also released in small quantities from the adrenal medulla along with epinephrine.

Epinephrine

Synthesis. Epinephrine is produced from norepinephrine in the adrenal medulla via the enzyme phenylethanolamine N-methyltransferase.

Degradation. Epinephrine is degraded by MAO and COMT.

Release. Epinephrine is released from the adrenal medulla along with some norepinephrine.

Dopamine

Synthesis. Dopamine is a precursor in the formation of both norepinephrine and epinephrine.

Degradation. Dopamine is degraded by MAO and COMT.

■ Coenzyme A

Pantothenic acid (vitamin B5) is a precursor of CoA. CoA participates in fatty acid synthesis and oxidation, as well as the oxidation of pyruvate in the citric acid cycle. A molecule of CoA that has an acetyl group is referred to as acetyl CoA. Acetate, which is derived from acetyl CoA, combines with choline to form the neurotransmitter acetylcholine.

■ Monoamine oxidase inhibitors

Monoamine oxidase inhibitors (MAOIs, e.g., isocarboxazid and phenelzine) inhibit both forms of the enzyme monoamine oxidase (MAO-A and MAO-B). In doing so, they prevent the degradation of norepinephrine, epinephrine, and dopamine. These drugs are used in the treatment of depression when tricyclic antidepressants are ineffective. It takes 2 to 3 weeks for the desired effects of these drugs to occur.

■ Hypertensive crisis with monoamine oxidase inhibitors

Hypertensive crisis may occur within hours of ingestion of tyramine-containing foods, including cheese, certain meats (liver and fermented or cured meats), cured or pickled fish, overripe fruits and vegetables, Chianti wine, and some beers. Hypertensive crisis is characterized by headache, palpitation, neck stiffness or soreness, nausea, vomiting, sweating (sometimes with fever or cold, clammy skin), photophobia, tachycardia or bradycardia, constricting chest pain, and dilated pupils. Potentially fatal intracranial bleeding may result from this crisis. Patients should avoid tyramine-containing foods while taking MAOIs and for 2 weeks after treatment with MAOIs is discontinued to avoid precipitating this condition, but if it does occur, then treatment is with intravenous phentolamine.

Release.   Dopamine acts as a neurotransmitter in the central nervous system (CNS), especially in the extrapyramidal motor system.

Glutamate

Synthesis.   Glutamate is synthesized from glucose via glutamine.

Degradation.   Glutamate is converted back to glutamine, and its action is terminated by reuptake into cells in the CNS.

Release.   Glutamate is the principal excitatory amino acid neurotransmitter in the CNS.

Gamma-Aminobutyric Acid

Synthesis.   Glucose is the principal in vivo source of gamma-aminobutyric acid (GABA). There is a GABA "shunt" of the Krebs cycle that results in the conversion of glutamate into GABA by the action of the enzyme glutamate decarboxylase.

Degradation.   GABA is converted back to glutamate, then to glutamine. Its action is terminated by reuptake into cells in the CNS.

Release.   GABA is the principal inhibitory amino acid neurotransmitter in the CNS.

Serotonin

Synthesis.   Serotonin (5-hydroxytryptamine [5-HT]) is synthesized from tryptophan by tryptophan hydroxylase.

Degradation.   Serotonin is degraded by MAO.

Release.   Serotonin acts as a neurotransmitter in the CNS.

Glycine

Synthesis.   Glycine is the simplest amino acid.

Degradation.   Glycine is broken down by glycine dehydrogenase.

Release.   Glycine is released by the inhibitory interneurons in the spinal cord that are activated by group Ia muscle afferents (see page 65). It acts by increasing Cl^- conductance in the postsynaptic membrane, hyperpolarizing it, and thus preventing action potential generation.

Histamine

Synthesis.   Histamine is synthesized from histidine by histidine decarboxylase.

Degradation.   Histamine is degraded by MAO.

Release.   Histamine acts as a neurotransmitter in the CNS.

Nitric Oxide

Synthesis.   Nitric oxide (NO) is not stored in vesicles. It is synthesized as required in the presynaptic terminal from arginine by the enzyme NO synthase.

Degradation.   NO has a half-life of only a few seconds.

Release.   NO acts as an inhibitory neurotransmitter in the CNS, GI tract, and blood vessels.

■ Carcinoid syndrome

Carcinoid tumors are neuroendocrine tumors of the gastrointestinal (GI) tract, urogenital tract, or pulmonary bronchioles. They can contain and secrete numerous autocoids, including prostaglandins and serotonin, causing symptoms such as flushing and diarrhea. Cardiac disease due to fibrosis of the endocardium and valves, along with asthma-like symptoms, are also common. Flushing may be precipitated by stress, alcohol, certain foods, or drugs, particularly serotonin-specific reuptake inhibitors (SSRIs), so these should be avoided. Heart failure, wheezing, and diarrhea are treated, respectively, with diuretics, a bronchodilator, and an antidiarrheal agent, such as loperamide or diphenoxylate. If patients remain symptomatic, serotonin receptor antagonists, antihistamines, and somatostatin analogues are the drugs of choice. $5-HT_3$ receptor antagonists (ondansetron, tropisetron, dolasetron, granisetron, palonosetron, ramosetron, alosetron, and cilansetron) can control diarrhea and nausea and occasionally ameliorate the flushing. A combination of histamine H_1 and H_2 receptor antagonists (diphenhydramine and cimetidine or ranitidine) may control flushing in patients with upper GI or pulmonary carcinoids. Synthetic analogues of somatostatin (octreotide and lanreotide) are the most widely used agents to control the symptoms of patients with carcinoid syndrome.

Table 2.1 provides examples that are predominantly excitatory or inhibitory.

Table 2.1 ► Examples of Excitatory and Inhibitory Neurotransmitters*	
Excitatory Neurotransmitters	**Inhibitory Neurotransmitters**
Acetylcholine	GABA
Norepinephrine	Glycine
Epinephrine	Nitric oxide
Dopamine	Histamine**
Glutamate	
Serotonin	
Histamine**	

* *Excitatory* and *inhibitory* refer to their predominant effect at postsynaptic membranes. These terms oversimplify the effects of individual neurotransmitters.
** Histamine has both excitatory and inhibitory effects.
GABA, gamma-aminobutyric acid.

2.2 Synaptic Transmission

Presynaptic Events

An action potential depolarizes the presynaptic terminal cell membrane, causing membrane Ca^{2+} channels to open and Ca^{2+} influx into the presynaptic terminal. This Ca^{2+} influx then stimulates the release of neurotransmitters from storage vesicles into the synaptic cleft (**Fig. 2.1**).

Fig. 2.1 ► **Synaptic signal transmission.**
An action potential (AP) arriving at the presynaptic membrane (**1**) causes voltage-gated Ca^{2+} channels to open (**2**). This increase in intracellular $[Ca^{2+}]$ triggers the release of neurotransmitters from their storage vesicles into the synaptic cleft (**3**). Neurotransmitter molecules then diffuse across the synaptic cleft (**4**) and bind with inotropic or metabotropic receptors on the postsynaptic membrane. Inotropic receptors are ligand-gated ion channels. Ligand (in this case, neurotransmitter) binding (**5**) causes the inflow of ions into the cell, resulting in either depolarization (inflow of cations) or hyperpolarization (inflow of anions). (**6**) Ligand binding to metabotropic receptors activates G proteins, which transduce a cellular response via second messenger molecules.

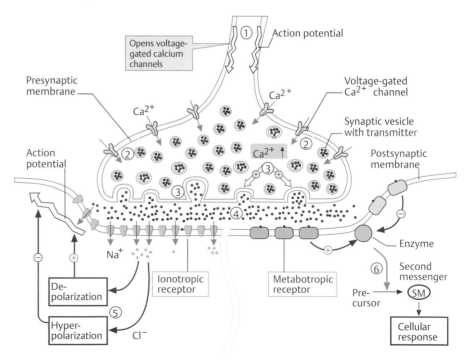

Postsynaptic Events

The neurotransmitters released diffuse across the synaptic cleft and bind to ligand-gated channels on the postsynaptic cell membrane, causing a change in conductance of ions.
— *Excitatory postsynaptic potentials* (EPSPs) are produced when excitatory neurotransmitters open Na^+ channels, resulting in depolarization of the postsynaptic membrane. K^+ channels also open, but the combined effect still depolarizes.
— *Inhibitory postsynaptic potentials* (IPSPs) are produced when inhibitory neurotransmitters open Cl^- channels, resulting in stabilization or hyperpolarization of the postsynaptic membrane.

An action potential is generated when the summated EPSPs and IPSPs bring the membrane to threshold. The neurotransmitter dissociates from the receptor and is removed from the synapse via enzymatic degradation, reuptake, or diffusion.

Summation of Postsynaptic Potentials. Postsynaptic neurons summate postsynaptic potentials spatially and temporally to integrate the total excitatory and inhibitory flow (**Fig. 2.2**).
— *Spatial summation* is the addition of synaptic depolarizations originating from various regions of the neuron.
— *Temporal summation* is the synchronicity of depolarizations, with additional depolarizations occurring before previous ones decay.

2.3 Neuromuscular Transmission

The neuromuscular junction is the synapse between motoneurons and skeletal muscle fibers.

Presynaptic Events

An action potential in the motoneuron depolarizes the membrane, causing membrane Ca^{2+} channels to open and Ca^{2+} influx into the presynaptic terminal. This Ca^{2+} influx then stimulates the release of acetylcholine from their storage vesicles into the synaptic cleft.

Postsynaptic Events

Acetylcholine diffuses across the synaptic cleft and interacts with nicotinic receptors on the postsynaptic membrane (motor end plate). The acetylcholine receptor is a Na^+ and K^+ ion channel. The binding of acetylcholine to this receptor causes an increased conductance of Na^+ and K^+ at the end plate, creating an *end plate potential* (EPP).

Myasthenia gravis

Myasthenia gravis is an autoimmune disease in which there are too few functioning acetylcholine receptors at the neuromuscular junction. Patients with this condition often present in young adulthood with easy fatiguability of muscles, which may progress to permanent muscle weakness. Treatment involves using neostigmine or similar agents to prolong the action of released acetylcholine.

Fig. 2.2 ► Summation of postsynaptic potentials: (A) spatial summation of stimuli; (B) temporal summation of stimuli.

Multiple presynaptic APs causing excitatory postsynaptic potentials (EPSPs) are summed to stimulate an action potential in the axon hillock of the postsynaptic neuron. The EPSPs may be summated spatially or temporally. In spatial summation (**A**), multiple APs arrive at the axon hillock simultaneously, and although none of them could generate an AP individually, their summated effect results in an AP. In temporal summation (**B**), presynaptic APs occurring close together in time generate a large summed EPSP.

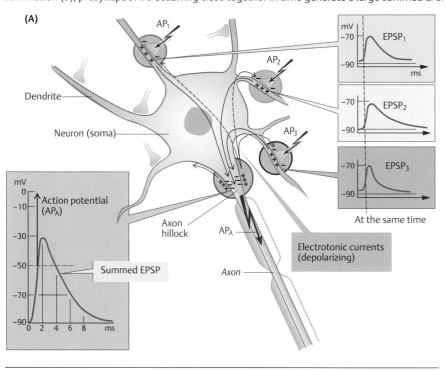

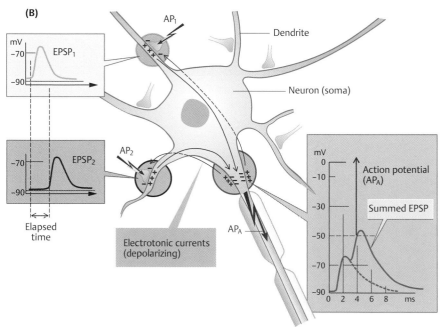

End Plate Potentials.   End plate potentials (EPPs) result from synchronous release of hundreds of vesicles of acetylcholine from the presynaptic terminal of the motoneuron, causing depolarization of the postsynaptic membrane (**Fig. 2.3**). This depolarization results in the influx of Na^+ through the postsynaptic membrane of the end plate. An action potential is generated when the summated EPPs bring the membrane region surrounding the end plate to threshold, causing the muscle to generate its own action potential. Miniature end plate potentials (MEPPs) occur upon spontaneous release of a single vesicle filled with ~10,000 molecules of neurotransmitter. They depolarize the postsynaptic membrane only 1 mV.

Fig. 2.3 ▶ **Motor end plate.**
Motor end plates are the contact between motor axon terminals and skeletal muscles fibers. The acetylcholine (ACh) vesicles release their contents into the synaptic cleft, where ACh binds with receptors on the sarcolemma. At the neuromuscular junction, motoneurons interface with muscle fibers. APs that travel along motoneurons will stimulate the muscle fibers to contract if the depolarization caused by the release of ACh from the presynaptic terminal reaches threshold and generates an AP at the end plate.

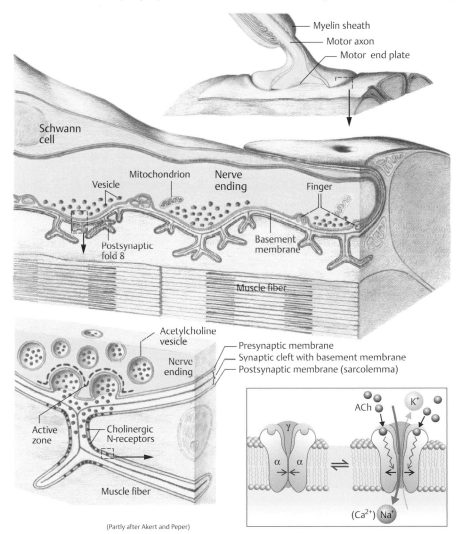

(Partly after Akert and Peper)

3 Muscle Cell Physiology

3.1 Skeletal Muscle

Structure

Skeletal muscle is composed of long, cylindrical, multinucleated cells called muscle fibers. Each fiber contains a bundle of myofibrils. Myofibrils contain myofilaments, overlapping thick and thin filaments. The filament arrays are arranged into sarcomeres, the functional unit of skeletal muscle cells. Sarcomere units, delimited by Z lines, are linked linearly.

Because the sarcomeres within the myofibrils are in register, the fibers of skeletal muscle have a striated appearance (**Fig. 3.1**).

Fig. 3.1 ▶ **Ultrastructure of striated muscle fibers.**
Skeletal muscle is composed of bundles of muscle fibers. Each muscle fiber is composed of myofibrils.

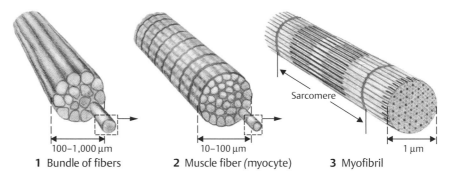

Sarcomere

100–1,000 μm   10–100 μm   1 μm
1 Bundle of fibers   **2** Muscle fiber (myocyte)   **3** Myofibril

Sarcomere Components

Thin Filaments.   Thin filaments are made of actin, with the regulatory proteins tropomyosin and troponin positioned along the surface.

Thick Filaments.   Thick filaments are made of myosin molecules assembled together with globular heads exposed.
— Titin is a giant protein that keeps the myosin filaments accurately lined up within the sarcomere.
— Cross-bridges are myosin heads that bind to actin filaments.

Sarcomere Organization

Z disks are at either end of a given sarcomere and anchor the thin filaments. The I band is the region near the Z line where there are thin filaments only, without overlap of thick filaments. The A band is the middle part of the sarcomere containing thick filaments. Thin filaments overlap into the A band. The H zone is the area in the center of the A band where the thin filaments do not reach, and there are thick filaments only. The M line is in the center of the H zone (**Figs. 3.2**, **3.3**).

Sarcotubular System

Sarcoplasmic Reticulum.   The sarcoplasmic reticulum (SR) is a modified endoplasmic reticulum found only in muscle cells. It consists of a flattened, irregular, saclike system that drapes around the myofibrils. Its membrane contains an active transport Ca^{2+} pump that is responsible for removing Ca^{2+} from the cytoplasm and its accumulation within the SR.

Fig. 3.2 ▶ Sarcomere structure.
Sarcomeres are bounded by Z disks. The I band contains only thin actin filaments. The A band contains thick filaments and is where the actin and myosin filaments overlap. The H zone solely contains myosin filaments, which thicken toward the middle of the sarcomere to form the M line. Actin is a globular protein molecule. Four hundred such molecules join to form F actin, a beaded polymer chain. Two of the twisted protein filaments combine to form an actin filament. Tropomyosin molecules joined end to end lie adjacent to the actin filaments, and a troponin molecule is attached every 40 nm or so. The sarcomere also has another system of filaments formed by the filamentous protein titin. Titin is anchored to the M and Z plates. Each myosin filament consists of bundles of myosin molecules (see **Fig. 3.3**).

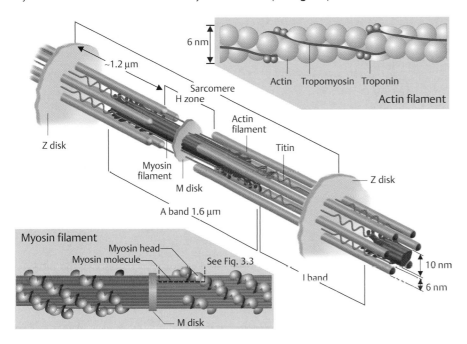

Fig. 3.3 ▶ Myosin II molecule.
Each myosin molecule has two globular heads connected by flexible necks. Each of the heads has a motor domain with a nucleotide binding pocket (for adenosine triphosphate [ATP], or adenosine diphosphate [ADP], + inorganic phosphate [P_i]) and an actin-binding site. The light protein chains are located on this heavy molecule; one is regulatory, the other, so-called essential. Conformational changes in the head–neck segment allow the myosin head to "tilt" when interacting with actin.

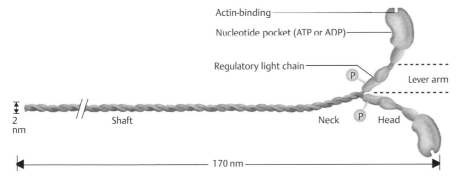

Transverse Tubules.   Transverse tubules are deep invaginations of the sarcolemma (the plasma membrane of the muscle fiber) that penetrate into the muscle fiber and allow the muscle action potential to be directed from the surface into the interior of the muscle fiber.

The sarcoplasmic reticulum adjacent to the transverse tubules is expanded to form terminal cisternae. A transverse tubule and its two adjacent terminal cisternae are linked into a triad.

Innervation

Skeletal muscle is innervated by the somatic nervous system.

Contraction Cycle

Excitation–Contraction Coupling

A motoneuron action potential, by release of acetylcholine at the neuromuscular junction, triggers a muscle action potential that is conducted along the sarcolemma and passes down the transverse tubules. The electrical impulse affects a voltage-sensing protein in the transverse tubule membrane (dihydropyridine receptor), which is linked to a Ca^{2+} release channel (ryanodine receptor) in the terminal cisterna membrane. Ca^{2+} is released through these channels into the cytoplasm to induce muscle contraction.

Action of Regulatory Proteins

Cross-bridge binding to thin filaments is controlled by the regulatory proteins troponin and the rodlike tropomyosin. Troponin is made up of three subunits that play specific roles:
— Troponin T keeps tropomyosin in the groove of actin.
— Troponin I inhibits the interaction of between the myosin cross-bridges and the actin thin filaments.
— Troponin C binds to Ca^{2+} shifting the position of tropomyosin to expose the binding sites on actin for the myosin cross-bridges.

Cross-Bridge Cycle (Fig. 3.4)

In the resting (relaxed) state, adenosine monophosphate (ADP) and inorganic phosphate (P_i) are bound to the myosin cross-bridge, which is energized, but the cross-bridge cannot attach to actin as the binding sites are blocked by tropomyosin.
— In step 1, due to arrival of an action potential, cytosolic Ca^{2+} rises and Ca^{2+} binds troponin, which shifts tropomyosin, exposing the binding sites on actin and allowing cross-bridge attachment.
— In step 2, P_i and then ADP are released from the cross-bridge, causing the cross-bridge to tilt (the power stroke) and the actin thin filament to slide along the myosin thick filament.
— In step 3, adenosine triphosphate (ATP) attaches to cross-bridge, which results in detachment of the cross-bridge from the actin thin filament.
— In step 4, the bound ATP is hydrolyzed to ADP and P_i, energizing the cross-bridge, and beginning the cycle again.

The cycle repeats if the cytosolic Ca^{2+} concentration remains high and Ca^{2+} continues to bind to troponin. It ends if reuptake of Ca^{2+} into the sarcoplasmic reticulum (via Ca^{2+}-ATPase) reduces the cytosolic Ca^{2+} concentration sufficiently so that Ca^{2+} is no longer bound to troponin, causing tropomyosin to shift and once again block the binding sites on actin.

Sliding Filament Mechanism

The rowing action of the myosin cross-bridges in each cycle cause the thin filaments to slide past the thick filaments toward the center of the sarcomere. Sarcomere shortening is manifested in reduction in the distance between Z lines, as well as reductions in the widths of the I band and the H zone. There is no change in the length of the A band.

Twitches and Tetani

The contraction of a muscle fiber following a single action potential is called a *twitch.*

Rapidly applied stimuli produce *tetani* (singular *tetanus*). Sustained contractions representing summation of individual twitches. The resulting tension (force) produced by the muscle is greater than that for a single twitch. The higher concentration of Ca^{2+} remaining in the cytoplasm during tetanic stimulation is responsible for the increased muscle tension (**Fig. 3.5**).

Fig. 3.4 ► **Cross-bridge cycle.**

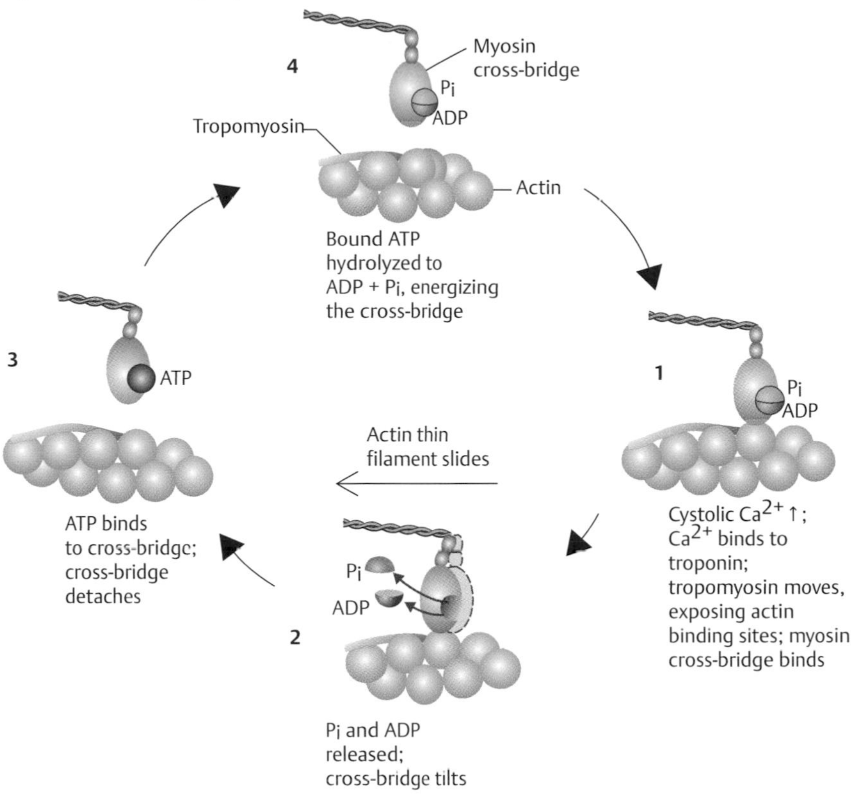

4
Myosin cross-bridge
Pi
ADP
Tropomyosin
Actin
Bound ATP hydrolyzed to ADP + Pi, energizing the cross-bridge
3
ATP
1
Pi
ADP
Actin thin filament slides
ATP binds to cross-bridge; cross-bridge detaches
Pi
ADP
2
Cystolic Ca2+ ↑; Ca2+ binds to troponin; tropomyosin moves, exposing actin binding sites; myosin cross-bridge binds
Pi and ADP released; cross-bridge tilts

Muscle Tension

Isometric, Isotonic, and Eccentric Contractions

- In *isometric contraction,* external muscle length is held constant.
- In *isotonic contraction,* the muscle length shortens.
- In *eccentric contraction,* the muscle exerts a pulling force but is actually getting longer because the external attachments exert more force than is developed within the muscle (**Fig. 3.6**).

Fig. 3.5 ► **Tetani.**
When individual twitches are elicited by rapidly repeating stimuli, the muscle does not relax between stimuli and continuously contracts. This is referred to as a tetanus. Muscle force is significantly greater in a tetanus than in a twitch because more cross-bridges are activated.

Stimulus
Range of summation
Single contractions
Tetanus
Isometric muscle force
0
2
4
6
8
10
Time (s)

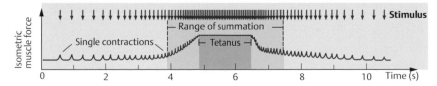

Fig. 3.6 ▶ **Isometric muscle tension relative to sarcomere length.**
The length (L) and force (F), or tension, of a muscle are closely related. Because the active force is determined by the magnitude of all potential actin–myosin interactions, it varies with sarcomere length. Skeletal muscle can develop maximum (isometric) force (F_0) at its maximum resting length (L_{MAX}). When the sarcomeres shorten ($L < L_{MAX}$), part of the thin filaments overlap, allowing only forces smaller than F_0 to develop. When L is 70% of L_{MAX}, the thick filaments make contact with the Z disks, and F becomes even smaller. In addition, at nonphysiological extensions ($L > L_{MAX}$), a muscle can develop only restricted force because the number of potentially available actin–myosin bridges is reduced.

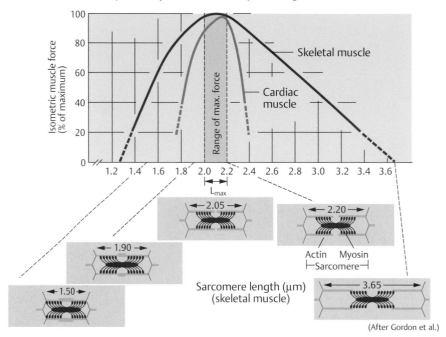

(After Gordon et al.)

The Length–Tension Relationship

The tension developed during isometric contractions at various muscle lengths demonstrates the mechanical properties of muscle tissue and of the sliding filament mechanism.

Passive tension is the force of the elastic elements of a muscle without involvement of the sliding filament apparatus. This is obtained from measurements made on resting, unstimulated muscle. This force is similar to that observed in other elastic structures, such as rubber bands.

Active tension is the force produced by the interaction between myosin and actin at various lengths. This unique curve demonstrates that there is a muscle length at which the sliding filaments will produce their maximum force output. The explanation for the existence of an optimum length is that this sarcomere length produces the maximum amount of overlap between actin and myosin and the largest number of active cross-bridges.

Total tension is the sum of passive and active tension (**Fig. 3.7**).

The Force–Velocity Relationship

During isotonic contraction, when a muscle lifts a load, the weight of the load affects the speed of shortening. Muscle contracts more slowly with a heavy load because cross-bridges cycle more slowly. When the cycling speed is slower, greater force is generated by the sustained attachment of the cross-bridges.

Types of Skeletal Muscle

The two main types of skeletal muscle are type I and type II.

Type I. Type I (red muscle) is employed for slow, sustained contractions as occurs in postural muscles. It can contract for long periods (minutes to hours) without fatigue.

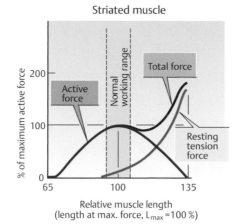

Fig. 3.7 ▶ **Length–tension curve for skeletal muscle.**
The total force, or tension, exerted by a muscle is the
sum of the active and passive (resting) forces. Active
tension is at its maximum when actin and myosin
filaments completely overlap. Skeletal muscle normally
functions in the plateau region of its length–tension
curve. When extended to 130% or more of the L_{MAX},
which is beyond the physiological range, the resting
tension becomes the major part of the total muscle
tension.

The isoform of myosin in this type of muscle has a slower rate of reaction with ATP. Red
muscle has more mitochondria, a greater capillary supply, and a high myoglobin content, resulting in a reliance on oxidative metabolism.

Type II.   Type II (white muscle) is used for rapid and/or strong muscle action. It fatigues rapidly (seconds).

Myosin reacts more rapidly in this type of muscle. It has fewer mitochondria and a lower
capillary density and is suited for anaerobic glycolysis.

3.2 Cardiac Muscle

Structure

Cardiac muscle is striated and organized into sarcomeres like skeletal muscle, except that each
myocyte is a single cell with a single nucleus. Like smooth muscle, cardiac muscle cells are connected by gap junctions and function in a coordinated fashion as a unit (functional syncytium).
Elongated mitochondria located close to myofibrils minimize the distance for diffusion of ATP
in cardiac muscle.

Cardiac tissue shows automaticity (spontaneous activity) and rhythmicity (the ability to
beat). This is due to the pacemaker activity at the sinoatrial (SA) node, which excites nearby
muscle cells. The contractile myocytes that make up the chamber walls are not automatic.

Innervation

Cardiac muscle is innervated by the autonomic nervous system.

Contraction Cycle

The contractile proteins and the molecular mechanism of contraction in cardiac muscle are
identical to that of skeletal muscle. Cardiac muscle cannot be tetanized: the long duration of
cardiac action potentials prevents reexcitation by high-frequency stimulation. The physiology
of cardiac muscle is discussed in more detail in Chapter 8 (**Fig. 3.8**).

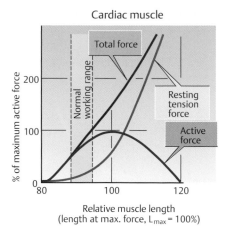

Fig. 3.8 ► **Length–tension curve for cardiac muscle.** The passive extension force, tension, of cardiac muscle is greater than that of skeletal muscle due to lower extensibility of noncontractile components. Cardiac muscle tends to operate in the ascending limb (below L_{MAX}) of its length–tension curve without a plateau. Hence, the ventricle responds to increased diastolic filling loads by increasing its force development (Frank–Starling mechanism). In cardiac muscle, extension also affects troponin's sensitivity to Ca^{2+}, resulting in a steeper curve.

3.3   Smooth Muscle

Structure

Smooth muscle differs from skeletal and cardiac muscle in several ways:
— Smooth muscle cells are much smaller.
— Smooth muscle lacks striations, and its filaments are not arranged in sarcomeres. Instead, actin and myosin are linked to dense bodies that are located on the sarcolemma and in the cytoplasm of the smooth muscle cell.
— The sarcoplasmic reticulum is not as extensive in smooth muscle as in striated muscle. Although they lack transverse tubules, the sarcolemma shows numerous invaginations, which allow influx of Ca^{2+} during an action potential.

Innervation

Smooth muscle, like cardiac muscle, is innervated by the autonomic nervous system. Some smooth muscles also have stretch-gated channels, and so stretch can also initiate contraction.

Contraction Cycle

Ca^{2+} plays a different role in regulating cross-bridge binding in smooth muscle than in skeletal muscle, as smooth muscle lacks the Ca^{2+} binding protein troponin. In smooth muscle, myosin cross-bridges cannot bind actin thin filaments unless activated by phosphorylation. Ca^{2+} regulates this process. When cytosolic Ca^{2+} rises in response to a stimulus, Ca^{2+} binds to calmodulin. This complex then binds to and activates the enzyme myosin light chain kinase, which phosphorylates the myosin light chain protein, forcing the myosin cross-bridge toward the actin thin filament and allowing for binding and muscle contraction (**Fig. 3.9**).

Cross-bridge cycling is the same process as in skeletal muscle. Cross-bridges go through repeated cycles as long as they are phosphorylated. Myosin is dephosphorylated by the enzyme myosin light chain phosphatase, which leads to detachment of the cross-bridge and muscle relaxation.

In smooth muscle there is an additional mechanism (latch bridge mechanism) for cross-bridge attachment that results in a long duration, contracted state that does not require ATP (**Fig. 3.9**).

Types of Smooth Muscle

Unitary smooth muscle.   Unitary smooth muscle is found in the walls of the hollow viscera of the body, such as the gastrointestinal (GI) tract, uterus, and blood vessels. Communication via gap junctions permits electrical potential changes to be communicated from cell to cell, allowing for coordinated contraction in which the muscle sheet acts as one unit.

Multiunit smooth muscle. Multiunit smooth muscle is less common than single unit. It is found in the iris and vas deferens.

Multiunit smooth muscle cells do not have gap junctions; thus, each smooth muscle cell operates independently of its neighbors. Each cell receives its own neural input, permitting finer control (**Fig. 3.10**).

Fig. 3.9 ▶ **Regulation of cross-bridge cycle in smooth muscle.**

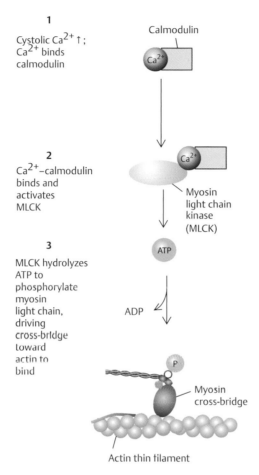

1

Cystolic Ca^{2+} ↑ ; Ca^{2+} binds calmodulin

Calmodulin

2

Ca^{2+}–calmodulin binds and activates MLCK

Myosin light chain kinase (MLCK)

3

MLCK hydrolyzes ATP to phosphorylate myosin light chain, driving cross-bridge toward actin to bind

ATP

ADP

Myosin cross-bridge

Actin thin filament

Fig. 3.10 ▶ **Single- and multiunit smooth muscle fibers.**
(**A**) Single-unit fibers have many gap junctions, permitting coordinated contraction. (**B**) In multiunit fibers smooth muscle cells are individually stimulated.

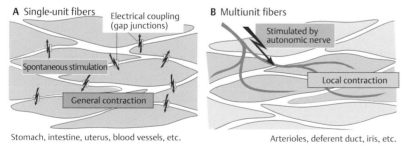

A Single-unit fibers   Electrical coupling (gap junctions)

Spontaneous stimulation

General contraction

Stomach, intestine, uterus, blood vessels, etc.

B Multiunit fibers

Stimulated by autonomic nerve

Local contraction

Arterioles, deferent duct, iris, etc.

Table 3.1 compares skeletal, cardiac, and smooth muscle.

Table 3.1 ► Comparison of Skeletal, Smooth, and Cardiac Muscle			
Feature	**Skeletal Muscle**	**Cardiac Muscle**	**Smooth Muscle**
Appearance	Striated	Striated	Nonstriated
Innervation	Somatic nervous system	Autonomic nervous system	Autonomic nervous system
Postsynaptic receptor	Nicotinic receptor at the neuromuscular junction	Adrenergic and muscarinic receptors on the cell surface	Adrenergic and muscarinic receptors on the cell surface
Neurotransmitter(s)	ACh	NE and EPI	ACh, NE, and EPI
Development of SR	Highly	Moderate	Little
Origin of Ca^{2+}	SR	SR and extracellular fluid	SR and extracellular fluid
Activation of contraction	Ca^{2+}–troponin C complex	Ca^{2+}–troponin C complex	Ca^{2+}–calmodulin leading to activation of MLCK
Contraction as a functional unit	No; action potentials travel along individual muscle cells	Yes; gap junctions permit coordinated contractions	Yes; gap junctions permit coordinated contractions
Determinants of the force of contraction	Regular array of thick and thin filaments	Regular array of thick and thin filaments	Scattered thick and thin filaments
Response to stimulus	Graded	All-or-none	Change in tone or rhythm frequency
Tetanizable	Yes	No	Yes
Work range	At peak of length–force curve	In rising length–force curve	Length–force curve is variable

Abbreviations: ACh, acetylcholine; EPI, epinephrine; MLCK, myosin light chain kinase; NE, norepinephrine; SR, sarcoplasmic reticulum.

Review Questions

1. What do antiporters in cell membranes do?
   A. Move single solutes out of a cell
   B. Move two solutes together out of a cell
   C. Move one solute in and another solute out of a cell
   D. Actively transport single solutes using the energy of adenosine triphosphate (ATP)
   E. Move single solutes in whichever direction is against their concentration gradient

2. A patient poisoned with cyanide cannot utilize ATP. Which of the following transport processes would be most directly affected by this loss?
   A. Transport of Na^+ out of cells
   B. Transport of glucose into red blood cells
   C. Transport of amino acids into cells
   D. Osmosis of water into cells
   E. Diffusion of CO_2 out of cells.

3. A neurotoxin is applied to resting skeletal muscle cells, irreversibly increasing sodium conductance but having no effect on potassium conductance. The transmembrane potential will
   A. hyperpolarize.
   B. depolarize.
   C. not change.
   D. transiently hyperpolarize, then return to what it was before applying the neurotoxin.
   E. transiently depolarize, then return to what it was before applying the neurotoxin.

4. After reaching the peak of the action potential, nerve fibers repolarize rapidly because
   A. K^+ channels are inactivated.
   B. Na^+ channels are activated.
   C. Na^+ channels are inactivated.
   D. there are no net ionic fluxes.
   E. the Na^+–K^+ ATPase removes the Na^+ that entered during the rising phase.

5. Before open heart surgery, a patient's heart might be stopped by injecting isotonic potassium chloride (KCl) into the coronary arteries. Which of the following mechanisms produces this effect?
   A. The cardiac muscle membrane potentials would fall to near zero.
   B. The Na^+–K^+ ATPase pump is inhibited.
   C. K^+ flows freely into cardiac myocytes.
   D. Cl^- equilibrates across cardiac myocyte membranes.
   E. Ca^{2+} leaks out of the cardiac myocytes.

6. Synaptic inhibition may be caused by
   A. increase in Cl^- conductance at the postsynaptic membrane.
   B. increase in Ca^{2+} conductance at the postsynaptic membrane.
   C. decrease in K^+ conductance at the postsynaptic membrane.
   D. increase in Na^+ conductance at the postsynaptic membrane.
   E. increase in Ca^{2+} conductance at the presynaptic membrane.

7. Stretch of muscle fibers
   A. is a direct cause of contraction in some smooth muscles.
   B. has no effect on contraction in visceral smooth muscle.
   C. induces contraction only when mediated by acetylcholine.
   D. causes a direct and sustained contraction in skeletal muscle.
   E. has no effect on the strength of contraction in skeletal muscle.

8. Before suturing a deep skin wound, the physician infiltrates the surrounding tissue with lidocaine. How does this block transmission in pain fibers?
   A. It facilitates Na^+ influx into nerve fibers.
   B. It promotes K^+ efflux from nerve fibers.
   C. It blocks Na^+ channel activation, which prevents generation of action potentials.
   D. It blocks the Na^+–K^+ ATPase pump.
   E. It depolarizes the nerve fibers' membrane potentials.

9. Why does a patient with myasthenia gravis have muscle weakness?
   A. Decreased presynaptic Ca^{2+}-binding sites
   B. Decreased postsynaptic acetylcholine receptor sites
   C. Decreased release of acetylcholine from motor nerve terminals
   D. Decreased myelin on motor nerve fibers
   E. Decreased Ca^{2+} in the muscle fiber sarcoplasmic reticulum

10. When the arm is fully outstretched, which of the following is true about the biceps muscle?
    A. The muscle is at its best mechanical advantage for flexing the forearm.
    B. The overlap between thick and thin filaments in the muscle fibers is at a maximum.
    C. The muscle can develop more force at this position than when the arm is flexed.
    D. The muscle cannot be stimulated tetanically.
    E. Shortening the muscle will allow more neuromuscular junctions to be activated.

11. Why do muscles undergo rigor mortis immediately after death?
    A. Most of the Ca^{2+} is sequestered into the sarcoplasmic reticulum.
    B. The cross-bridges are cycling in a tetanic contraction.
    C. Myosin and actin are separated and cannot interact.
    D. Myosin heads are unable to detach from actin thin filaments.
    E. ATP molecules remain bound to myosin heads

12. What would be the effect on the heart of treating a cardiac patient with a calcium channel–blocking drug?
    A. Increased preload
    B. Increased Ca^{2+} stored within the sarcoplasmic reticulum of cardiac myocytes
    C. Increased contractility
    D. No change in contractility
    E. Decreased contractility

Answers and Explanations

1. **C** Antiporters, also called exchangers, move two solutes in opposite directions (therefore B is incorrect) using the energy from moving one solute with its electrochemical gradient to move the other solute against its electrochemical gradient (p. 5).
    A,E Antiporters always move two solute species.
    D Although some primary active transporters move solutes in opposite directions (e.g., the Na^+–K^+ ATPase), they use the energy of ATP rather than the electrochemical gradient of one of the solutes to provide the energy.

2. **A** Although all of these processes would ultimately be affected, the Na^+–K^+ pump uses ATP directly and would be affected first (p. 4).
    B Glucose is transported into red blood cells by a uniporter.
    C Amino acids are transported into cells by secondary active transport.
    D Osmosis is passive and depends only on the concentration of particles in the cells.
    E CO_2 and O_2 move across cell membranes by free diffusion.

3. **B** Any increase in Na^+ conductance relative to K^+ conductance will depolarize the membrane (p. 13).
    A,D An increase of K^+ conductance relative to Na^+ conductance will hyperpolarize the membrane.
    C,E The depolarization continues as long as the high Na^+ conductance is maintained by the toxin.

4. **C** Na^+ channel inactivation lowers the inward Na^+ current and allows the outward K^+ current to repolarize the membrane (p. 13).
    A K^+ channel inactivation would depolarize, not repolarize.
    B Na^+ channel activation generates the rising phase of the action potential.
    D There is a net outward ionic flux.
    E Although the Na^+–K^+ ATPase contributes a small outward current at all times, it is the transiently high outward K^+ current that drives the falling phase of the action potential.

5. **A** The membrane potential in cardiac myocytes depends primarily on a K^+ concentration difference, which would be near zero with isotonic KCl outside the muscle cells (p. 12). With a zero membrane potential, action potentials cannot be generated.
    B The pump is not affected.
    C K^+ concentration is the same inside and outside of the cells because they are bathed in an isotonic solution, and so K^+ will not flow into the myocytes.
    D Cl^- passively follows the cations and does not affect membrane potential differences.
    E Ca^{2+} would still remain within depolarized cells.

6. **A** An increase in chloride conductance hyperpolarizes and/or clamps the membrane potential and inhibits depolarization of the postsynaptic membrane by an excitatory postsynaptic potential (EPSP) (p. 19).
    B–D An increase in Ca^{2+} or Na^+ conductance or a decrease in K^+ conductance would depolarize the postsynaptic membrane.
    E An increase in Ca^{2+} conductance at the presynaptic membrane would cause the release of a neurotransmitter.

7. **A** Stretch can cause depolarization and generation of action potentials in some smooth muscle, including visceral muscles (B), due to the opening of stretch-gated channels.
    C Stretch acts directly on the cell membrane without a neurotransmitter.
    D,E Stretch does not depolarize skeletal muscle fibers, but stretch will change the overlap of thick and thin filaments and alter the strength of contraction.

8. **C** Lidocaine blocks Na^+ channels, which prevents the generation of action potentials and thus blocks transmission of sensory information in pain fibers.
    A,B Lidocaine does not increase Na^+ influx into or K^+ efflux from nerve fibers.
    D,E Lidocaine does not affect the pump or the membrane potential.

9. **B** A reduced number of acetylcholine receptor sites makes the muscle fibers less sensitive to released acetylcholine. When some of the fibers fail to respond to neural stimulation, this produces muscle weakness (p. 19).
A Presynaptic Ca^{2+} binding is normal.
C The amount of acetylcholine released declines somewhat with normal use but is adequate except when there is a deficit in receptors at the postjunctional membrane.
D Myasthenia gravis does not affect myelination.
E Ca^{2+} within muscle fibers is unaffected.

10. **C** As described by the length–force curve, the greatest force can be developed at the maximum physiological length (p. 25).
A Although muscle force is maximized, the fully extended position offers the least mechanical advantage.
B Overlap between thick and thin filaments increases as the muscle shortens. With the arm fully outstretched, the overlap is at a minimum.
D A muscle can be stimulated tetanically at any length.
E Muscle length does not affect neuromuscular junctions.

11. **D** It requires binding of ATP to detach cross-bridges from actin-binding sites. This cannot occur once ATP becomes depleted (pp. 24–25).
A The depletion of ATP allows Ca^{2+} to escape from the sarcoplasmic reticulum.
B The cross-bridges are fixed, not cycling.
C Myosin and actin are attached.
E ATP is depleted and is not available to bind.

12. **E** Calcium channel–blocking drugs reduce the influx of Ca^{2+}. This reduces the stimulating effect of Ca^{2+} influx, so contractility falls (p. 28).
A Preload is a function of filling pressure, not ionic flux.
B Over time, the decreased Ca^{2+} entry would lead to lower Ca^{2+} stored in the sarcoplasmic reticulum.
C,D Contractility is proportional to intracellular Ca^{2+} concentration during systole, so it is reduced.

4 Autonomic Nervous System

The autonomic nervous system (ANS) is a branch of the peripheral nervous system (PNS). ANS neurons innervate smooth muscle, cardiac muscle, and glands to maintain homeostasis. The ANS is not under voluntary control.

The other branches of the PNS are the somatic nervous system and the enteric nervous system.

— The *somatic nervous system* innervates skeletal muscle and is discussed in Chapter 6.
— The *enteric nervous system* innervates nerve plexuses in the gut and is discussed in Chapter 19.

4.1 Organization of the Autonomic Nervous System

The ANS has a parasympathetic and sympathetic division. Preganglionic neurons of both divisions originate in the central nervous system (CNS) but emerge from different regions. Preganglionic neurons are lightly myelinated B fibers; postganglionic neurons are unmyelinated C fibers with slower conduction velocity. **Table 4.1** summarizes the main features of each division, and they are illustrated in **Fig. 4.1**.

■ Quadriplegia

Quadriplegia is caused by spinal cord injury at a high level. A ventilator may be required if the injury involves C3–C5, as these spinal nerves control the diaphragm (via the phrenic nerve), which is the major muscle that allows us to breathe. However, quadriplegic patients can survive because the cranial nerves (ANS preganglionic parasympathetic nerves) remain intact and can coordinate vital bodily functions despite the patient's having no voluntary control from below the level of injury.

■ Horner syndrome

Horner syndrome may occur due to an interruption of the sympathetic supply to the face through disease of the brainstem or thoracolumbar region of the spinal cord, for example, by stroke, tumors, carotid artery dissection (tearing on the lining of the artery), or spinal cord injury; or it may be idiopathic (cause not known). It causes pupillary constriction (miosis), a sunken eye (enophthalmos), drooping of the upper eyelid (ptosis), and ipsilateral loss of sweating (anhydrosis) on the affected side of the face. Treatment for this disease is directed at the underlying cause.

Table 4.1 ▶ Summary of the Parasympathetic and Sympathetic Divisions of the Autonomic Nervous System

	Parasympathetic Division	Sympathetic Division
Region of spinal cord from which preganglionic neurons emerge	Cranial and sacral: – Cell bodies of preganglionic neurons are in the midbrain, pons, and medulla, giving rise to the autonomic components of cranial nerves III, VII, IX, and X – Cell bodies of preganglionic neurons are in S2–S4	Thoracolumbar: – Cell bodies of preganglionic neurons are in T1–L3
Length of preganglionic neurons	Long	Short
Length of postganglionic neurons	Short	Long
Location of ganglia	Near target organs	Sympathetic chain ganglia (located parallel to the spinal cord on both sides) Abdominal prevertebral ganglia Adrenal medulla*
General functions	Principally concerned with maintenance, conservation, and protection of body resources (anabolic)	Principally involved with expenditure of body resources or energy (catabolic)
Comments	A functioning parasympathetic system is necessary to sustain life, as it maintains essential bodily functions. Parasympathetic nerves can act in isolation from the system as a whole, producing discrete effects at specific end organs.	The sympathetic system is not strictly necessary to maintain life. – It is capable of a mass response, the emergency "flight-or-fight" response. The neuronal basis of this widespread response lies in the wide divergence of preganglionic axons within the sympathetic chain ganglia.

* The adrenal medulla is innervated by preganglionic sympathetic nerves and is physiologically similar to a sympathetic ganglion. However, because the chromaffin cells of the adrenal medulla lack axons, it responds to preganglionic secretion of acetylcholine by secreting the hormones epinephrine (and norepinephrine to a lesser extent) into the bloodstream, producing a response at target tissues.

Fig. 4.1 ▶ **The autonomic nervous system.**
The autonomic nervous system has a parasympathetic division and a sympathetic division.
Parasympathetic preganglionic neurons arise in the cranial nerve nuclei or in the sacral region of the spinal
cord. They are relatively long compared with the parasympathetic postganglionic neurons. Sympathetic
preganglionic neurons, which arise in the thoracolumbar region of the spinal cord, are relatively short
compared with the postganglionic neurons. Both pre- and postganglionic parasympathetic neurons
release acetylcholine as their neurotransmitter substance. Sympathetic preganglionic neurons also release
acetylcholine, but postganglionic neurons release norepinephrine. The exception to this is sweat glands,
which are innervated by sympathetic postganglionic fibers that release acetylcholine.

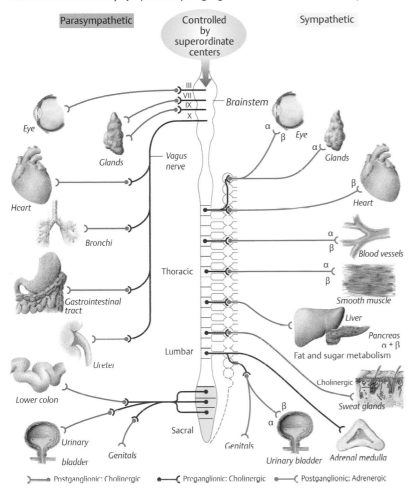

4.2 Neurotransmitters

Neurotransmitters that act in both the CNS and ANS are also discussed in Chapter 2.

Acetylcholine

Synthesis.   Acetylcholine is synthesized via the combination of acetate from acetyl coenzyme
A (acetyl-CoA) and choline and is catalyzed by the enzyme choline acetyltransferase. The uptake
of choline into the presynaptic terminal is the rate-limiting step in acetylcholine synthesis.

Release.   Depolarization of the preganglionic membrane by an action potential causes increased Ca^{2+} influx into the presynaptic terminal and the release of acetylcholine (**Fig. 4.2**).

Degradation.   The breakdown of acetylcholine is rapid via a specific enzyme, acetylcholinesterase, to produce acetate and choline. Acetylcholinesterase is located in neuronal membranes
and red blood cells. Pseudocholinesterases (nonspecific) and butyrylcholinesterases, which are
more widely distributed, can also hydrolyze acetylcholine.

Fig. 4.2 ► **Acetylcholine: release and degradation.**
Acetylcholine is stored in vesicles in the axoplasm of presynaptic nerve terminals. These vesicles are anchored via the protein synapsin to the cytoskeletal network, thus allowing for the concentration of vesicles near the presynaptic membrane while preventing fusion with the membrane. An action potential causes Ca^{2+} influx into the axoplasm of the presynaptic terminal through voltage-gated channels. Ca^{2+} then activates protein kinases that phosphorylate synapsin. This causes the vesicles to become free of the cytoskeleton, dock at the active zone, fuse with the presynaptic membrane, and release acetylcholine into the synaptic gap. Acetylcholine attaches to receptors on the postsynaptic membrane and exerts its effects. Acetylcholine is then hydrolyzed by acetylcholinesterase, with reuptake of the choline component into the presynaptic terminal. See page 38 for discussion of acetylcholine's effects on muscarinic cholinergic receptors (M_1, M_2, M_3).

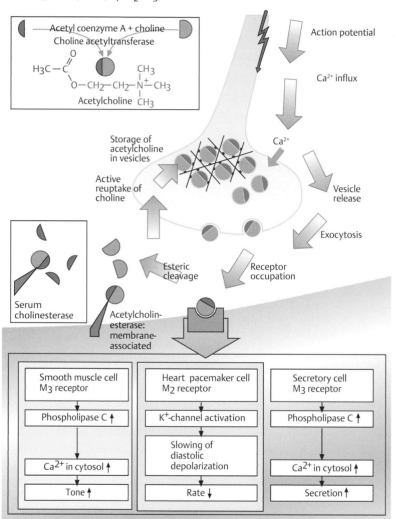

■ **Ascorbic acid redox reactions**

Ascorbic acid (vitamin C) is involved in many processes in the body, including collagen and bile acid synthesis, activation of neuroendocrine hormones (e.g., gastrin, corticotropin-releasing hormone [CRH], and thyrotropin-releasing hormone [TRH]), iron absorption, and detoxification (via stimulation of cytochrome P450 enzymes in the liver. Ascorbic acid is oxidized to dehydroascorbic acid via an extremely reactive intermediate, semidehydro-L-ascorbate. The hydrogen ions that are liberated in this oxidation are able to act as donors in hydroxylation reaction throughout the body (which accounts for some of its effects). One such reaction is when ascorbic acid acts as a cofactor for dopamine-β-hydroxylase in the synthesis of norepinephrine and epinephrine.

Norepinephrine

Synthesis. Norepinephrine is synthesized from the precursor amino acid tyrosine by hydroxylation to dihydroxyphenylalanine (dopa) in postganglionic neurons of the sympathetic division. Dopa is decarboxylated to dopamine, which is oxidized to norepinephrine and packaged in vesicles.

Release. When stimulated by an action potential, the presynaptic membrane is depolarized, Ca^{2+} influx increases levels in the presynaptic terminal, and vesicles of norepinephrine are released. Norepinephrine then binds to adrenergic receptors on target cells (**Fig. 4.3**).

Fig. 4.3 ► **Synthesis and termination of norepinephrine (NE) and adrenergic transmission.**
Postganglionic sympathetic nerve terminals possess varicosities that enable the nerves to lie in close proximity to effector organs. NE synthesis and storage in vesicles occur in these varicosities. An action potential at the nerve terminal causes the influx of Ca^{2+} and the subsequent release of NE into the synaptic cleft. NE then binds to adrenergic receptors on effector organs, thus exerting its physiological effect. Note that NE has little effect on the β_2 adrenergic receptor, whereas epinephrine, synthesized in the adrenal medulla, acts at all adrenergic receptors. Approximately 70% of NE is taken back up into the presynaptic nerve terminal and repackaged in vesicles or it is inactivated by monoamine oxidase (MAO). In the heart, NE is inactivated by MAO or catecholamine-O-methyltransferase (COMT). See pages 38 and 39 for discussion of norepinephrine's effects on α- and β-adrenergic receptors. (cAMP, cyclic adenosine monophosphate; PKA, protein kinase A)

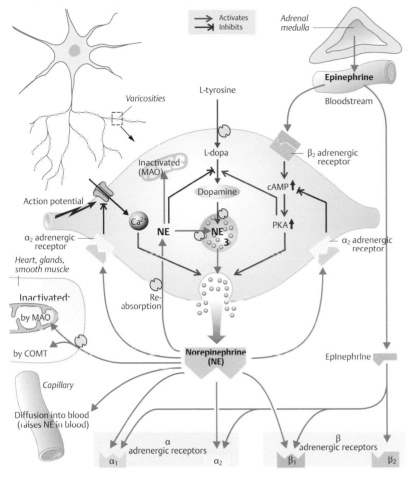

Degradation. Termination of action of norepinephrine is primarily by reuptake (60–90%) into the nerve terminal. Secondary degradation is by monoamine oxidase (MAO) and catechol-O-methyltransferase (COMT).

4.3 Neurotransmitter Receptors

Neurotransmitter receptors and signal transduction are also discussed in Chapter 1.

Cholinergic Receptors

Acetylcholine receptors fall into two classes: nicotinic receptors and muscarinic receptors.

Nicotinic Receptors

Nicotinic receptors are ligand-gated ion channels that are activated by acetylcholine or nicotine. Receptor activation leads to opening of Na^+ and K^+ channels (**Fig. 4.4**) and excitation.

Fig. 4.4 ▸ Acetylcholine receptors.
The nicotinic receptor consists of five protein subunits. Binding of acetylcholine to the two α subunits is thought to change its conformation, permitting the central pore to open and allowing the influx of ions into the cell. The muscarinic receptor is coupled to intracellular G proteins, which may then transduce excitatory effects via phospholipase C or inhibitory/excitatory effects via adenylate cyclase.

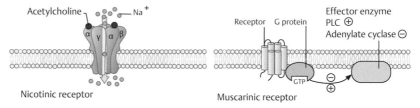

Locations:
— Postganglionic neurons in autonomic ganglia of both the parasympathetic and sympathetic divisions
— Motor end-plates of skeletal muscles at the neuromuscular junction (NMJ)
— Chromaffin cells in the adrenal medulla

Muscarinic Receptors

Muscarinic receptors are metabotropic and are coupled to G proteins. They are activated by acetylcholine and muscarine.
— They are present on the targets of all postganglionic parasympathetic neurons (see each receptor subtype for the location where they are predominantly found and their effects).

M_1 receptors
— *Location*: CNS
— *Signal transduction mechanism*: G_q activation followed by ↑ inositol 1,4,5-triphosphate (IP_3) and diacylglycerol (DAG)
— *Effect*: Excitatory

M_2 receptors
— *Location*: Heart
— *Signal transduction mechanism*: G_i activation followed by ↓ cyclic adenosine monophosphate (cAMP)
— *Effect*: Inhibitory (decreased heart rate and velocity of contraction)

M_3 receptors
— *Location*: Smooth muscle, glands
— *Signal transduction mechanism*: G_q activation followed by ↑ IP_3 and DAG
— *Effect*: Excitatory (increased contraction of smooth muscle; increased secretion from glands)

Adrenergic Receptors

Adrenergic receptors are divided into α receptors (α_1 and α_2) and β receptors (β_1, β_2, and β_3). Both norepinephrine and epinephrine activate adrenergic receptors, but the sensitivity varies within the receptor subtypes.

α_1 receptors
— *Location*: Vascular smooth muscle
— *Neurotransmitter sensitivity*: Equal sensitivity to norepinephrine and epinephrine.
— *Signal transduction mechanism*: G_q activation followed by ↑ IP_3 and DAG
— *Effect*: Excitatory (contraction of vascular smooth muscle)

α₂ receptors

- *Location*: Autoreceptors at presynaptic terminals of sympathetic neurons
- *Neurotransmitter sensitivity*: More sensitive to epinephrine than norepinephrine
- *Signal transduction mechanism*: G_i activation followed by ↓ cAMP
- *Effect*: Inhibitory (block the release of norepinephrine from presynaptic terminals)

β₁ receptors

- *Location*: Sinoatrial (SA) node, atrioventricular (AV) node, and cardiac ventricular muscle
- *Neurotransmitter sensitivity*: Equally sensitive to both norepinephrine and epinephrine
- *Signal transduction mechanism*: G_s activation followed by ↑ cAMP
- *Effect*: Excitatory (increased heart rate, force, and velocity of contraction)

β₂ receptors

- *Location*: Bronchial smooth muscle
- *Neurotransmitter sensitivity*: More sensitive to epinephrine than norepinephrine
- *Signal transduction mechanism*: G_s activation followed by ↑ cAMP
- *Effect*: Relaxation (dilation) of bronchial smooth muscle

β₃ receptors

- *Location*: Adipose tissue
- *Neurotransmitter sensitivity*: More sensitive to epinephrine than norepinephrine
- *Signal transduction mechanism*: G_s activation followed by ↑ cAMP
- *Effect*: Excitatory

Table 4.2 summarizes ANS receptors and their signal transduction mechanisms.

Table 4.2 ▶ Summary of Autonomic Nervous System Receptors			
Receptor Classification	**Location**	**Neurotransmitter**	**Signal Transduction Mechanism**
Cholinoceptors			
Nicotinic	Postganglionic neurons in autonomic ganglia Motor end-plate of skeletal muscle at the NMJ Postganglionic neurons that innervate the adrenal medulla	ACh	Nicotinic receptors open Na^+ and K^+ channels at autonomic ganglia and at the end-plate.
Muscarinic	Targets of postganglionic parasympathetic neurons: M_1: CNS M_2: Heart M_3: Smooth muscle* and glands	ACh	M_1: G_q leading to ↑ IP_3 and DAG M_2: G_i leading to ↓ cAMP M_3: G_q leading to ↑ IP_3 and DAG
Adrenergic receptors			
α	α_1: Vascular smooth muscle α_2: Autoreceptors at presynaptic terminals of sympathetic neurons	NE and EPI	α_1: G_q leading to ↑ IP_3 and DAG α_2: G_i leading to ↓ cAMP
β	β_1: SA node, AV node, and ventricular muscle β_2: Bronchial smooth muscle β_3: Adipose tissue	NE and EPI	β-receptors: G_s leading to ↑ cAMP

Abbreviations: ACh, acetylcholine; AV, atrioventricular; cAMP, cyclic adenosine monophosphate; CNS, central nervous system; DAG, diacylglycerol; EPI, epinephrine; IP3, inositol 1,4,5-triphosphate; NE, norepinephrine; NMJ, neuromuscular junction; SA, sinoatrial
* Effects on smooth muscle are the most important in terms of autonomic responses.

4.4 Physiologic Effects of the Autonomic Nervous System

Dual and Single Innervation

Dual Innervation.   Most organs receive innervation from both the parasympathetic and sympathetic neurons. The two divisions usually work in opposition to each other. For example, sympathetic activity increases heart rate, whereas parasympathetic activity reduces heart rate.

Single Innervation.   Some autonomic effectors receive input from only one division of the ANS. For example, sweat glands receive only cholinergic innervation from the sympathetic division, and ciliary muscles of the eye receive only parasympathetic innervation.

Fig. 4.5 ► **Physiologic responses of the autonomic nervous system.**
This figure shows the effects of parasympathetic and sympathetic stimulation on organs throughout the body. Most organs are innervated by both systems with opposing effects. However, blood vessels, for example, are only innervated by sympathetic postganglionic neurons. (CNS, central nervous system; VIP, vasoactive intestinal peptide.)

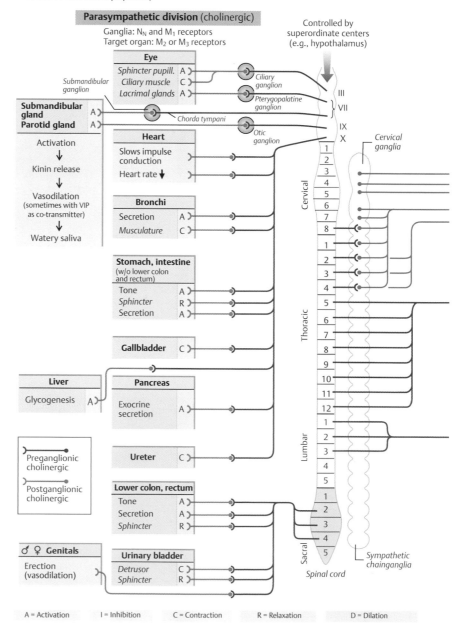

A = Activation   I = Inhibition   C = Contraction   R = Relaxation   D = Dilation

Autonomic Tone

Under normal conditions, there is a low-level tonic firing in both the parasympathetic and sympathetic divisions. This *autonomic tone* is determined by higher autonomic centers in the brainstem. This continuous tonic firing allows the ANS to produce a response by reducing background neuronal activity to an organ. For example, vascular smooth muscle is controlled by sympathetic outflow. Therefore, a reduction in tonic sympathetic outflow results in vasodilation. Tonic activity is overridden in the emergency response when the entire sympathetic nervous system is activated as a unit.

Figure 4.5 provides a comprehensive chart of the effects of the parasympathetic and sympathetic nervous systems on organs throughout the body.

Fig. 4.5 ▶ (Continued) **Physiologic responses of the autonomic nervous system.**

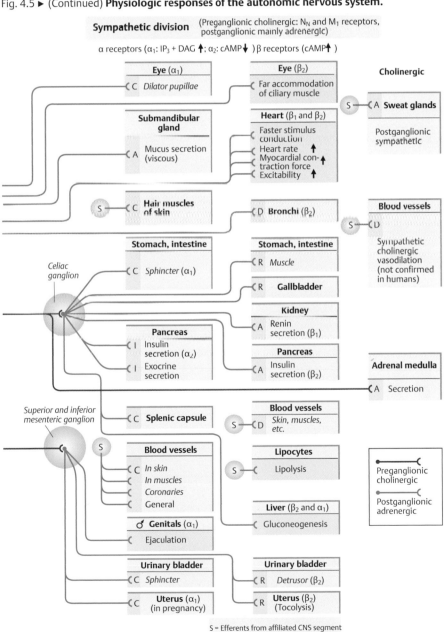

S = Efferents from affiliated CNS segment

4.5 Drugs Affecting the Autonomic Nervous System

Cholinergic agonist drugs are termed *parasympathomimetics,* as they mimic the actions of acetylcholine.

— Direct-acting parasympathomimetics bind directly to cholinergic receptors.
— Indirect-acting parasympathomimetics inhibit the enzyme acetylcholinesterase and thus increase the concentration of acetylcholine.

Cholinergic antagonist drugs block neuronal and muscle nicotinic receptors, along with muscarinic receptors.

— Ganglionic blocking agents block nicotinic receptors at autonomic ganglia (but not nicotinic receptors at the neuromuscular junction).
— Depolarizing neuromuscular blocking agents persistently depolarize nicotinic receptors at the NMJ, leading to receptor desensitization.
— Nondepolarizing neuromuscular blocking agents block nicotinic receptors at the NMJ without depolarization.
— Muscarinic receptor antagonists block muscarinic receptors throughout the body and therefore have low organ specificity.

Table 4.3 provides a summary of cholinergic agonist and antagonist drugs.

Table 4.3 ▶ **Summary of Cholinergic Agonists and Antagonists**		
Type	**Mechanism**	**Agents**
Cholinergic Agonists (parasympathomimetics)		
Direct	Mimics ACh directly at cholinergic receptors	Carbachol, pilocarpine
Indirect	Inhibits acetylcholinesterase that breaks down ACh in synaptic cleft	Neostigmine, physostigmine
Cholinergic Antagonists		
Ganglion-blocking agents	Blocks nicotinic receptors at autonomic ganglia	Hexamethonium, trimethaphan
Depolarizing neuromuscular blockers	Persistently activates nicotinic receptors at the NMJ, leading to initial excitation, then desensitization	Succinylcholine
Nondepolarizing neuromuscular blockers	Competitive antagonists at NMJ	Vecuronium, curare
Muscarinic antagonists	Competitive antagonists at muscarinic receptors	Atropine, scopolamine, ipratropium
Abbreviations: ACh, acetylcholine; NMJ, neuromuscular junction.		

Adrenergic agonist drugs are termed *sympathomimetics,* as they mimic the actions of norepinephrine and epinephrine.

— Catecholamines are endogenous sympathomimetics that directly stimulate specific adrenergic receptors.
— Direct sympathomimetics are synthetic agents that directly stimulate specific adrenergic receptors.
— Indirect sympathomimetics stimulate the release of stored norepinephrine from presynaptic nerve terminals or block the reuptake of norepinephrine (many do both).

Adrenergic antagonist drugs block specific adrenergic receptors.

— Alpha blockers block specific α receptors.
— Beta blockers block specific β receptors.

Table 4.4 lists the adrenergic agonist and antagonist drugs and the receptors where they are active.

Table 4.4 ▶ Adrenergic Agonists and Antagonists		
Type	**Mechanism**	**Agents and Receptors (Stimulated/Blocked)**
Adrenergic Agonist (Sympathomimetics)		
Catecholamines	Endogenous agents that directly stimulate specific α or β receptors	Norepinephrine (α_1, α_2, β_1) Epinephrine (α_1, α_2, β_1, β_2) Dopamine (α_1, β_1)
Direct sympathomimetics	Synthetic agents that directly stimulate specific α or β receptors	Isoproterenol (β_1, β_2) Dobutamine (β_1) Phenylephrine (α_1, α_2) Clonidine (α_2) Metaproterenol, terbutaline, and albuterol (β_2)
Indirect sympathomimetics	Stimulate the release of stored NE from nerve terminals or block the reuptake of NE	Tyramine (↓ release of NE) Amphetamine, ephedrine (↓ release of NE and ↓ its reuptake)
Adrenergic Antagonists (Sympatholytics)		
α-blockers	Block specific α receptors	Phenoxybenzamine and phentolamine (α_1, α_2) Prazosin, terazoin, and doxazosin (α_1)
β-blockers	Block specific β receptors	Propranolol, nadolol, and timolol (β_1, β_2) Acebutolol, atenolol, esmolol, and metoprolol (β_1)

Abbreviation: NE, norepinephrine.

4.6 Central Autonomic Control

The hypothalamus selectively activates components of the endocrine system and the autonomic and somatic nervous systems to maintain homeostasis. It initiates appropriate behavioral responses to stimuli such as stress, reproduction, exercise, heat, and cold. It also monitors body water and plasma glucose. The neuroendocrine functions of the hypothalamus are described in Chapter 23.

Temperature Regulation

The preoptic region and adjacent anterior nuclei of the hypothalamus contain a thermostat for establishing the *set point temperature* (~37°C/98.6°F), the core temperature that the system attempts to maintain.

Mechanisms of temperature regulation

If the core temperature falls below the set point, then the following mechanisms may be induced by the posterior hypothalamus (**Fig. 4.6**):

— Somatic nervous system activation induces shivering. Shivering generates heat by causing adenosine triphosphate (ATP) hydrolysis in the contractile apparatus of skeletal muscle.
— Thyroid hormones may be released, which generates heat by increasing the activity of $Na^+–K^+$ ATPase.
— Sympathetic nervous system activation causes vasoconstriction of blood vessels to the skin, resulting in heat conservation.

If the core temperature rises above the set point, then the following mechanisms may be induced by the anterior hypothalamus:

— Sympathetic cholinergic activation of sweat glands (via muscarinic receptors) increases heat loss by evaporation of water from the skin.
— Lowered sympathetic adrenergic activity causes dilation of blood vessels to the skin, resulting in heat loss by convection and radiation.

Countercurrent exchange of heat

In countercurrent exchange of heat, the fluid within two tubes flows in opposite directions. Because a temperature gradient is present in all parts of the tube, heat is exchanged along the entire length. The result of countercurrent heat exchange is thermal equilibrium. The body uses this property to minimize heat loss from the surface of the skin, as heat is transferred from arterial blood to venous blood. This facilitates maintenance of a stable core temperature. Countercurrent exchange of heat is analogous to the countercurrent exchange of water and the countercurrent multiplier in the loop of Henle (see pp. 178-181, **Fig. 17.4**).

Fig. 4.6 ▶ Neural factors affecting thermoregulation.
Sensors within the hypothalamus, the skin, and the spinal cord sense temperature. The hypothalamus regulates body temperature by coordinating the activity of heat loss and heat-generating/conservation mechanisms, which stimulate adjustments in the body to maintain an appropriate temperature.

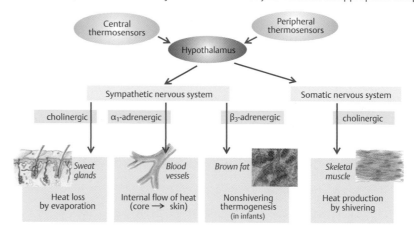

Pathophysiology of Temperature Regulation

Fever.   Fever is produced by endogenous pyrogens (e.g., interleukin-1) released by cells of the immune system in response to infective bacteria. These pyrogens act on the anterior hypothalamus to increase prostaglandin synthesis, which in turn stimulates the thermoregulatory center to reset the set point to a higher temperature. Because body temperature is cooler than the set point, body temperature increases (by heat production and conservation of heat loss) until it stabilizes at the new, elevated set point temperature. When the fever breaks and the set point returns to 37°C (98.6°F), the patient vasodilates and sweats to lose heat until the body temperature returns to normal.

For suppressing fever, aspirin is effective therapy because it inhibits cyclooxygenase and therefore inhibits prostaglandin synthesis. In doing so, aspirin lowers the set point temperature and will cause activation of the heat loss mechanisms. Steroids may also be used because they block the release of arachidonic acid (the precursor of prostaglandins) from membrane phospholipids.

Heat exhaustion.   Heat exhaustion occurs when the body overheats, causing profuse sweating. This may result in a drop in blood pressure and fainting. Treatment for this condition involves rehydration and resting in a cool place.

Heat stroke.   Heat stroke represents a failure of heat loss mechanisms but not a change in set point. In this case, a person in a hot environment fails to adequately mobilize cutaneous vasodilation and sweating. People experiencing heat stroke must be drastically cooled (placed in a bathtub of ice water), or the core temperature will continue to rise, resulting in death.

Hypothermia.   Hypothermia results from exposure to cold temperatures when the capacity to generate body heat is inadequate to maintain the core temperature. Death can occur by myocardial fibrillation (uncoordinated, chaotic contractions of cardiac muscle).

Feeding Behavior and Satiety

Feeding behavior is controlled by a reciprocal interaction between the lateral feeding center of the hypothalamus, which controls hunger, and the ventromedial satiety center of the hypothalamus.

— The lateral nucleus of the hypothalamus contains glucose receptors, which are sensitive to changes in blood glucose. These "glucostats" probably initiate feeding behavior.
— Gastric distention activates the ventromedial nucleus satiety center to terminate feeding behavior.

Inflammatory mediators

Cytokines, which are secreted primarily by activated macrophages, are important mediators of inflammation. The most important inflammatory mediators are interleukin-1 (IL-1), interleukin-6 (IL-6), and tumor necrosis factor-α (TNF-α). Local effects include induction of adhesion molecule expression on vascular endothelium (promoting cell adherence), increased vascular permeability (promoting influx of serum components), and activation of lymphocytes. Systemic effects include increased leukocyte production, fever, and induction of an acute phase response. Other important cytokines are interleukin-8 (IL-8), a potent chemotactic factor for recruitment of neutrophils, basophils and T cells, and interleukin-12 (IL-12), which activates NK cells and promotes differentiation of T cells into T-helper (TH) cells.

Nonsteroidal antiinflammatory drugs

Nonsteroidal antiinflammatory drugs (NSAIDs), for example, aspirin and ibuprofen, inhibit the cyclooxygenase enzymes, COX-1 and COX-2. These enzymes catalyze the formation of prostaglandin H_2, which is the precursor for prostaglandin, prostacyclin, and thromboxane synthesis. Aspirin inhibits the cyclooxygenase enzymes by acetylating a single serine residue. This is an irreversible covalent modification that inactivates both COX-1 and COX-2. Other NSAIDs are competitive inhibitors of the cyclooxygenases. COX-1 maintains the normal lining of the stomach and is involved in kidney and platelet function. Inhibition of COX-1 is responsible for many of the side effects of NSAIDs. COX-2 is induced by inflammation. COX-2 inhibition is thought to lead to the analgesic, antipyretic, and antiinflammatory effects of aspirin and the other NSAIDs.

Water Balance

The anterior hypothalamus controls water balance by controlling the excretion of water by the kidney and creating the sensation of thirst.

— In water deficit, hypothalamic osmoreceptors are activated by the increased osmolality of extracellular fluid. Pressure-sensitive receptors in the great veins are also activated by the decrease in osmotic pressure. These stimulate the supraoptic and paraventricular nuclei of the hypothalamus to release antidiuretic hormone (ADH) from the posterior pituitary. ADH causes salt and water retention in the kidney (see Chapter 17).

— The activation of osmoreceptors in the subfornical organ, located in the diencephalon but outside the blood−brain barrier, causes the release of central angiotensin II. Angiotensin II activates the lateral nucleus of the hypothalamus, which initiates thirst.

Table 4.5 summarizes the functions of the hypothalamus and the effects of lesions in certain regions.

Table 4.5 ▶ Summary of Hypothalamic Functions		
Region or nucleus	**Function**	**Lesion**
Anterior preoptic region	Maintains constant body temperature	Central hypothermia
Posterior region	Responds to temperature changes (e.g., sweating)	Hypothermia
Midanterior and posterior regions	Activate sympathetic nervous system	Lack of sympathetic outflow
Paraventricular and anterior regions	Activate parasympathetic nervous system	Lack of parasympathetic outflow
Supraoptic and paraventricular nuclei	Regulate water balance by stimulation or inhibition of ADH release from the posterior pituitary	Diabetes insipidus
Lateral and ventromedial nuclei	Regulate appetite, food intake, and satiety	Hypothalamic obesity syndrome Anorexia and emaciation
Lateral nuclei	Regulate thirst by stimulation or inhibition of angiotensin II release from the subfornical organ in the diencephalon	Lack of thirst response, resulting in hyponatremia (low Na+)
Abbreviation: ADH, antidiuretic hormone.		

Febrile convulsions

Febrile convulsions are seizures associated with elevated body temperature. They are the most common type of seizure in children, affecting 2 to 5% between the ages of 6 months and 5 years, with the peak incidence at 18 months. These seizures are not associated with trauma, infection, metabolic disturbances, or history of seizures, and most last < 10 minutes. More serious illnesses must be ruled out, but the treatment of simple febrile seizures with anticonvulsants is generally not recommended, as the potential drug toxicities associated with these medications outweigh the relatively minor risks associated with the convulsion.

Malignant hyperthermia

Malignant hyperthermia is a rare complication of anesthesia with any volatile anesthetic, particularly halothane. The anesthetic produces a substantial increase in skeletal muscle oxidative metabolism, which consumes oxygen (O_2) and causes a buildup of carbon dioxide (CO_2). The body also loses its capacity to regulate temperature, which rises rapidly (e.g., 1°C [1.8°F]/5 min). This can lead to circulatory collapse and death. Signs include muscular rigidity with accompanying acidosis, increased O_2 consumption, hypercapnia (increased CO_2), tachycardia (increased heart rate), and hyperthermia. Malignant hyperthermia may be treatable if dantrolene (a drug that reduces muscular contraction and the hypermetabolic state) is given promptly.

5 Sensory Systems

5.1 Sensory Receptors and Signal Transduction

Sensory Receptors

Sensory receptors are specialized cells or neurons that are dedicated to the task of transducing stimuli into nerve action potentials.

Classification of Receptors

Sensory receptors are classified as *mechanoreceptors, chemoreceptors, photoreceptors,* or *thermoreceptors,* depending on the type of energy to which they respond.

They can also be categorized according to the location from which the stimulus induces a response:
— *Exteroceptors* respond to stimuli originating outside the body.
— *Interoceptors* respond to stimuli from within the body (e.g., visceroceptors, equilibrium receptors of the inner ear, and proprioceptors that sense the positions and movement of muscles, bones, and joints).

Morphology of Receptors

Sensory receptors may be one of the following:
— Free nerve endings
— Nerve endings associated with connective tissue capsules
— Sensory endings synaptically linked to receptor cells. When these receptor cells are depolarized, synaptic vesicles containing neurotransmitter are released onto sensory afferent nerve endings.

The variety of skin receptors are shown in **Fig. 5.1**.

Fig. 5.1 ▶ **Skin receptors.**
The various sensory receptors in hair-bearing skin and in hairless skin are shown. Nociceptors, and receptors for heat and cold are free nerve endings. *Pain receptors*, or *nociceptors*, constitute about half of all receptors.
From *Thieme Atlas of Anatomy, Head and Neuroanatomy*, © Thieme 2007, Illustration by Markus Voll.

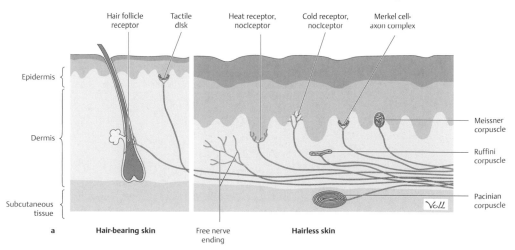

Electrophysiology of Receptors and Signal Transduction

Application of a stimulus to a sensory nerve ending produces an electronic potential across the cell membrane known as a *generator potential*. In receptor cells, the electronic potential resulting from a stimulus is termed a *receptor potential*.

Receptor and generator potentials are generally depolarizations caused by ion influx into the nerve ending or receptor cell.

Note: An important exception is light acting on the photoreceptors in the retina, leading to a decreased influx of ions and hyperpolarization.

Electronic potentials are graded; that is, as the intensity of the applied stimulus is increased, the magnitude of the potential change increases concomitantly.

— In the case of generator potentials, when threshold is reached, an action potential is propagated along the afferent nerve. The trigger point for the action potential is the first node of Ranvier.

— In the case of receptor potentials, the quantity of synaptic transmitter released by receptor cells is proportional to the magnitude of the receptor potential. If threshold is reached, an action potential is propagated along the afferent nerve.

Generator and receptor potentials can exhibit temporal summation in response to repetitive stimuli.

Coding of Stimulus Intensity

Information about the intensity of an applied stimulus detected by a sensory receptor is carried to the central nervous system (CNS) by the afferent nerve fiber as a *frequency code*. The magnitude of the stimulus is coded by the firing frequency of the nerve fiber (spikes per second). This, in turn, determines the frequency of release of neurotransmitter at the synapse (**Fig. 5.2**).

The number of afferent fibers recruited by the applied stimulus is another indicator of stimulus intensity.

Adaptation

Adaptation is the term used to describe the steadily reduced rate of firing of some afferent nerves associated with sensory receptors that occurs when a constant stimulus is applied to the sensory receptor for a prolonged period. The psychological perception of the intensity of the stimulus is concurrently reduced.

Fig. 5.2 ▸ **Stimulus processing and information coding.**
The original stimulus is coded by frequency for transmission along nerve fibers and decoded by bursts of neurotransmitters from presynaptic nerve terminals. Such neurotransmitters initiate action potentials (APs) in the next neuron that imitates the original stimulus.

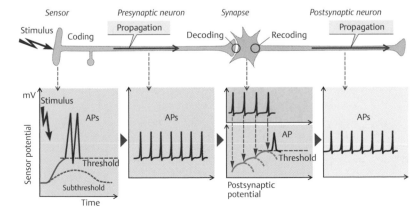

Table 5.1 ▶ Adaptation of Somatosensory Receptors		
Receptor	**Sensation Detected**	**Adaptation**
Pacinian corpuscles	Vibration	Very rapid
Hair follicles	Touch	Rapid
Meissner corpuscles	Touch	Rapid
Merkel disks	Pressure	Slow
Ruffini endings	Stretch	Slow

Slowly adapting (tonic) receptors
— The afferent nerves associated with these receptors will continue to fire action potentials at the same rate when a constant stimulus is applied.
— Slowly adapting receptors signal the magnitude of a stimulus.

Rapidly adapting (phasic) receptors
— The afferent nerves associated with these receptors will rapidly reduce their rate of firing of action potentials when a constant stimulus is applied.
— Rapidly adapting receptors signal onset and offset of a stimulus.

The adaptation rates of somatosensory receptors are listed in **Table 5.1**.

5.2 Somatosensory System

Somatosensory Sensations

Somatosensory sensations are grouped into two broad divisions:
— The *discriminative touch system* is responsible for touch, pressure, and vibration sensibility. The mechanoreceptors for this system are included in **Table 5.1**
— The *pain and temperature system*

Sensory information from each of these systems is conveyed to the cortex by different somatosensory tracts.

Somatosensory Tracts

Sensory afferent nerves travel to the spinal cord in the posterior (dorsal) roots.

Dorsal Column System

This system conveys discriminative touch information in the spinal cord to higher centers. It also carries information concerning position sense from proprioceptors at skeletal joints.
— The sensory afferent nerves of this system are myelinated Aβ fibers.

Course of the dorsal column system
— Primary afferent neurons (first-order neurons) ascend in the spinal cord in the dorsal (posterior) columns without synapsing. They synapse in the dorsal column nuclei (nucleus gracilis and nucleus cuneatus) of the medulla.
— The second-order neurons decussate (cross the midline) and pass to the relay nuclei of the thalamus (ventral posterolateral [VPL]) via the medial lemniscus. The information is arranged topographically in the thalamus.

▪ Two-point discrimination

Two-point discrimination is the ability to discriminate one mechanical stimulus touching the skin from two stimuli. Sensitivity varies from one region of the body surface to another. The thresholds for discriminating two stimuli are lower on the tongue, fingers, and lips than they are at other locations, such as the back. This is correlated with the greater density of sensory receptors and, hence, their smaller receptive fields in these sensitive areas.

▪ Dermatomes

The strip of skin innervated by the posterior (dorsal) root of a single spinal segment is known as a dermatome. A dermatomal map is important for localizing injuries to peripheral sensory nerves.

— Thalamocortical projections (third-order neurons) carry the information to the postcentral gyrus of the parietal lobe.

Course of sensation from the head and face

— Sensation arising from tactile stimulation of the face and head is carried by the trigeminal nerve to the principal sensory nucleus of the trigeminal in the pons.
— Second-order fibers decussate and ascend via the trigeminothalamic tract to the ventral posteromedial (VPM) nucleus of the thalamus.
— Thalamocortical projections relay the information to the face area of the postcentral gyrus of the parietal lobe.

Anterolateral System

This system conveys pain and temperature information in the spinal cord to higher centers.
— The spinothalamic pathway of the anterolateral system is responsible for the discriminative aspects of pain perception, such as localization and magnitude estimation.
— The spinoreticular pathway of the anterolateral system is responsible for the arousal, autonomic, and emotional responses to pain.
— The sensory afferents of this system are myelinated Aδ and unmyelinated C fibers.

Course of the anterolateral system

— Primary afferent neurons (first-order neurons) enter into the dorsal horn of the spinal cord gray matter and synapse with interneurons and projection neurons of the substantia gelatinosa.
— Second-order neurons decussate in the spinal cord and ascend in the anterolateral tract to terminate in the VPL nucleus of the thalamus or in the brainstem reticular formation.
— Third-order neurons project from the VPL nucleus to the somatosensory cortex of the parietal lobe. The reticular formation projects diffusely to the forebrain and brainstem areas, which are believed to be involved in the affective responses to pain (amygdala and anterior cingulate gyrus).

Course of pain sensations from the head and face

— Pain originating from the head and face is carried first by pain afferent fibers in the trigeminal nerve that synapse in the descending (spinal) nucleus of the trigeminal nerve.
— Second-order fibers decussate and ascend to the thalamus and reticular formation.

Table 5.2 summarizes the features of dorsal column and anterolateral systems. Their course is illustrated in **Fig. 5.3**.

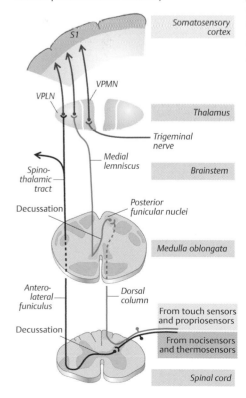

Fig. 5.3 ▶ **Somatosensory tracts.** (VPLN, ventral posterolateral nucleus; VPMN, ventral posteromedial nucleus)

Table 5.2 ▶ Features of the Dorsal Column and Anterolateral System				
Somatosensory Tract	**Sensations Projected**	**Types of Afferent Fibers**	**Course**	**Characteristics**
Dorsal column system	Fine touch Pressure Two-point discrimination Vibration Proprioception	Aα and Aβ	Crosses in medulla	Preservation of modality specificity Precise mapping of body surface carried through all relays and onto the cortical surface High synaptic security
Anterolateral system	Temperature Pain	Aδ and C	Crosses in the spinal cord	Imprecise mapping of body surface Cross-modality (e.g., cutaneous and muscle) convergence Low synaptic security

Primary Sensory Cortex

The primary sensory cortex (S1) is located on the postcentral gyrus of the parietal lobe. The body surface is mapped onto the cortical surface in a strict, somatotopic fashion in the form of a *sensory homunculus*. The map reflects the density of skin innervation rather than the actual area, with greater cortical area devoted to the index finger and lips than to the back (**Figs. 5.4** and **5.5**).

Afferent Input to the Primary Sensory Cortex

— Somatic sensory area 2 (S2) adjoins the lower posterior portion of area 1 on the parietal cortex. It receives information from both sides of the body with low synaptic security. It is particularly sensitive to direction of stimulus movement on the body surface.

— The secondary somatosensory area is located on the parietal cortex (areas 5 and 7) directly posterior to the postcentral gyrus. It mediates stereognosis (the ability to determine form via touch) and complex perceptions (e.g., spatial orientation). It forms associations between the somatic, auditory, and visual systems.

Fig. 5.4 ► **Sensory centers of the brain.**
The contralateral side of the body is represented on somatic sensory area 1 (S1) of the parietal cortex. S2 receives sensory information from both sides of the body. The cortical areas that process hearing and vision are also shown.

Fig. 5.5 ► **Somatotopic organization of S1.**
S1 has somatotopic organization, meaning that each part of the body is represented at a particular cortical area. The size of the cortical area devoted to each part is determined by the density of its sensory receptors.

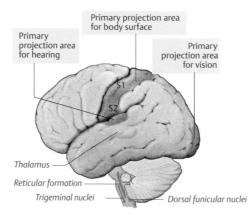

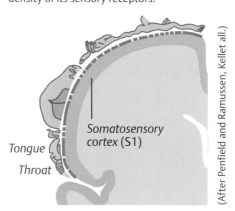

(After Penfield and Ramussen, Kellet all.)

5.3 Pain

Pain is an unpleasant sensory and emotional experience associated with actual or potential tissue damage or perception of such damage. It serves a protective function in signaling tissue injury or processes that might damage tissue. The responses to pain may be autonomic, psychological, emotional, or behavioral.

Pain Receptors

— Morphologically, pain receptors, or *nociceptors*, are simple free nerve endings in the skin or other tissues.

— They respond specifically to intense mechanical, thermal (hot or cold), or chemical stimulation. Some nociceptors are unimodal, responding to only one energy form. Others are polymodal and are activated by two types of stimulus energy.

Afferent Pain Fibers

— Myelinated Aδ fibers rapidly convey "first" pain, characterized as short, sharp pain that is easily localized. This type of pain usually originates in the skin.
— Unmyelinated C fibers slowly convey "second" pain, or diffuse, aching, burning, or throbbing pain that is poorly localized. This type of pain may originate in skin, muscles, joints, or viscera.
— The main transmitter released by the afferent pain fibers is glutamate. Many fibers also release peptide transmitters, particularly substance P.

Sensitization of Afferent Pain Fibers

Tissue injury or inflammation can lower the threshold for activation of nociceptor afferents. Inflammatory mediators such as bradykinin, prostaglandins, cytokines, and histamine released by leukocytes, platelets, and mast cells, as well as adenosine triphosphate (ATP) and H^+ ions released by damaged cells, act on the nociceptor terminals to increase firing rates and lower the threshold for activation.
— *Allodynia,* painful perception produced by normally innocuous stimuli, is one manifestation of sensitization. An example of this phenomenon is the temporary sensitization of the skin to light touch following sunburn.
— *Hyperalgesia* refers to the increased intensity of the pain perception resulting from sensitization.

Referred Pain

Referred pain occurs because somatic (relating to the body) and visceral (relating to the internal organs) afferent fibers converge onto the same projection neuron of the dorsal horn. Referred pain originates from the viscera but is perceived as if coming from an overlying or nearby somatic structure within the same dermatome (**Fig. 5.6**).

Fig. 5.6 ► **Mechanism of referred pain.**
The convergence of somatic and visceral fibers at the same relay neuron confuses the relationship between perceived and actual sites of pain. The pain is typically perceived at the somatic site, as somatic pain is well localized, whereas visceral pain is not.
From *Thieme Atlas of Anatomy, Head and Neuroanatomy,* © Thieme 2007, Illustration by Karl Wesker.

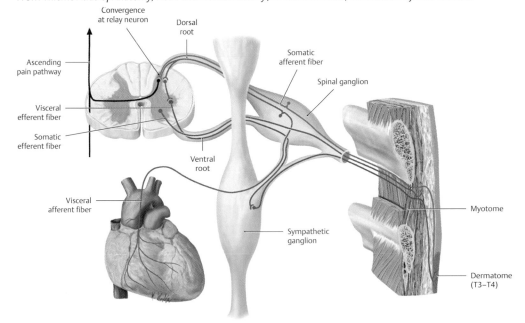

Convergence at relay neuron
Dorsal root
Ascending pain pathway
Somatic afferent fiber
Spinal ganglion
Visceral efferent fiber
Somatic efferent fiber
Ventral root
Visceral afferent fiber
Sympathetic ganglion
Myotome
Dermatome (T3–T4)

Inhibition of Pain

— Pain responses can be suppressed by descending pathways that synaptically inhibit the neurons of the ascending pain pathways. The periaqueductal gray matter (PAG) of the midbrain gathers information from higher centers and uses an indirect pathway to the dorsal horn that synaptically inhibits projection neurons.
— The neurons of the dorsal horn of the spinal cord, PAG, and amygdala are rich in opiate receptors. Stimulation of these receptor molecules by morphine induces presynaptic inhibition or postsynaptic hyperpolarization and results in analgesia.

5.4 Vision

The structure of the eye is shown in **Fig. 5.7**.

Fig. 5.7 ▶ **Eye structure.**
Sagittal section of the right eye viewed from the left.

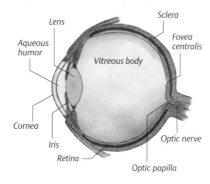

Physical Optics

Parallel rays coming from a point source of light at a distance > 20 feet are refracted (bent) by a convex glass lens so that the image is focused at a point behind the lens called the *principal focus*. The field of vision is inverted, and left to right is reversed on the retina.
— The *focal length* is the distance in meters from the lens to the principal focus.
— *The refractive power of a lens* is calculated in diopters (D):

$$D = 1/\text{focal length}$$

If the object is brought closer to the lens, the light rays are focused at progressively greater distances behind the principal focus.

Physiological Optics

When the ciliary muscle is relaxed in the normal (emmetropic) eye, parallel light rays are focused on the retina; that is, the retina lies at the principal focus of the eye. The diopteric power of the eye is ~60 diopters. Most of that refractive power is the result of light refraction at the interface of the air and the cornea. The remainder is contributed by the lens of the eye.

■ Primary open-angle glaucoma

Glaucoma refers to a group of eye diseases that cause damage to the optic nerve. Primary open-angle glaucoma is the most common form of glaucoma. In this case, drainage of the aqueous humor is prevented due to blockage of the drainage channels between the cornea and the iris. The buildup of aqueous humor results in raised intraocular pressure and subsequent damage to the optic nerve. Symptoms of this condition include gradual loss of peripheral vision that progresses to tunnel vision. Drug treatment is aimed at reducing intraocular pressure by decreasing the production of aqueous humor and/or increasing the drainage of the aqueous humor.

■ Adaptation to changes in light intensity

The eye adapts to changes in light intensity through pupillary aperture variation and by changes in retinal photoreceptor cells. Pupillary aperture variation alters the amount of light reaching the retina over a 30-fold range. It is mediated by the pupillary reflex with increased parasympathetic outflow to the iris constrictor muscle causing the pupil to constrict (miosis). Increased sympathetic outflow to the iris dilator muscle causes the pupil to dilate (mydriasis). With high levels of illumination, rods gradually become less sensitive to light. This is why it takes several minutes to adapt when entering a darkened theater. They are inactivated at high light intensities.

Accommodation

Accommodation allows the eye to focus on objects closer than 20 feet by increasing the curvature of the lens, thereby shortening the focal length. Lens curvature is controlled by the ciliary muscles, which receive parasympathetic innervation (fibers are derived from cranial nerve [CN] III). Muscle contraction counteracts a continuous stretch on the lens from the supporting ligaments (zonular fibers), allowing the lens to bulge more spherically for near vision (**Fig. 5.8**).

Fig. 5.8 ▶ Accommodation.
During accommodation for near vision, the ciliary muscles contract, allowing zonular fibers to relax and the lens to bulge. During accommodation for far vision, the zonular fibers are stretched, and the lens becomes flatter.

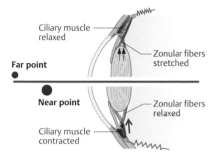

Optical Defects

Optical defects occur due to refractive errors related to the curvature of the cornea or loss of accommodation. **Table 5.3** summarizes optical defects, which are also illustrated in **Figs. 5.9** and **5.10**.

Table 5.3 ▶ Optical Defects			
Defect	Mechanism for Defect	Result	Method of Correction
Myopia	Far images (> 20 ft) focus in front of the retina	Nearsightedness	Use of a biconcave (negative) lens
Hyperopia	Near images focus behind the retina	Farsightedness	Use of a convex (positive) lens
Presbyopia	With aging, the lens gradually loses its elasticity, allowing for less accommodation.	Inability to focus on close objects Distant vision is maintained	Use of a convex (positive) lens for near vision
Astigmatism	The curvature of the cornea or lens is asymmetric	The refractive power of the eye varies in different planes parallel to the optical axis. This causes different parts of the images to be focused at different focal lengths	Use of a cylindrical lens

Amblyopia

Amblyopia (lazy eye), common in preschool children, causes them to squint in an attempt to see more clearly. When they accommodate, their convergence is faulty, so they see two images (diplopia). The stronger eye suppresses the image from the weaker eye. If this is not corrected by surgery on the extraocular muscle and/or patching the dominant eye, the patient will develop cortical blindness in the weaker eye.

Nystagmus

Nystagmus is a slow movement of the eyes in a particular direction (usually laterally) alternating with a quick recovery movement in the opposite direction. It is an alternation of a smooth pursuit movement and a saccade.
— Optokinetic nystagmus is a following reflex of the eyes to movement of the visual field. If the visual field moves to the right, then the eyes follow the field to the right (slow movement) and then recover to the left (fast movement).

Types of eye movements

The following are normal eye movements:
— Saccades are voluntary rapid flicks, or fast movements to shift the gaze.
— Smooth pursuit movements track or follow visual stimuli.
— Vestibular movements compensate for change in head orientation with body movements.
— Microtremors (microsaccades) are constant tiny involuntary movements during fixation, shifting the target image to unadapted receptors. If the target image were not shifted, it would fade by adaptation of receptors.

Atropine and the eye

Atropine is a competitive antagonist of acetylcholine (ACh) at muscarinic receptors. It causes sedation (by blocking M_1 receptors), tachycardia and mild vasodilation (by blocking M_2 receptors), and decreased gastrointestinal (GI) motility, urinary retention, and cycloplegia (paralysis of the ciliary muscle of the eye) with mydriasis (by blocking M_3 receptors). Many of these effects are unwanted and are due to the low organ specificity of muscarinic antagonist drugs. Atropine analogues (homatropine or tropicamide) are used as ophthalamic solutions to produce mydriasis to allow retinal examination. Atropine is also used in the treatment of urinary incontinence, as a pre-anesthetic agent (for autonomic stability), to treat bradycardia in emergency situations, and as an antidote for cholinesterase inhibitor poisoning (e.g., insecticides).

The macula

The macula (fovea centralis) is the central region of the retina. It is extremely sensitive due to the following characteristics: it has a high concentration of cones that are closely packed; the overlying retinal layers diverge, permitting direct access of light to outer segments of the cones; and blood vessels detour around the fovea in order not to obscure reception.

Fig. 5.9 ▶ **Presbyopia, myopia, and hyperopia.**
In presbyopia, the lens becomes less flexible, and the eye cannot focus on near objects. This can be corrected with a convex lens, but distant vision does not require correction. In myopia, the image is focused in front of the retina; this can be corrected with a biconcave lens. In hyperopia, the image is focused behind the retina; this can be corrected with a convex lens.

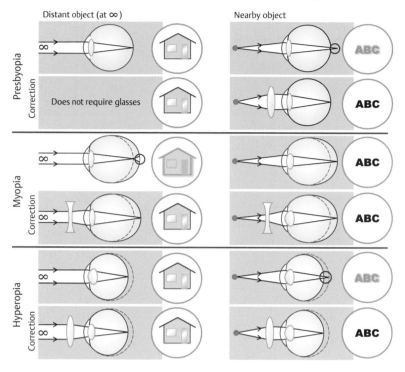

Fig. 5.10 ▶ **Astigmatism.**
In regular astigmatism, the curvature of the lens is asymmetrical in the horizontal and vertical axes. This abnormality can be corrected by a cylindrical lens.

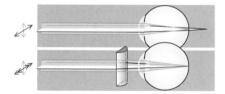

The Retina

Retinal Circuit

The retinal circuit consists of three synaptically connected neurons: photoreceptor cells (rods or cones), bipolar cells, and ganglion cells (**Fig. 5.11**). Horizontal cells and amacrine cells are interneurons that link adjacent circuits for generating surround inhibition. All retinal cells, other than ganglion cells, are small neurons that do not transmit action potentials. Ganglion cells, whose axons form the optic nerve, are the only retinal cells that transmit action potentials.

Photoreceptor Cells: Rods and Cones. The eye contains 6 million cones that are clustered in the central region of the retina, known as the *fovea centralis* or *macula*, and 120 million rods that are mainly in the peripheral retina (**Fig. 5.12**).

Fig. 5.11 ▶ **Layers of the retina.**

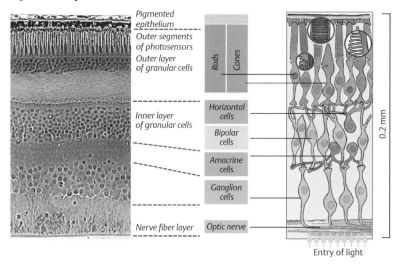

Fig. 5.12 ▶ **Photoreceptor cells.**
Light-absorbing visual pigments and a variety of enzymes and transmitters in retinal rods and cones (**1**) mediate the conversion of light stimuli into electrical stimuli (phototransduction). The membranous disks of the retinal rods contain rhodopsin (**2**), a photosenstitive purple-red chromoprotein. Rhodopsin consists of the integral membrane protein opsin and 11-*cis*-retinal. Photic stimuli trigger a primary photochemical reaction in rhodopsin in which 11-*cis*-retinal is converted to all-*trans*-retinal (**3**). See also **Fig. 5.13**.

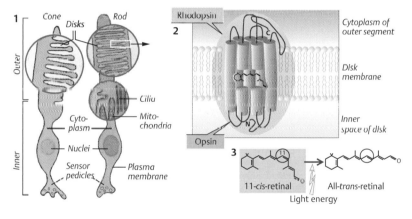

— Rods contain the light-absorbing pigment rhodopsin. They are responsible for vision at low levels of illumination and are not color sensitive. Black and white vision mediated by rods is called *scotopic* (or night) vision.
— Cones are responsible for color vision. There are three types of cones—red (long wave), green (medium wave), and blue (short wave), which each have different photopigments. Cones function at high levels of illumination and are responsible for *photopic* (or daytime) vision.
Photoreceptors in the dark have a membrane potential of about – 30 mV and tonically release the transmitter glutamate.

Phototransduction in Rods

The disks in the outer segment of rods are studded with molecules of rhodopsin. Rhodopsin consists of a protein, opsin, linked to a chromophore, 11-*cis*-retinal. The cytoplasmic end of the rhodopsin molecule is coupled to a G-protein complex, transducin.
— In the dark, plasma membrane Na⁺ channels of the rod cell are held in the open position by cyclic guanosine monophosphate (cGMP).

Fig. 5.13 ▶ **Process of phototransduction.**
cGMP, cyclic guanosine monophosphate; 5′ GMP, 5′ guanosine monophosphate.

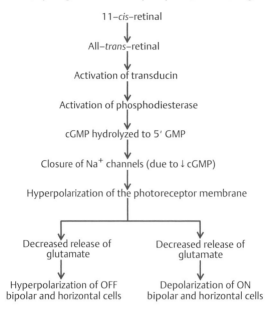

— A collision between a photon of light and a molecule of rhodopsin leads to the stereoisomerization (change in configuration) of 11-*cis*-retinal to all-*trans*-retinal (**Figs. 5.12** and **5.13**). This activates transducin, which in turn induces phosphodiesterase to become active. Phosphodiesterase breaks down the cGMP that is holding open the membrane Na^+ channels, and the channels shut.

— As a result, the influx of Na^+ ions is interrupted, and the membrane hyperpolarizes. This reduces the tonic release of glutamate at the synaptic ending of the photoreceptor.

— Each photoreceptor synapses with two types of bipolar cells: ON cells and OFF cells. Reduced release of glutamate by photoreceptors induces depolarization (disinhibition) of ON bipolar cells. Subsequently, a neurotransmitter is released from these bipolar cells, which causes the firing of ON ganglion cells (**Fig. 5.14**). A parallel network involving OFF bipolar cells causes inhibition of OFF ganglion cells.

Fig. 5.14 ▶ **Potentials of photoreceptor cells, ON bipolar and ON ganglion cells.**

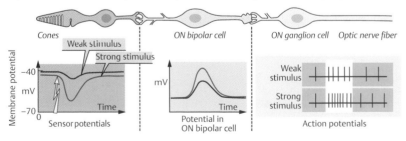

Receptive Fields of Ganglion Cells and Surround Inhibition

Horizontal and amacrine cells connect bipolar cells to surrounding retinal circuits, resulting in surround inhibition. Thus, ganglion cell receptive fields consist of a center with an antagonist surround (e.g., ON center, OFF surround or vice versa) (**Fig. 5.15**).

Fig. 5.15 ▶ **Receptive fields of ON ganglion cells and OFF ganglion cells.**
ON ganglion cells (**1,2**) increase their firing when the center of their receptive field is illuminated and slow their firing when light strikes their surrounding periphery. OFF ganglion cells (**3,4**) fire more slowly when their center is illuminated and fire more rapidly when light strikes their periphery.

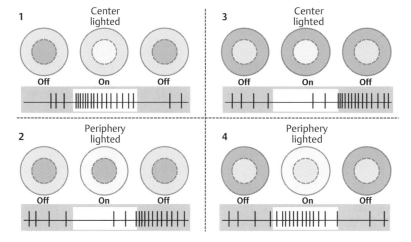

Optical Pathways

- The axons of ganglion cells form the optic nerve and optic tract. They convey information to the lateral geniculate body of the thalamus for processing.
- The left and right optic nerves undergo partial decussation at the optic chiasm, which is just anterior to the pituitary stalk. Optic nerve fibers arising from the nasal hemiretinas (central visual field) of each eye cross, whereas fibers of the temporal hemiretinas (nasal visual field) pass through the chiasm but remain ipsilateral.
- From the lateral geniculate, the information is relayed to the occipital cortex.

Lesions of Optical Pathways

Lesions of optical pathways will cause varying symptoms, depending on their location. The lettering in the following description refers to **Fig. 5.16**.
- Cutting the left optic nerve (**A**): This will cause blindness in the ipsilateral (left) eye.
- Cutting the optic chiasm (**B**): This will cause blindness in the temporal (nonnasal) halves of the visual fields of both eyes.
  Cutting the left optic tract (**C**): This will cause blindness in the right halves of the visual fields of both eyes.
- Cutting the left optic radiation (**D**): This will also cause blindness in the right halves of the visual fields of both eyes, but the foveal fibers remain intact so central vision is spared.

Visual Processing by the Cortex

Visual information is transformed to visual perceptions in the occipital lobe of the cortex. Foveal receptors occupy < 1% of total retinal area but project to ~50% of the primary visual cortex.

Fig. 5.16 ▶ Visual pathways and visual field deficits.
Cutting the left optic nerve causes blindness in the ipsilateral (left) eye **(A)**. Cutting the optic chasm causes blindness in the temporal (nonnasal) halves of the visual fields of both eyes. **(B).** Cutting the left optic track causes blindness in the right halves of the visual fields of both eyes. **(C).** Cutting the left optic radiation will also causes blindness in the right halves of the visual fields of both eyes, but the foveal fibers remain intact so central vision is spared **(D).**

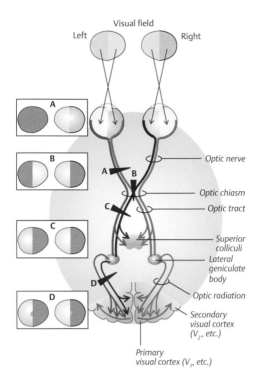

Receptive Fields of Cells in the Visual Cortex

There are three types of cells in the visual cortex. Each of these cells has a different receptive field pattern that causes them to respond preferentially to different types of visual stimuli:

— *Simple cells* have elliptically shaped receptive fields. An excitatory region may lie alongside or within the long axes of the ellipse parallel to an inhibitory region. Maximal activation occurs with stationary stimuli at the proper angle or orientation.
— *Complex cells* respond best to an edge stimulus in the receptive field with the edge at the preferred orientation, usually perpendicular to the long axis of the receptive field. Moving stimuli also have a preferred direction for optimal activation.
— *Hypercomplex cells* have a central excitatory field flanked by one or two inhibitory fields. They respond best to bars of optimum length moving in one direction along a preferred orientation.

This hierarchy may develop in a cascade of inputs from simple to more complex cells.

5.5 Hearing (Audition)

Physics of Sound

The sensation of sound is produced by vibrating air molecules striking the ear. The simplest sound waves are sinusoidal pressure waves with bands of condensation and rarification. They have defined amplitudes (measured in decibels [dB]) and defined frequencies (measured in cycles/sec [Hz]). The greater the amplitude of a sound wave, the greater the loudness. The greater the frequency, the higher the pitch.

Structure of the Ear

Outer Ear

— Sound waves are channeled through the air-filled external auditory canal and cause the tympanic membrane (eardrum) to oscillate.

Middle Ear

- The middle ear contains the three auditory ossicles (bones): the malleus, incus, and stapes (**Fig. 5.17**).
- The *malleus* inserts into the tympanic membrane.
- The *incus* connects the malleus to the stapes.
- The *stapes* inserts into the membrane covering the oval window of the cochlea, thereby conducting sound vibrations into the inner ear.

Inner Ear

- The inner ear, or *cochlea,* is a bony tubular structure wound into a spiral of 2.5 turns (**Fig. 5.17**). Within the bony tube is an elongated, triangular membranous tube. The base of the triangle is the basilar membrane where the sensory receptor cells, of the ear, the hair cells, reside.
- These hair cells are innervated by afferent fibers of the auditory division of the vestibulocochlear nerve (CN VIII).

Fig. 5.17 ▶ **Structures of the middle and inner ear and sound conduction.**
Sound waves are transmitted to the organ of hearing via the external ear and the auditory canal, which terminates at the tympanic membrane (1). Vibration of the tympanic membrane is conducted by the ossicular chain (malleus, incus, and stapes) to the membrane of the oval window, where the inner ear begins. The inner ear consists of the vestibular organ and the cochlea. Inside the cochlea is an endolymph-filled duct called the scala media. The scala media is accompanied on either side by two perilymph-filled cavities: the scala vestibuli and the scala tympani (2). These cavities merge at the apex of the cochlea to form the helicotrema. The scala vestibuli arises from the oval window, and the scala tympani terminates on the membrane of the round window. The outer and inner hair cells that both sit upon the basilar membrane are the sensory cells of the hearing organ (3).

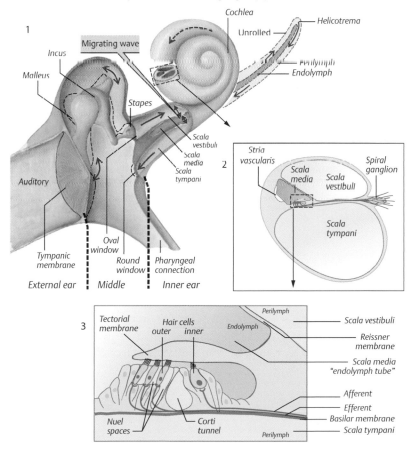

Mechanisms of damage protection in the ear

The linkage of ossicles is protected from damage due to loud sounds by an attenuation reflex mediated through an efferent inhibitory system. This reflexively contracts the tensor tympani and stapedius muscles when a loud sound is heard. However, the reflex has a 40 msec latency, so there is no protection for fast transient sounds of high intensity (e.g., a thunder clap). The eustachian tube is a pressure relief system that maintains equal pressure on both sides of the eardrum. Abrupt changes in atmospheric pressure cause a popping sensation due to air moving in or out of the eustachian tube.

Ototoxic drugs

Ototoxic drugs are those that are harmful to the auditory nerve, cochlea, and vestibular system. Examples of ototoxic drugs include gentamicin, streptomycin, aspirin (in high doses), and furosemide. Damage caused by these drugs may be temporary and will subside when the drug is stopped, or it may be permanent. Symptoms of ototoxicity may be tinnitus (ringing in the ears), hearing loss, and vertigo.

— The spiraling is unrelated to auditory perception but is a convenient configuration for supplying blood vessels and nerves to the cells of the inner ear.

— The space within the membranous tube is the scala media. It is filled with endolymph, which has a high K⁺ concentration.

— The space above the membranous tube is the scala vestibuli, and the space below the basilar membrane at the base of the tube is the scala tympani. They are filled with perilymph, which has a high Na⁺ concentration.

— The cell bodies of hair cells are fixed in the basilar membrane and extend into the tectorial membrane. The differences in elastic properties between these two membranes are important in auditory transduction.

Auditory Transduction

The steps involved in auditory transduction are as follows:

1. Sound waves cause the tympanic membrane to vibrate.
2. Vibration of the tympanic membrane causes the ossicles to vibrate, which pushes the stapes against the oval window, generating pressure waves in the cochlea.
3. Pressure waves in the cochlea cause the basilar membrane to vibrate.
4. Vibration of the basilar membrane causes a shearing action between it and the tectorial membrane. The cilia of the hair cells connected to these membranes are bent to one side or the other as the tectorial membrane moves across the basilar membrane. This stimulates the opening of ion channels, increasing K⁺ permeability and initiating a receptor potential. Hair cell depolarization causes Ca²⁺ entry, which releases the neurotransmitter glutamate (**Fig. 5.18**).

Fig. 5.18 ► **Stimulation of hair cells by membrane deformation.**
Vibration of the cochlea causes a discrete shearing of the tectorial membrane against the basilar membrane, causing bending of the stereocilia of the outer hair cells **(1)**. This bending causes mechanosensitive cation channels in the stereocilia membrane to open, allowing K⁺ to enter and depolarize the outer hair cells. This causes the outer hair cells to shorten. Repolarization is achieved by the opening of K⁺ channels on the perilymph side of the hair cell. The outflowing K⁺ is taken up by K⁺-Cl⁻ cotransporters in the supporting cells and recirculated via gap junctions to the stria vascularis. The shortening of the outer hair cells changes the amplitude of the traveling sound wave, which then additionally bends the stereocilia of the inner hair cells. Depolarization occurs in the same way as for outer hair cells and causes the opening of basolateral Ca²⁺ channels, increasing the cytostolic Ca²⁺ concentration. This leads to the release of the neurotransmitter glutamate and the subsequent conduction of impulses in afferent neurons to the central nervous system **(2,3)**.
APs, action potentials.

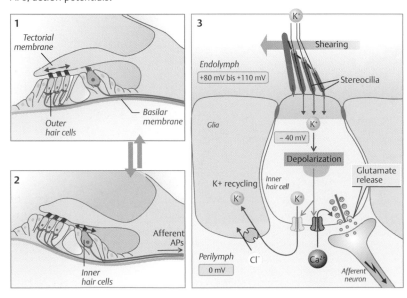

Note:

- The large size of the tympanic membrane compared with the oval window has the effect of amplifying the pressure input into the cochlea.
- Inner hair cells detect sounds; outer hair cells sharpen the frequency response of the system, improving clarity.
- High sound frequencies cause maximum displacement of the basilar membrane at the basal end of the cochlea near the oval window. Low frequencies cause maximum displacement at the apical end.
- There is tonotopic organization of frequencies at all levels of the central auditory pathway. This occurs as a result of the displacement of a particular region of the basilar membrane by a particular range of frequencies. Each region is innervated by particular neurons, which preserve the tonotopic map.

Auditory Pathways

- Auditory nerve fibers from the spinal ganglion terminate in the anterior and posterior cochlear nuclei.
- Secondary neurons cross, or project ipsilaterally, to the olivary nucleus. From there the neurons in the pathway continue to the lateral lemniscus, the medial geniculate, and finally the auditory cortex.
- Auditory signals use four to six neurons to project from the cochlea to the auditory cortex.
- Projections are mainly contralateral but also unilateral. Consequently, binaural hearing is still possible in the presence of unilateral lesions to higher relays or the auditory cortex.

Auditory Processing by the Cortex

The auditory cortex processes information from neural inputs:

- *Discriminate pitch (frequency):* the auditory system can discriminate 1 dB of sound pressure difference at low intensities and even lower sound pressure differences at higher intensities. A particular auditory neuron responds best to a limited range of frequencies. Surround and lateral inhibition produce sharper pitch discrimination.
- *Localize sound sources in space:* the input to auditory neurons is often inhibitory from one ear and excitatory from the other. The amount of lag time can be used for low frequencies, whereas intensity is used for high frequencies.
- *Recognize patterns of sounds:* neurons are more sensitive to changes of pitch and intensity at higher relay centers. They respond only to increasingly complex sound patterns as activity ascends toward the cortex.

5.6 The Vestibular System

Structure and Function of the Vestibular System (Fig. 5.19)

- The sensory ridges (ampullary crests) of three endolymph-filled semicircular canals, which are oriented in different planes, detect angular acceleration (e.g., nodding, tilting, and rotating of the head).
- The sensory maculae of the saccule and utricle in the petrous bone detect linear acceleration and changes in direction of gravity relative to the head.
- The sensory cells of both the ampullary crests and the maculae are hair cells. Two types of cilia, a long kinocilium and many shorter stereocilia, extend from these hair cells. These hair cells are spontaneously active. Movement of the cilia in response to movement of the endolymph (due to angular or linear acceleration) is either excitatory or inhibitory depending on direction.

Fig. 5.19 provides greater detail on how endolymph movement results in ciliary movement.

■ Hearing tests

Rinne test: With normal hearing, air conduction (as determined by a tuning fork held laterally to the external auditory meatus) is better than bone conduction (tuning fork placed on the mastoid process). Rinne-positive results (air conductance > bone conductance) occurs with normal ears and sensorineural (perceptive) deafness. Rinne-negative results (bone conductance > air conductance) are seen in conduction deafness.

Weber test: The foot of the tuning fork is placed in the middle of the patient's forehead, and the patient is asked in which ear the sound is heard. Sound localizes to the affected ear with conduction deafness, to the contralateral ear in sensorineural deafness, and is perceived as equally loud in both ears if both ears are normal.

■ Vertigo

Vertigo is the illusion of movement. It is most commonly caused by disorders of the inner ear such as Meniere disease (a syndrome characterized by vertigo, tinnitus, and deafness), vestibular neuronitis, lesions involving CN VIII, head injury causing vestibular damage, benign postural vertigo (vertigo occurs when certain positions are adopted or movements made), and drugs (e.g., gentamicin, barbiturates, and alcohol). Other causes of vertigo are migraine, epilepsy, multiple sclerosis, and tumors. Treatment depends on the cause, but anticholinergic drugs and antihistamines are often used to prevent nausea and vomiting.

Nystagmus is a slow movement of the eyes in a particular direction (usually laterally) alternating with a quick recovery movement in the opposite direction. It is an alternation of a smooth pursuit movement and a saccade.

—*Caloric nystagmus* is elicited by placing cool or warm water in the external ear. The resultant thermal convection currents stimulate the semicircular canals. Cool water produces nystagmus with slow recovery to the opposite side as the ear being tested. Warm water produces nystagmus with a quicker recovery to the same side as the ear being tested.

The integrity of the vestibular system is tested by inducing caloric nystagmus and looking for the results described, which demonstrate that the vestibular neural pathways are intact.

Fig. 5.19 ► **The vestibular system.**
(1) This system consists of three semicircular canals with sensory ridges (ampullary crests) and the saccule and utricle with sensory maculae. (2) Ampulla, dilated portions of the canals (shown in cross-section), contain connective-tissue ridges, the ampullary crests, each with sensory cells that are encased in a gelatinous cupula, which extends from the crest and attaches to the roof of the ampulla. Sensory cells on the crest extend into the cupula and bear on them one kinocilium and many shorter stereocilia. When the head is rotated in the plane of one of the canals, the lag of the endolymph causes a deflection of the cupula, which causes bending of the cilia. (3) The sensory cells are either hyperpolarized (inhibitory) or depolarized (excitatory) depending on the direction of ciliary motion. (4) The saccular and utricular maculae are areas of the epithelial lining that contain arrays of sensory cells. These cells also bear a kinocilium and stereocilia which project to an otolithic membrane. The latter consists of a gelatinous layer similar to the cupula but with otoliths, calcium carbonate crystals, embedded in it. The crystals exert traction on the gelatinous mass in response to linear acceleration, inducing shearing motions in the cilia with similar results to those found for the sensory cells on the ampullary crests (3). The afferent fibers that synapse with the sensory cells carry the messages to the vestibular ganglion.

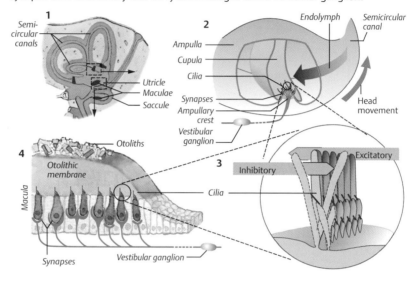

Vestibular Pathways

Afferent fibers from utricle, saccule, and semicircular canals synapse in vestibular nuclei in the brainstem; they also project to the cerebellum. The vestibular nuclei have major connections to the oculomotor system for stabilizing the visual image, to the neck muscles for stabilizing the head, and to the postural muscles for body balance.

5.7 Taste (Gustation)

Taste Receptor Cells

On the dorsal surface of the tongue, there are four types of lingual papillae: filiform, fungiform, vallate, and foliate. The latter three types have taste buds on their surface (**Fig. 5.20**). Taste receptor cells are located within taste buds. The microvilli at the apical surface of the receptor cells come in contact with taste molecules dissolved in saliva. On their basal surface, they synapse with gustatory afferent fibers.

Taste Qualities

The five primary taste qualities are as follows:
— *Salty taste:* evoked by cations such as Na^+
— *Sour (acid) taste:* evoked by H^+ ions
— *Bitter taste:* evoked by alkaloids

Fig. 5.20 ▶ **Structure of a taste bud.**
Nerves induce the formation of taste buds in the oral mucosa. Nerve fibers (from cranial nerves VII, IX, and X) grow into the oral mucosa at the basal side and induce the epithelium to differentiate into taste cells. These cells have microvilli that extend to the gustatory pore.

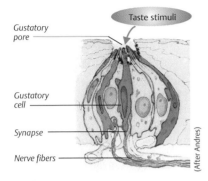

— *Sweet taste:* evoked by certain organic compounds (sugars and amino acids)
— *Umami (meaty) taste:* evoked by amino acids

It is believed that every taste can be duplicated by proper mixing of the five primary taste qualities.

Taste Transduction

— Salty taste results from the potential changes that follow entry of Na^+ through Na^+ channels on the microvilli of taste receptor cells. Sour taste also results from the direct effects of protons on membrane channels.
— Sweet taste is the result of activation of second messengers by a G protein (gustducin) coupled to a protein receptor in the membrane. Bitter taste likewise uses a second messenger cascade.
— The mechanism for tasting umami is unknown.

Membrane potential changes on taste receptor cells result in the release of transmitter and subsequent generation of action potentials in taste afferents.

Taste Pathways

Three cranial nerves carry taste information to the CNS:
— The *chorda tympani of the facial nerve (CN VII)* innervates taste buds on the anterior two thirds of the tongue.
— The *glossopharyngeal nerve (CN IX)* innervates taste buds on the posterior one third of the tongue.
— The *vagus nerve (CN X)* innervates scattered taste buds of the oropharynx and epiglottis.

Taste information from these cranial nerves is gathered at the solitary tract nucleus of the medulla. From there information flows to the ventral posteromedial nucleus of the thalamus and subsequently to the face area of the postcentral gyrus. Other targets are the hypothalamus, amygdala, and insular cortex (**Fig. 5.21**).

Taste Processing in the Central Nervous System

Each taste afferent can respond to more than one of the primary taste qualities. Although individual fibers respond preferentially to one or two of these qualities, gustatory coding is not a strictly labeled line system. Instead, the CNS must analyze taste stimuli based on a cross-fiber pattern code.

Fig. 5.21 ▶ **Gustatory pathways.**

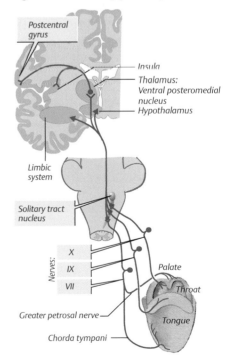

5.8 Olfaction

Olfactory Receptor Neurons

— Olfactory receptor neurons (ORNs) are bipolar neurons located in the olfactory epithelium. In contrast to taste receptor cells, these cells are themselves neurons whose axons extend into the CNS.
— The basal cells of the olfactory epithelium are adult stem cells that continuously divide and differentiate into ORNs. Because the olfactory epithelium is exposed to harmful gases and airborne pathogens, resulting in cell death, there must be continual cell turnover in this epithelium.

Olfactory Transduction

Odorant molecules bind to ORNs on 1 of roughly 400 different sensory cell types. The activated receptor stimulates G_s proteins, causing an increase in cyclic adenosine monophosphate (cAMP) and opening of Na^+ or Ca^{2+} channels. This, in turn, causes membrane depolarization (**Fig. 5.22**).

Olfactory Pathways

Axons of the olfactory sensor cells penetrate the bony cribriform plate in the nose to synapse on mitral and bristle cells and inhibitory interneurons in the olfactory bulb. The olfactory bulb also receives descending efferent input from the olfactory cortex. Axons from mitral and bristle cells project as the olfactory tract to the prepiriform cortex directly or via the thalamus and to the amygdala and hypothalamus.

Fig. 5.22 ► **Transduction of olfactory stimuli.**
(ATP, adenosine triphosphate; cAMP, cyclic adenosine monophosphate; GTP, guanosine triphosphate; ICF, intracellular fluid)

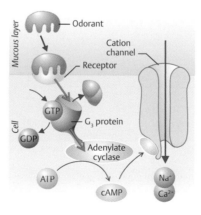

6 Motor Systems

6.1 Organization of the Motor System

The system for control of movement is hierarchically organized. At the lowest level, the spinal cord contains motoneurons and elemental reflex circuits. Superimposed on these basic circuits are descending tracts arising from the brainstem, which control basic postural functions, and from the motor cortex, which control more skillful movements.

The cerebellum is responsible for coordinating motor activity, and the basal ganglia are responsible for initiating motor programs.

The Motoneuron and the Motor Unit

- Skeletal muscle fibers, termed *extrafusal fibers*, form the bulk of muscle and generate its force and contraction.
- Skeletal muscle is innervated by α motoneurons, whose cell bodies are situated in the ventral horn of the spinal cord gray matter and in cranial nerve motor nuclei in the brainstem.
- The *motor unit* is the functional unit of motor control. It is composed of a single α motoneuron and the skeletal muscle fibers it innervates.
- The number of muscle fibers innervated by an α motoneuron ranges from about three in the ocular muscles to several hundred in some postural muscles.
- Small motor units are employed where fine movements are necessary. The cell bodies of motoneurons of small motor units are smaller in diameter than those of large motor units.

The Size Principle

The force of muscle contraction is graded primarily by the number of motor units activated. The size principle states that small diameter motoneurons are more excitable than large motoneurons. Therefore, as the excitatory input to a pool of motoneurons increases, the smaller motor units generating less tension are recruited first. Later the larger motor units, generating greater muscle tension, are brought into play.

Muscle Sensory Receptors

Muscle Spindles

Muscle spindles, termed *intrafusal fibers*, are stretch receptors embedded within muscles to monitor muscle length and its rate of change. They consist of about a dozen modified skeletal muscle fibers surrounded by a capsule (**Fig. 6.1**) innervated by afferent nerve fibers.
- Intrafusal fibers do not contain actin and myosin except near the polar ends of the spindle.
- Contraction of intrafusal fibers does not contribute to the force exerted by the mass of the muscle.

Types of intrafusal muscle fibers.   There are two types of intrafusal muscle fibers: *nuclear bag fibers* and *nuclear chain fibers*. Nuclear bag fibers bulge in the middle due to a cluster of cellular nuclei in the central part of the fiber. Nuclear chain fibers have nuclei linearly arranged in the central region.

Muscle spindle afferent output.   Intrafusal fibers are innervated by two types of afferent nerve fibers: group Ia and group II.
- Group Ia fibers (annulospiral endings) innervate both nuclear bag and nuclear chain fibers. They are dynamic, rapidly adapting sensory afferents and as such are sensitive to the rate of change in length of the muscle.
- Group II endings innervate nuclear chain fibers. They are slowly adapting sensory afferents that signal muscle length rather than the rate of change of length.

Fig. 6.1 ▶ Muscle spindles and Golgi tendon organs.
Muscle spindles are sensors that function to regulate muscle length. Group Ia afferent neurons coil around the nuclear chain and nuclear bag fibers (intrafusal fibers), whereas group II coil around nuclear chain fibers only. These annulospiral endings detect longitudinal stretching of intrafusal fibers and transmit information about their length (II) and rate of change in length (Ia) to the spinal cord. Efferent γ motoneurons innervate both types of intrafusal fibers, allowing variation of their length and stretch sensitivity.

Golgi tendon organs are sensors found in tendons that function to regulate muscle tension. Group Ib afferent fibers that originate in Golgi tendon organs transmit information regarding muscle tension to inhibitory interneurons in the spinal cord. These interneurons inhibit α motoneurons of the muscle from which the type Ib afferent impulse originated. They also activate antagonistic muscles via excitatory mechanisms. These factors combine to adjust muscle tension.
From *Thieme Atlas of Anatomy, Head and Neuroanatomy*, © Thieme 2007, Illustration by Markus Voll.

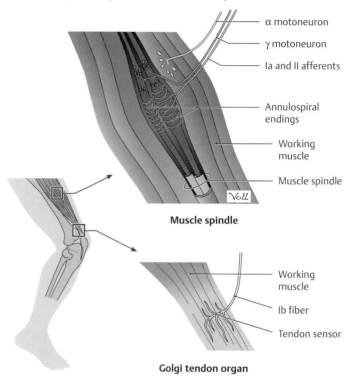

Muscle spindle

Golgi tendon organ

Muscle spindle efferent input. Gamma (γ) motoneurons are also present in the ventral horn and innervate only the polar ends of the spindle fibers, which can contract and thereby stretch the central regions of the intrafusal fibers.

— Gamma motoneurons increase the sensitivity of the spindle to overall muscle stretch.
— Gamma motoneurons also maintain the sensory function of the spindle when the muscle shortens during contraction. Shortening would collapse the spindle, terminate firing on the spindle afferents, and deprive the central nervous system (CNS) of information concerning muscle length. Therefore, there must be *coactivation* of α and γ motoneurons during voluntary movement.

Golgi Tendon Organs

Golgi tendon organs are sensory receptors located in muscle tendons that are activated by muscle tension (**Fig. 6.1**).

Golgi tendon organs are oriented in series with skeletal extrafusal muscle fibers while muscle spindles are in parallel. This means that although muscle lengthening will activate both receptors, muscle shortening will suppress spindle receptor activity but activate tendon organ receptors.

Interneurons in the Ventral Horn

The neuronal assemblies in the ventral horn contain, in addition to motoneurons, small interneurons that are either excitatory or inhibitory to motoneurons.

Renshaw Cells

Renshaw cells are inhibitory interneurons that receive input from collateral axons of α motoneurons. They synapse both with the motoneurons that activate them and with adjacent α motoneurons.

— Renshaw cells are activated by acetylcholine, the neurotransmitter synthesized by α motoneurons. They release glycine, an inhibitory transmitter. Strychnine, a rodenticide, blocks the receptors for glycine.
— By inhibiting motoneurons, Renshaw cells dampen firing and provide stability to motoneuron output.

6.2 Spinal Reflexes

Myotatic (Stretch) Reflex

The myotatic (stretch) reflex is a monosynaptic reflex (**Fig. 6.2**).

Mechanism

Stretching a muscle (and the spindles within it) induces firing of spindle afferents, which synapse directly on the α motoneurons that innervate the same (homonymous) muscle. Excitation of α motoneurons results in contraction (shortening) of the muscle.

Function

The stretch reflex is a negative feedback system to maintain the constancy of muscle length. It automatically and rapidly resists perturbations of muscle position that might be produced by momentary fatigue or other disturbances. It is particularly important in postural and other muscles that maintain preset positions.

The stretch reflex is also responsible for generating muscle tone (basal level of contraction). It is activated by the slight stretching of the muscle by the force of gravity.

Fig. 6.2 ▶ **Myotatic (stretch) reflex.**
The example given here is the patellar reflex. Tapping the patellar tendon by a reflex hammer causes stretching of the muscle spindles within the quadriceps muscle. This initiates group Ia afferent impulses, which enter the spinal cord via the posterior (dorsal) root and terminate in the anterior (ventral) horn on the α motoneuron of the same muscle. Activation of the α motoneuron causes the muscle to contract.

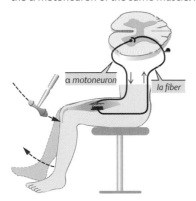

Inverse Myotatic Reflex

The inverse myotatic reflex (Golgi tendon organ reflex) is a disynaptic reflex.

Mechanism

Contraction of a muscle activates group Ib afferent fibers from Golgi tendon organs. The group Ib afferent fibers synapse on inhibitory interneurons, which inhibit α motoneurons to the homonymous muscle. Inhibition of the α motoneuron reduces the tension being produced by the muscle.

> **Clinically important monosynaptic reflexes**
>
> Some clinically important deep tendon reflexes include the biceps reflex, triceps reflex, patellar reflex, and Achilles reflex. In each case, tapping the tendon with a reflex hammer causes the attached muscle to contract, if the reflex arc is intact. Even though the test involves a single muscle, the muscle may be innervated by motoneurons in more than one spinal cord segment. The biceps reflex corresponds to the spinal cord segment of C5–C6; the triceps reflex corresponds to C6–C7; the patellar reflex corresponds to L3–L4; and the Achilles reflex corresponds to S1–S2. It is clinically necessary to routinely test the reflexes on both the right and the left sides to allow for comparison.

Function

The reflex acts to dampen the force of muscle contraction and thereby participates in ongoing regulation of muscle force.

Flexor Reflex, Antagonist Inhibition, and Crossed Extensor Reflex

The flexor (withdrawal) reflex is multisynaptic and is accompanied by antagonist inhibition and the crossed extensor reflex (**Fig. 6.3**).

Mechanism

In this protective reflex, activation of nociceptors causes firing in group II, III, and IV afferent fibers. These afferents make multiple synapses via interneurons with α motoneurons, resulting in withdrawal of the limb from the painful stimulus (usually by flexion) and inhibition of motoneurons that project to antagonist muscles (*antagonist inhibition*). Thus, limb movement is facilitated and not impeded by the action of an opposing muscle. At the same time, the *crossed extensor reflex* extends the contralateral limb to support the body following flexor withdrawal.

Fig. 6.3 ► **Withdrawal reflex.**
A painful stimulus in the sole of the right foot leads to flexion of all joints of that leg (flexor reflex). Action potentials from nociceptive afferents are conducted via stimulatory interneurons (**1**) in the spinal cord to motoneurons of ipsilateral flexors leading to their contraction and via inhibitory interneurons (**2**) to motoneurons of ipsilateral extensors (**3**), leading to their relaxation (antagonist inhibition). The crossed extensor reflex consists of contraction of the extensor muscles (**5**) and relaxation of the flexor muscles in the contralateral leg (**4, 6**). Action potentials from nociceptive afferents are also conducted to other segments of the spinal cord (**7, 8**) because different flexors and extensors are innervated by different segments.

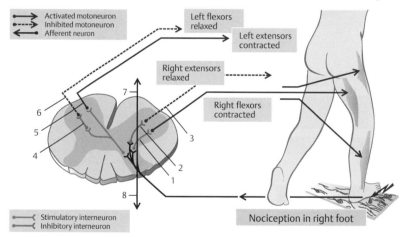

6.3 Cortical Motor Control

Primary Motor Cortex

The primary motor cortex is located directly anterior to the central sulcus on the precentral gyrus (**Fig. 6.4**). It is somatotopically organized so that a map of the muscles of the body can be plotted on the surface of the precentral gyrus (**Fig. 6.5**). A disproportionately large area of the primary motor cortex is dedicated to the fingers and thumb and the muscles used in speech.

Afferent Input to the Primary Motor Cortex

The neurons of the motor cortex are influenced by inputs from several sources:
- The premotor cortex, located anterior to the primary motor area, is thought to play a role in planning movements.

Fig. 6.4 ▶ **Motor cortex.**
Lateral view of the left hemisphere. The primary motor cortex (M1; area 4) is located in the precentral gyrus and is involved in the execution of voluntary movement. The premotor cortex is located anterolaterally adjacent to the primary motor cortex and is involved in the planning and initiation of movement. The supplementary motor cortex is located on the medial surface of the precentral gyrus and is involved in controlling axial and girdle muscles on both sides of the body. The motor cortex is functionally related to the somatosensory cortex on the postcentral gyrus, as sensory information is required for the cortical representation of space, which allows for precision of movement. This is the reason these cortical areas are referred to as the sensorimotor system.
From *Thieme Atlas of Anatomy, Head and Neuroanatomy*, © Thieme 2007, Illustration by Markus Voll.

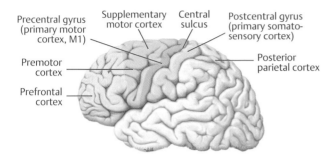

- The supplementary motor area, located on the medial surface of the premotor cortex, controls larger muscle groups than the primary motor area and is also involved in preparation for complex movements.
- The ventral anterior (VA) and ventral lateral (VL) nuclei of the thalamus are essential relays for excitatory input to the motor cortex.

Pyramidal Tracts

The pyramidal tracts, comprising mainly the corticospinal but also the corticonuclear and corticoreticular tracts, originate in the motor cortex. The corticonuclear and corticoreticular tracts innervate motor cranial nerve nuclei and the reticular formation in the brainstem, respectively. The corticospinal tracts pass through the pyramids of the medulla (**Fig. 6.6**). Most of the corticospinal fibers cross at the decussation of the pyramids. Because of the pyramidal decussation, the motor cortex of one hemisphere controls movement on the contralateral side of the body.
- The corticospinal fibers terminate directly on α or γ motoneurons and on interneurons that then synapse with α or γ motoneurons.
- The main function of these corticospinal tracts is control of the fine, skilled movements performed by the distal musculature of the limbs.

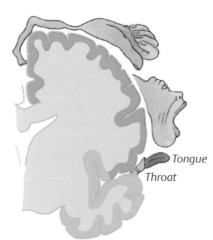

Tongue
Throat

Fig. 6.5 ▶ **Somatotopic organization of the primary motor cortex.**
The primary motor cortex exhibits somatotopic organization with respect to the target muscles it controls. Muscles that are involved in fine or complex tasks have more area of the primary motor cortex dedicated to them.

Fig. 6.6 ► Course of the pyramidal tracts.
The pyramidal tracts originate in the motor cortex. The fibers of the main part, the corticospinal tract, travel through the pyramids of the medulla, where 80% cross and descend in the spinal cord as the lateral corticospinal tract. The remaining 20% descend without crossing, forming the anterior corticospinal tract. Most fibers of the anterior corticospinal tract ultimately cross at the segmental level. The axons of the corticospinal tract terminate either directly or via interneurons on α and γ motoneurons. The fibers of the corticonuclear tract innervate motor cranial nerve (CN) nuclei in the brainstem (CN III–VII and IX–XII). The corticoreticular tract fibers pass to the reticular formation in the brainstem (not shown).
From *Thieme Atlas of Anatomy, Head and Neuroanatomy,* © Thieme 2007, Illustration by Markus Voll.

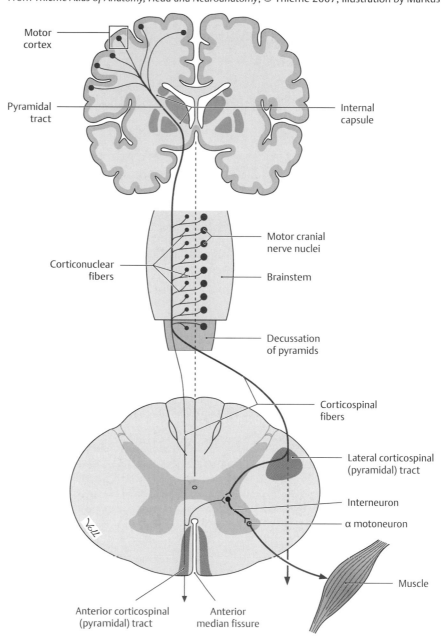

6.4 Brainstem Motor Control

Descending Brainstem Motor Tracts

— These tracts, called the *extrapyramidal tracts*, originate from nuclei in the brainstem and do not pass through the medullary pyramids (**Fig. 6.7**). Their projections are concentrated on the neurons that control proximal and axial muscles — for postural control — rather than distal limb muscles.

— Their primary function is to control posture and positioning of the body and limbs.

Fig. 6.7 ▶ **Course of the extrapyramidal tracts.**
The extrapyramldal tracts originate in various regions of the brainstem and terminate on interneurons that synapse with α and γ motoneurons, which they control. They also receive input (*blue*) from the cortex and the cerebellum.
From *Thieme Atlas of Anatomy, Head and Neuroanatomy*, © Thieme 2007, Illustration by Markus Voll.

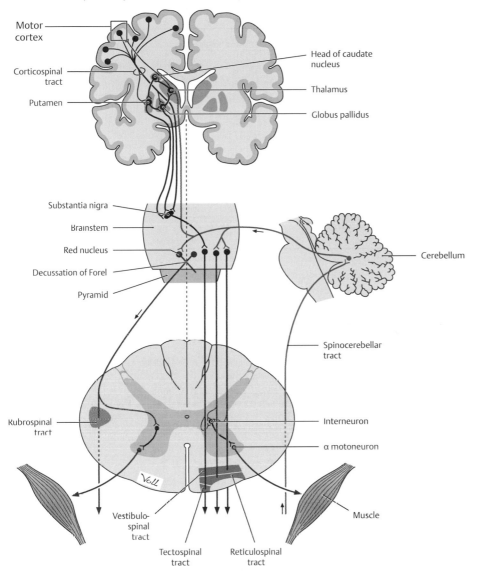

Lateral and Medial Reticulospinal Tracts

— The lateral and medial reticulospinal tracts originate from the medullary and pontine reticular formation and have a strong influence on γ motoneurons.
— They are involved mainly in the control of axial and girdle muscles.

Lateral and Medial Vestibulospinal Tracts

— The lateral and medial vestibulospinal tracts originate from the vestibular nuclei.
— The lateral vestibulospinal tract projects to ipsilateral spinal cord neurons and is involved in the activation of extensor muscles in the maintenance of balance.
— The medial vestibulospinal tract innervates motoneurons of neck muscles and is involved in the control of head movements.

Tectospinal Tract

— The tectospinal tract originates in the superior colliculus and innervates motoneurons of neck muscles.
— It is involved in the control of head movements in response to moving visual stimuli.

Rubrospinal Tract

— The rubrospinal tract originates in the red nucleus of the midbrain.
— It is involved in activating flexor and inhibiting extensor motoneurons. It collaborates with the corticospinal tract in the control of motoneurons.

6.5 Upper and Lower Motoneurons

Upper and Lower Motoneuron Pathways

The projections that control muscle movement are broadly divided into lower and upper motoneuron pathways.

— *Lower motoneurons* are the α motoneurons that directly synapse with muscle at the neuromuscular junction.
— *Upper motoneurons* refer to descending motor pathways, principally the pyramidal tract, the most important of the motor control pathways.

Motoneuron Lesions

Lesions of both lower and upper motoneurons result in paralysis (lack of voluntary movement), but the physiological characteristics of the paralysis differ noticeably.

Lower Motoneuron Lesions

Lower motoneuron lesions sever the connection between the muscle and the CNS. Denervation induces accelerated atrophy of the muscle, possibly by removal of signaling molecules from the nervous system that maintain and nourish muscle cells. Denervation also provokes increased synthesis of acetylcholine receptor molecules by muscle cells. These receptor proteins are inserted all along the plasma membranes of the denervated muscle and are not confined to the neuromuscular junction as in normal muscle. As a result, the muscle membrane becomes supersensitive to acetylcholine and exhibits fasciculation (spontaneous rippling contraction of groups of muscle fibers).

The loss of lower motoneuron input also means that reflex circuits are interrupted, and the muscle shows no reflex responses and no tone. Thus, lower motoneuron lesions result in *flaccid paralysis*.

Upper Motoneuron Lesions

With upper motoneuron lesions, there is no denervation supersensitivity, much less muscle atrophy, and no loss of segmental reflexes. In fact, lesions of the motor cortex cause hypersensitivity of the stretch reflexes, resulting in elevated muscle tone. This type of paralysis is termed *spastic paralysis*.

Table 6.1 summarizes the signs of lower and upper motoneuron lesions.

Table 6.1 ▶ **Upper and Lower Motoneuron Lesions**	
Site of Lesion	**Signs**
Lower motoneuron lesions	Flaccid paralysis Muscle atrophy Fasciculations (spontaneous twitches) of single muscles Absence of reflexes
Upper motoneuron lesions	Increased muscle tone (tone sometimes decreased, depending on the region damaged) Derangements of groups of muscles Enhanced tendon reflexes Positive Babinski sign

6.6 Basal Ganglia

Structure

The basal ganglia, a group of nuclei situated deep to the cortex, receive and/or provide input to a number of brain areas. The main components of the basal ganglia are the following:

— the striatum, composed of the caudate and the putamen
— the globus pallidus, composed of the lateral globus pallidus and the medial globus pallidus
— the substantia nigra, composed of the substantia nigra pars reticulata and the substantia nigra pars compacta
— the subthalamic nucleus

Functions

The basal ganglia play a key role in voluntary motor control, by stimulating some motor circuits while inhibiting others. They do this as part of a circuit with three main participants, the cortex, the basal ganglia, and the thalamus.

Pathways

Cortical neurons involved in movement planning project to the striatum of the basal ganglia releasing excitatory glutamate. Once excited, the cells of the striatum project to two pathways, the direct and the indirect **(Fig. 6.8)**.

— In the *direct pathway*, the neurons of the striatum project inhibitory GABAergic neurons onto the cells of the substantia nigra reticulata-medial globus pallidus complex, which unless inhibited, acts to inhibit the thalamus from releasing excitatory glutaminergic input to the cortex. So the overall result of this pathway is that the striatum inhibits the inhibitory SNr-GPm complex freeing the thalamus from inhibition. The thalamus stimulates the cortex via glutaminergic input resulting in muscle action.

Fig. 6.8 ► **The pathways within the basal ganglia.**
The excitatory glutaminergic pathways are green, the inhibitory GABAergic pathways are red, and the modulatory dopaminergic pathways are black. The GPi-SNr complex when stimulated is inhibitory to the thalamus. For the thalamus to excite the cortex to action, the GPi-SNr complex must be inhibited so that is cannot inhibit the thalamus. (GPm, medial globus pallidus; GPl, lateral globus pallidus; SNr, substantia nigra pars reticulata; SNc, substantia nigra pars compacta; STN, subthalamic nucleus)

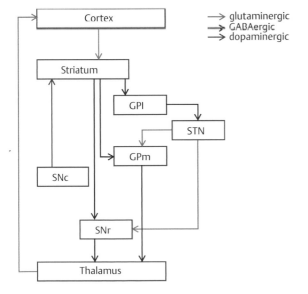

glutaminergic
GABAergic
dopaminergic

> **GABA synthesis and degradation**
>
> Gamma-aminobutyric acid (GABA) is the principal inhibitory neurotransmitter in the CNS. It is released by interneurons in the cerebral cortex and cerebellum, by neurons of the basal ganglia, and by neurons in the spinal cord that control muscle tone. GABA is mainly synthesized from glucose via the GABA "shunt" of the Krebs cycle. This "shunt" converts ketoglutarate to glutamic acid, which is then converted to GABA. Thus, glutamate, the principal excitatory CNS neurotransmitter, is converted to GABA. Termination of the action of GABA is by reuptake into neurons and glial cells

— In the *indirect pathway*, the neurons of the striatum project inhibitory GABAergic neurons to the lateral globus pallidus, which then project inhibitory GABAergic neurons to the subthalamic nucleus. This inhibition of inhibitory outputs (disinhibition) of the lateral globus pallidus, results in the subthalamic nucleus projecting excitatory input to the SNr-GPm (which then inhibits the thalamus). The net result is a reduction in thalamic stimulation of the cortex, resulting in inhibition of muscle action.

— The antagonistic functions of the two pathways are modulated by the substantia nigra pars compacta that produces dopamine and which stimulates either D_1 or D_2 receptors in the striatum, which favor activation of the direct or indirect pathway, respectively.

Via these pathways the body maintains a balance of excitation and inhibition of motion.

Basal Ganglia Disorders

Parkinsonism

Parkinsonism, a hypokinetic disorder, results from a loss of dopaminergic input to the striatum from the substantia nigra. Dopamine normally activates the direct pathway. Withdrawal of this activation gives free rein to the indirect pathway, resulting in excessive inhibition of the thalamus and a slowing of movement.

Symptoms of parkinsonism usually start between 60 and 70 years of age and include a "pill-rolling" tremor, "lead pipe" rigidity (limbs stay where they are placed when they are passively moved), and bradykinesia (slow execution of movement and speech), resulting in a masklike face and shuffling gait.

Treatment is with the dopamine precursor levodopa (L-dopa) and anticholinergic drugs, which are given as an adjunct for the tremor.

Ballismus

Ballismus and hemiballismus are hyperkinetic disorders caused by damage, usually vascular, to the subthalamic nucleus, which interrupts the indirect pathway. The thalamus is thus freed from inhibition generating excessive activity of the motor cortex.

Ballismus is characterized by irregular, flinging movements of the limbs.

Treatment, when necessary, involves the use of dopamine-blocking agents (e.g., pimozide, haloperidol, and chlorpromazine), despite the fact that dopamine has not been definitively linked to the disorder.

Huntington Chorea

Huntington chorea is a genetic (autosomal dominant) hyperkinetic disorder associated with degeneration of striatal neurons that project via the indirect pathway.

Symptoms start insidiously in middle age with chorea (involuntary, continuous jerky movements) that involve the face and extremities and make the patient seem fidgety or jumpy. Neuronal degeneration progresses to involve the frontal cortex, resulting in dementia.

There is no cure nor any treatment to prevent progression of this disease.

Other Basal Ganglia Disorders

Dystonia (abnormal twisting movements), Tourette syndrome, and tics are also believed to involve abnormal activity in basal ganglia circuits.

6.7 Cerebellum

Functions

The cerebellum provides a mechanism for rapid feedback control of movement. It receives a copy of motor commands from the cerebral cortex, as well as ongoing information about the progress of movements from the periphery. After comparing the command and feedback information for errors, it issues correction signals to the motor cortex (via the thalamus) and other motor centers. It is also involved in planning, initiating, and learning movement sequences.

Dopamine: release, degradation, and uses

Dopamine is released from dopaminergic neurons following an action potential. It then binds to two major types of G-protein coupled receptors: D_1 and D_2. D_1 receptors increase cyclic adenosine monophosphate (cAMP), whereas D_2 receptors decrease cAMP, so the differing effects of dopamine binding depend on signal transduction. Dopamine's action is terminated by reuptake into neurons, where it is stored in vesicles for reuse, or it is degraded by catechol *O*-methyltransferase (COMT) or monoamine oxidase (MAO). D_2 agonists are used to treat Parkinson disease and to inhibit prolactin, whereas D_2 antagonists are used as antiemetics and in the treatment of schizophrenia.

Lewy body dementia

Lewy body dementia (LBD) is the second most common form of dementia after Alzheimer disease. It occurs as a result of abnormal proteins (Lewy body proteins) being deposited throughout the cortex of the brain. If this deposition occurs in the substantia nigra, dopamine stores become depleted, causing parkinsonian symptoms. LBD manifests with cognitive impairment and increasing difficulty performing tasks, as well as memory problems and visual hallucinations. There is no cure for this disease, so treatment aims to reduce symptoms by using cholinesterase inhibitors, l-dopa, and neuroleptics.

Functional Subdivisions

Although there is considerable overlap of functions, the cerebellum can be divided into three general subdivisions: the vestibulocerebellum, the spinocerebellum, and the cerebrocerebellum (**Table 6.2**).

Vestibulocerebellum
- The vestibulocerebellum (floculonodular lobe) functions to maintain balance and positioning of the body in space.
- It receives input from the vestibular organ and vestibular nuclei.
- Its output is directed at the vestibular nuclei, which project to spinal cord motoneurons innervating antigravity extensor muscles and to ocular motor centers responsible for positioning of the eyes.

Spinocerebellum
- The spinocerebellum (anterior lobe and vermis) monitors and controls ongoing movements, ensuring their accuracy in direction and force level.
- It receives its principal input from the ascending spinocerebellar tracts that convey ongoing information from muscle and joint receptors.
- The outflow of the spinocerebellum is to the red nucleus and, most importantly, to the VA and VL nuclei of the thalamus. These thalamic nuclei are the major inputs to the motor cortex whose outflow, via the pyramidal tract, directly controls spinal motoneurons responsible for movement.

Cerebrocerebellum
- The cerebrocerebellum (lateral hemispheres), the largest subdivision of the cerebellum in humans, has responsibilities in the planning and initiation of movements, particularly complex learned movements.
- It receives its major input from the cerebral cortex via the pontine nuclei.
- Its output is directed to the VA and VL nuclei of the thalamus.

Table 6.2 summarizes the afferent and efferent connections of the divisions of the cerebellum and the symptoms associated with damage to each section.

Table 6.2 ▶ Summary of the Divisions of the Cerebellum			
Division of Cerebellum	**Afferent Source**	**Deep Cerebellar Nucleus**	**Efferent Target**
Vestibulocerebellum	Vestibular organ, vestibular nuclei	Fastigial nucleus	Lateral vestibular nucleus
Spinocerebellum	Muscle spindles, tendon organs via spinocerebellar tracts	Emboliform and globose nuclei (interposed nuclei)	Motor cortex via VA and VL nuclei of the thalamus
Cerebrocerebellum	Cerebral cortex via pontine nuclei	Dentate nucleus	Motor cortex via VA and VL nuclei of the thalamus
Abbreviations: VA, ventral anterior; VL, ventral lateral.			

Neuronal Circuitry

The cerebellum of all three subdivisions consists of a cortex of gray matter and an underlying accumulation of white matter (**Fig. 6.9**). The deep cerebellar nuclei lie within this white matter. The only projection neurons from the cerebellar cortex, the Purkinje cells, project their axons to these nuclei (**Fig. 6.10**). It is the deep cerebellar nuclei that convey instructions from the cerebellum to the motor nuclei of the brainstem and thalamus.

Note: Purkinje cells are GABAergic and act solely by inhibitory modulation of the rapidly firing cells of the deep cerebellar nuclei.

Fig. 6.9 ▶ **Cerebellar cortex.**
The inner granular layer of the cerebellar cortex mostly contains granule cells (*blue*) plus mossy fibers (*green*) and climbing fibers (*orange*). Golgi cells are also present (not shown). The Purkinje layer contains the cell bodies of Purkinje cells (*purple*). The molecular layer contains parallel fibers (axons of granule cells) that synapse with the dendrites of Purkinje cells. The molecular layer also contains axons from the inferior olive and its accessory nuclei (climbing fibers) and a small number of inhibitory interneurons (basket and stellate neurons).
From *Thieme Atlas of Anatomy, Head and Neuroanatomy*, © Thieme 2007, Illustration by Markus Voll.

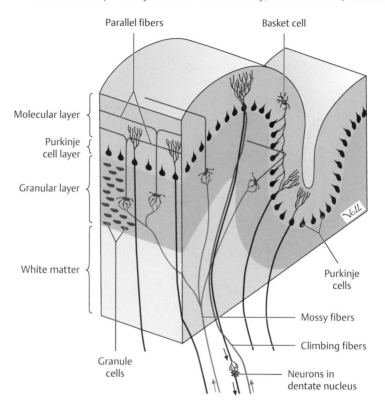

Fig. 6.10 ▶ **Synaptic circuitry of the cerebellum.**
The afferent fibers to the cerebellum are mossy and climbing fibers (origins shown). Climbing fibers form multiple excitatory synapses on the cell bodies and dendrites of Purkinje cells. Mossy fibers form excitatory synapses with granule cells. Collateral axons of both mossy and climbing fibers excite inhibitory interneurons and neurons of cerebellar nuclei. The parallel fibers of granule cells form excitatory synapses with the dendrites of Purkinje cells, which, in turn, make inhibitory synapses with cerebellar and vestibular nuclei. Cerebellar efferent neurons arise from the cerebellar nuclei. This complex circuit provides a means for feedforward and feedback control of motor function. (Asp, aspartate; Glu, glutamate)
From *Thieme Atlas of Anatomy, Head and Neuroanatomy*, © Thieme 2007, Illustration by Markus Voll.

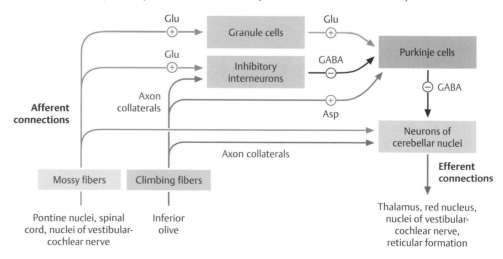

Cerebellar Disorders

Damage to the cerebellum from tumors or ischemia affects muscles on the ipsilateral side, resulting in the loss of coordination and balance. The underlying defect is the loss of feedback control and disturbed accuracy of motion.

Table 6.3 lists signs that may occur in cerebellar disorders.

Table 6.3 ► **Signs that May Occur in Cerebellar Disorders**		
Sign	**Description**	**Division of Cerebellum Responsible**
Gait ataxia	Wide stance and unsteady walking	Vestibulocerebellum
Nystagmus	Alternate fast and slow eye movements	
Vertigo	Dizziness with a sense of rotational movement	
Intention tremor	Tremor during voluntary movement	Spinocerebellum
Dysmetria	Difficulty moving limbs to a precise position, including overshoot and undershoot	
Dysdiadochokinesia	Difficulty with rapid pronation and supination of the hands	Cerebrocerebellum
Dysarthria	Slurred speech	

7 Higher Cortical Functions

7.1 The Electroencephalogram

The electroencephalogram (EEG) is used to measure spontaneous electrical activity arising from the cerebral cortex via electrodes placed on the scalp. This electrical activity results from temporal summation of cortical postsynaptic potentials (not action potentials) in many neurons and is represented by waves (**Fig. 7.1**).

Fig. 7.1 ▶ Recording the electroencephalogram (EEG).
An EEG measures electrical activity in the brain using multiple electrodes placed on the scalp. Alpha (α) waves are typical of an awake, relaxed patient (with eyes closed) and are generally detected in multiple electrodes (synchronization). Beta (β) waves are typical of an awake, alert patient (with eyes open). The frequency and amplitude of β waves varies greatly from different electrodes (desynchronization). Theta (θ) waves are seen in drowsiness or very relaxed states, and delta (δ) waves occur during deep sleep.

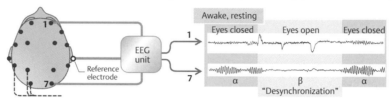

EEG Waves

EEG waves are described by their amplitude and frequency.

— EEG wave amplitude is determined by the following:

— The number of active synapses
— Synchronization of postsynaptic potentials
— The distance of these potentials from recording electrodes

— EEG wave frequency is determined by the state of central nervous system (CNS) arousal.

Table 7.1 lists the types of waves that are present normally in an EEG. Their appearance is shown in **Fig. 7.2**.

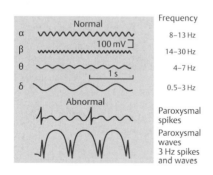

Fig. 7.2 ▶ EEG waves.
EEG waves have different frequencies and amplitudes. Localized or generalized paroxysmal spikes and waves are used to diagnose epilepsy.

Table 7.1 ▶ Waves Seen on the Electroencephalogram		
Wave	**Frequency**	**Comments**
Beta (β)	High	β waves are typical of awake, mentally active individuals (with eyes open)
Alpha (α)	Low	α waves are seen in awake individuals (with eyes closed)
Theta (θ)	Low	θ waves are seen in drowsy or very relaxed states
Delta (δ)	Low	δ waves are seen in deep sleep when the cortex and thalamus are highly synchronized

Clinical Use of the EEG

The EEG is used to determine the foci of localized or generalized seizure (convulsive) activity. Derangement of the EEG is also seen with brain lesions, such as tumors, abscesses, and subdural or extradural hematomas, and with metabolic disorders, infections (e.g., encephalitis), and hypoxia.

Table 7.2 summarizes the EEG changes that occur in certain diseases or conditions.

Table 7.2 ▶ Electroencephalogram (EEG) Changes that Occur in Disease States and Conditions	
Disease/Condition	**EEG Changes**
Epilepsy	EEG changes vary with the type of seizure and between phases of the seizure
Tumors, abscesses, subdural or extradural hematomas	Slows frequency, alters amplitude
Metabolic disorders, encephalitis, hypoxia (use of general anesthetics)	Slowing and synchrony of the EEG
Brain death	Flat EEG

7.2 Sleep

Sleep is a periodic and reversible state of decreased ability to interact with the external environment. It is accompanied by a significant reorganization of the neural, endocrine, and somatic systems. It is known that sleep follows a circadian rhythm, but the physiological causes of sleep onset, sleep maintenance, and waking are unknown, as is the need for sleep.

Sleep Cycle

The daily sleep–wake cycle is thought to be controlled by a rhythm generator in the suprachiasmatic nucleus of the hypothalamus.
- As light levels are reduced in the evening, firing of cells in the nucleus increases, leading ultimately to the release of melatonin by the pineal gland. Melatonin is a sleep-inducing substance.
- The alert state is activated by nonspecific ascending activating impulses from the reticular formation of the brainstem via the thalamus to wide areas of the cerebral cortex. This activity involves both cholinergic and monoaminergic (norepinephrine, 5-hydroxytryptamine [5-HT], and dopamine) systems.

Stages of Sleep

There are five sleep stages: four synchronized slow-wave sleep stages and rapid eye movement (REM) sleep. These five sleep stages are identified by polygraphic recordings: EEG, EMG (electromyogram), and EOG (electro-oculogram), and by behavioral criteria.

Table 7.3 shows the EEG changes, eye movements, and other notable occurrences during wakefulness and each stage of sleep. In normal sleep, a person descends through stages 1 to 4 in order, ascends back to stage 2, and enters REM sleep. REM sleep is also exited through stage 2. Repeated cycles are made through the stages, and REM sleep is entered four to six times per night at an average inter-REM period of 90 minutes.

■ Epilepsy

Epilepsy is the tendency to have recurring seizures. These seizures take many forms, ranging from brief cessations of responsiveness without loss of conciousness to convulsions with accompanying loss of consciousness. It occurs in almost 17% of adults. In one-quarter of those the EEG and MRI finds a focal lesion on the brain. Antiepileptic drugs (eg., phenytoin, valproate) are usually able to prevent further seizures. The 10–20% of patients in whom drugs are insufficient are candidates for surgical removal of the small area of the brain in which the seizures originate.

■ Sleep apnea

Sleep apnea is a sleep disorder characterized by episodic cessation of breathing during sleep. It may be central, in which breathing cessation occurs due to a lack of respiratory effort; obstructive, in which breathing cessation occurs due to airflow obstruction; or a mixture of the two (complex sleep apnea). Symptoms of sleep apnea include snoring, insomnia (difficulty staying asleep), hypersomnia (sleepiness in the daytime), morning headaches, awakening with a dry mouth or sore throat, memory problems, mood swings, and depression. Sleep apnea sufferers tend to be unaware of their difficulty breathing, except in cases of central sleep apnea, in which the patient can awaken abruptly with shortness of breath (dyspnea). Sleep apnea can place the cardiovascular system under great strain due to hypoxia, resulting in cor pulmonale (right heart failure due to pulmonary hypertension), sudden cardiac death, and stroke. Risk factors for this condition include excess weight, smoking, use of alcohol/sedatives/tranquilizers, a positive family history of the condition, and narrowed airway. It is more common in men and tends to occur after the age of 65. Treatment depends on the cause and severity and may include reducing risk factors, oral splints to help keep the airway open, CPAP (continuous positive airway pressure), or surgery (e.g., uvulopalatopharyngoplasty or tracheostomy).

Somnambulism

Somnambulism (sleepwalking) is an arousal disorder that occurs in stage 4 sleep. The person may simply sit up in bed, or he or she may walk around and perform tasks such as getting dressed, cooking, or even driving. Enuresis (involuntary urination), when it occurs, is often correlated with the transition from stage 4 to 2 and REM sleep. This condition is usually not dangerous unless it causes the person to adopt dangerous behaviors (e.g., driving while asleep). Somnambulism is common in preteen children and is not indicative of any underlying pathology; however, in adults, it is often diagnostic of psychological disturbances. Usually, no treatment is required for this condition unless there is causative underlying pathology. Medication such as benzodiazepines may be useful in some cases.

Table 7.3 ▶ Stages of Sleep and Their Accompanying Features

Stage	EEG Activity	Eye Movements	Comments
Stage W	α activity associated with quiet wakefulness or low-voltage desynchronized activity of active wakefulness	There are frequent voluntary, conjugate eye movements (EOG)	Stage W is wakefulness The EMG has a high level of activity
1	α waves	Occasional slow, rolling eye movements occur	This stage is drowsiness.
2	Theta waves	No conjugate eye movements	
3	EEG shows further slowing and greater amplitude waves	No conjugate eye movements	Increased parasympathetic activity; lowered sympathetic activity Heart rate and blood pressure are reduced Pupils are miotic Monosynaptic reflexes are slightly depressed
4	Delta waves	No conjugate eye movements	This is deep sleep. Autonomic changes similar to Stage 3 Monosynaptic reflexes are moderately depressed. Growth hormone is secreted in young children.
REM sleep	EEG resembles that of an alert, awake patient with high-frequency, low-amplitude waves (despite patient's being in deep sleep)	Rapid conjugate eye movements	REM is deep sleep, patients in this stage are hard to awaken. It occurs ~90 minutes after sleep onset. **Continuous events:** –Desynchrony of the EEG –Atonia (no muscle tone) of all skeletal muscle, except those supporting eye and respiration-related movements, –Difficulty in being awakened –Penile or clitoral erection and lubrication **Phasic events:** –Abrupt release of atonia, allowing twitches, especially in distal limb musculature –Large variations in blood pressure and heart and respiration rates –Many people report dreams when awakened during REM sleep

Abbreviations: EEG, electroencephalogram; EMG, electromyogram; EOG, electro-oculogram; REM, rapid eye movement.

7.3 Language

— The dominant hemisphere as defined by speech and language capabilities is the left hemisphere in 90 to 95% of right-handed people (who represent 90% of people) and in 70% of left handed people (who represent 10% of people).

— Damage to the Broca area causes an expressive aphasia in which the patient is unable to produce language (written or spoken), but comprehension of language is intact.

— Damage to the Wernicke area causes a receptive aphasia in which the patient is unable to comprehend written and spoken language, but speech may be fluent and well articulated.

— The nondominant right hemisphere has executive control for spatial abilities and is more adept at tasks requiring spatial orientation.

— The corpus callosum provides communication between hemispheres, allowing information received on one side to be shared by both sides.

7.4 Memory

— *Short-term memory* involves synaptic changes. It can retain about seven units of information (e.g., groups of numbers) for a few seconds.
— *Long-term memory* involves neural changes. It has a very large capacity to store information for minutes to years.

7.5 Barriers in the Brain and Cerebrospinal Fluid

Blood–Brain Barrier

Anatomy and Functions

— The blood–brain barrier is the barrier between cerebral capillary blood and cerebrospinal fluid (CSF). It is composed of the tight junctions between capillary endothelial cells and glial cells.
— The blood–brain barrier regulates the osmolarity of brain tissue and CSF and thereby the intracranial pressure and volume. It also maintains a constant environment for neurons in the CNS.

Disruption to the Blood–Brain Barrier

The blood–brain barrier may be disrupted by inflammation (e.g., meningitis), irradiation, and tumors.

Cerebrospinal Fluid

Formation and Composition

— CSF is produced in the choroid plexus and occupies the subarachnoid space and ventricles of the brain.
— CSF is an ultrafiltrate of serum and very similar in composition. It has less Ca^{2+}, glucose, protein, and immunoglobulin G (IgG) compared with serum, but more Cl^- and lactate.

Functions

— The primary function of CSF is to impart buoyancy to the brain. Decreased CSF production therefore increases pressure on the spine and renders the brain more susceptible to injury (less cushioning).
— CSF also transports nutrients and hormones into the brain and carries waste products out of it.
— Lipid-soluble substances (O_2 and CO_2) are able to freely diffuse across the blood–brain barrier and enter CSF.
— Proteins cannot pass through the blood–brain barrier due to their large molecular size.
— Lipophilic (nonionized) drugs will pass through the blood–brain barrier more readily than water-soluble (ionized) drugs.

■■■ Alzheimer disease

Alzheimer disease is a progressive neurodegenerative disorder producing marked atrophy of the cerebral cortex. It is the most common cause of dementia. It produces impairment of short-term memory, cognition, and language; increasing difficulty performing the activities of daily living; personality changes (e.g., anxiety, depression, aggression, and social withdrawal); and immobility leading to death. Treatment involves drugs such as tacrine, donepezil, rivastigmine, and galantamine, which are centrally acting, reversible inhibitors of cholinesterase. They prevent the hydrolysis of acetylcholine, thus increasing the concentration of acetylcholine available to neurons. This improves cognition in mild to moderate Alzheimer disease. Memantine, an N-methyl-D-aspartate (NMDA) receptor antagonist, may also be used in moderate to severe Alzheimer disease to protect neurons from damage caused by glutamate.

■■■ Meningitis

Meningitis is inflammation of the meninges of the brain (pia mater and arachnoid), usually due to a viral infection, but it can also be caused by a bacterial or fungal infection. Symptoms include headache, stiff neck on passively moving the chin toward the chest, photophobia (sensitivity to light), irritability, drowsiness, vomiting, fever, seizures, and rashes (viral or meningococcal meningitis). Predisposing factors for meningitis include head injury (especially basal skull fracture), otitis media, sinusitis, mastoiditis, and a compromised immune system (e.g., carcinoma, acquired immunodeficiency syndrome [AIDS], diabetes, splenectomy, and immunosuppressant drugs). A lumbar puncture often provides a definitive diagnosis of meningitis. Blind treatment with a broad-spectrum antibiotic or prompt treatment with IV antibiotics that are sensitive to the causative organism is required for bacterial meningitis. For viral meningitis, treatment includes bed rest and fluids, but this normally resolves on its own in a week or two.

■■■ Lumbar puncture

Lumber puncture is a procedure for sampling CSF. It allows for the diagnosis of conditions such as meningitis, subarachnoid hemorrhage, and tumor metastases. To perform a lumbar puncture, the patient is asked to lie on his or her side with hips flexed. A needle is inserted precisely in the midline between the spinous process of L3 and L4 and advanced into the dural sac (lumbar cistern). A sample of CSF is then aspirated. CSF pressure may also be measured by connecting the needle to a manometer.

Review Questions

1. Characteristics of the sympathetic division of the autonomic nervous system (ANS) include
   A. effector organs that have muscarinic receptors.
   B. norepinephrine as the transmitter at preganglionic nerve terminals.
   C. acetylcholine as the transmitter at preganglionic nerve terminals.
   D. little divergence between pre- and postganglionic neurons.
   E. primary innervation of ciliary muscles.

2. Activation of the parasympathetic division of the autonomic nervous system may produce
   A. increased release of epinephrine from the adrenal medulla.
   B. increased basal metabolic rate.
   C. cutaneous vasodilation.
   D. decreased heart rate.
   E. thick, viscous saliva.

3. An anxious patient presents with dilated pupils. How do you explain this sign?
   A. The patient is trying to avoid squinting to see more clearly.
   B. The patient is overly sensitive to bright lights.
   C. There is subconscious relaxation of radial muscle fibers.
   D. There is parasympathetic activation to contract circular muscle fibers of the pupil.
   E. There is sympathetic activation to contract radial muscle fibers of the pupil.

4. How does the body respond when a patient starts to develop a fever?
   A. By shivering
   B. By sweating
   C. By blocking the response to pyrogens
   D. By decreasing the body temperature set point
   E. By inhibiting prostaglandin production

5. Thermal stress causes a person to collapse. The patient's core body temperature is 105°F (40.6°C), and the skin is warm and dry. What is this condition?
   A. Accidental hypothermia
   B. Heat stroke
   C. Heat cramps
   D. Heat exhaustion
   E. Not a medical emergency

6. Which of the following remains after removal of the human cerebral cortex?
   A. Postural control
   B. Conditioning to patterned tonal stimuli
   C. The capacity to regulate body temperature
   D. Patterned vision
   E. Hypotonia in the extensor muscles

7. Chronic stimulation of the hypothalamic satiety center would
   A. produce the hypothalamic obesity syndrome.
   B. produce anorexia.
   C. cause hyperphagia.
   D. activate the feeding center.
   E. produce bulimia.

8. Adaptation to a prolonged constant stimulus in a sensory receptor
   A. occurs when the receptor receives a subthreshold stimulus.
   B. is characterized by a decreasing frequency of action potentials.
   C. is characterized by decreasing amplitude of action potentials.
   D. is identical to accommodation in neurons.

9. Optimal discrimination between closely spaced tactile stimuli, as in two-point discrimination, is found in skin regions where the
   A. receptive field sizes are largest.
   B. receptive field sizes are smallest.
   C. density of receptors is lowest.
   D. receptor thresholds are lowest.
   E. sensory axons leaving the region have uniform conduction velocities.

10. A 45-year-old man is diagnosed with a heart attack partly due to pain in his left upper arm. Why does he experience pain in his arm?
    A. The heart attack limits blood flow to the arm.
    B. Afferent fibers from the heart and arm converge onto the same spinal neuron.
    C. Sensory input from the heart projects to relays in the thalamus.
    D. Sympathetic efferent fibers innervate both the heart and blood vessels in the arm.
    E. Afferents from the heart cause reflex muscle contractions in the arm.

11. What mediates analgesia produced by brain stimulation?
    A. Prostaglandins
    B. Histamine
    C. Serotonin
    D. Substance P
    E. Opiate receptors

12. If the near point is closer to the eye than normal, then
    A. correction to normal vision requires a cylindrical lens.
    B. correction to normal vision requires a positive lens.
    C. the patient has presbyopia.
    D. the patient has hyperopia.
    E. the patient has myopia.

13. The macula (fovea centralis) of the retina
    A. is the area of most acute vision.
    B. contains the highest concentration of rods.
    C. has the lowest threshold for excitation in the retina.
    D. is specialized for excitation in the dark.
    E. is not used during routine vision.

14. What would you suspect was the cause of weak vision bilaterally and temporally (the outside half of both visual fields)?
    A. Macular degeneration
    B. Early stages of cataracts
    C. A pituitary tumor
    D. Damage to the fovea centralis of both eyes
    E. Trauma to the occipital cortex

15. A 62-year-old woman has a tumor in the nucleus of her left lateral lemniscus. Which hearing loss would you expect her to have?
    A. Deafness in her left ear
    B. Deafness in her right ear
    C. Slight hearing loss only in her left ear
    D. Slight hearing loss only in her right ear
    E. Slight hearing loss in both ears

16. A patient presents with dizziness and disorientation. The caloric nystagmus test is normal. Which of the following do you observe?
    A. Quick flick of the eyes to the left from warm water in the left ear
    B. Quick flick of the eyes to the right from warm water in the left ear
    C. Quick flick of the eyes to the left from warm water in the right ear
    D. Slow movement of the eyes to the left from cold water in the right ear
    E. Slow movement of the eyes to the right from cold water in the left ear

17. A lesion in which location would produce deafness limited to high tones?
    A. Tympanic membrane
    B. Ossicular chain
    C. Apical portion of the basilar membrane
    D. Basal portion of the basilar membrane
    E. Medial geniculate body

18. What would cause sensorineural deafness?
    A. Blocked eustachian tubes
    B. Punctured eardrums
    C. Destruction of cochlear hair cells by antibiotics
    D. Immobilized stapes
    E. Caloric nystagmus

19. The taste of sweet
    A. is associated with some organic chemicals.
    B. is associated with acids.
    C. is associated with alkaloids and some plant-derived poisons.
    D. may stimulate a sneeze.
    E. is a combination of two basic tastes, minty and salty.

20. The action potential frequency in 1a afferent fibers from the muscle spindle signals
    A. muscle length only.
    B. the rate of change of muscle length only.
    C. both muscle length and the rate of change of muscle length.
    D. muscle tension only.
    E. rate of change of muscle tension only.

21. Muscle relaxation evoked by the inverse myotatic reflex
    A. is initiated monosynaptically.
    B. is initiated by annulospiral endings in the relaxing muscle.
    C. requires inhibition of motoneurons supplying the relaxing muscle.
    D. supports the weight of the body.
    E. occurs simultaneously with a crossed flexor reflex in the opposite limb.

22. "Voluntary" movements are produced by
    A. the simultaneous activation of both α and γ motoneurons.
    B. impulses traveling via "upper" motoneurons directly from the brain to muscle extrafusal fibers.
    C. direct activation of the γ loop and reflex activation of α motoneurons.
    D. direct activation of α motoneurons with little or no response of γ motoneurons.
    E. selective activation of γ motoneurons.

23. Postural tone in antigravity muscles depends *most* upon sustained activity of
    A. cerebellar Purkinje cells.
    B. neurons of the reticulospinal tracts.
    C. neurons of the vestibulospinal tracts.
    D. neurons in the red nucleus.
    E. corticospinal neurons.

24. Which of the following characterizes lower motoneuron disease?
    A. Exaggerated reflexes
    B. Enhanced recurrent activation of Renshaw cells
    C. Being a later stage from the development of upper motoneuron disease
    D. Atrophy of the affected muscles
    E. Spasticity

25. The alpha rhythm of the electroencephalogram (EEG)
    A. results from cortical desynchronization during mental excitation.
    B. is in the frequency range of 4 to 7 Hz, seen in drowsy states.
    C. is indicative of cortical hypoxia.
    D. is the intrinsic cortical frequency in deep sleep.
    E. is prominent during supine relaxation with eyes closed.

26. Rapid eye movement (REM) sleep is characterized
    A. by an EEG pattern of low-amplitude desynchronized activity.
    B. by a cortical EEG pattern of delta range (2–4 Hz) activity.
    C. by a significant increase in muscle tone.
    D. as paradoxical because it occurs only once during the night, whereas other sleep stages occur several times.
    E. by a low arousal threshold.

27. During which stage of sleep is there the greatest variability of autonomic excitability?
    A. Stage 1
    B. Stage 2
    C. Stage 3
    D. Stage 4
    E. Stage REM

28. An obese 37-year-old man complains of falling asleep frequently during the day. A sleep laboratory study shows frequent airway blockage during the night. What is your diagnosis?
    A. Narcolepsy
    B. Hypersomnia
    C. Sleep apnea
    D. Somnambulism
    E. Enuresis

29. Why will a 5-year-old boy with amblyopia (lazy eye) become blind in one eye unless his convergence is corrected?
    A. Patching his weaker eye will preserve his vision.
    B. The lazy eye will suppress the stronger eye.
    C. The visual pathways from the stronger eye will degenerate.
    D. The visual cortex suppresses the image from the weaker eye to avoid diplopia (double vision).
    E. The plasticity of the brain will compensate for the loss of vision.

30. Which sign would you expect a right-handed patient with a tumor in the right parietal lobe to exhibit?
    A. Speech defects
    B. Right homonymous hemianopsia
    C. Psychomotor seizures
    D. Neglect of the left extremities
    E. Recent memory loss

31. A 23-year-old man presents with spastic limb movements after being poisoned with strychnine, which was added to his methamphetamine. What is the mechanism of this symptom?
    A. Excess excitation onto spinal motoneurons
    B. Hyperactivity of upper motoneurons
    C. Hypoactivity of upper motoneurons
    D. Block of excitatory afferent fibers
    E. Block of inhibitory transmitters acting on spinal motoneurons

Answers and Explanations

1. **C** Acetylcholine (not norepinephrine [B]) is the transmitter for all preganglionic nerve terminals and parasympathetic postganglionic nerve terminals (p. 35).
   A,D,E The parasympathetic division has muscarinic receptors, little divergence, and is the primary innervation of ciliary muscles.

2. **D** Activity of the vagus nerve slows the heart (p. 40).
   A,B,E Sympathetic, not parasympathetic, activity releases epinephrine, increases metabolism, and produces viscous saliva.
   C Sympathetic, not parasympathetic, cholinergic fibers produce cutaneous vasodilation.

3. **E** Anxiety would produce sympathetic activation, which dilates the pupils by contracting the radial muscle dilator pupillae (p. 41).

A Squinting is characteristic of patients with diplopia (double vision) or a response to bright lights.

B–D The pupils are constricted in response to bright lights, by relaxation of radial muscle fibers, or by parasympathetic activation of the cicular muscle sphincter pupillae.

4. **A** The pyrogens that cause a fever act on the anterior hypothalamus to increase prostaglandin synthesis. These prostaglandins stimulate the thermoregulatory center to reset the set point temperature to a higher temperature (not D). The body tries to raise its temperature to this new set point temperature by shivering (pp. 43–44).

B,C,E Sweating decreases body temperature, as would blocking the response to pyrogens or inhibiting prostaglandin production.

5. **B** Heat stroke occurs with prolonged hyperthermia (elevated body temperature) and no sweating (p. 44).

A Hypothermia is abnormally low core body temperature (< 95°F [35°C]).

C Heat cramps are muscular spasms.

D Heat exhaustion is associated with excessive sweating.

E Heat stroke is a medical emergency.

6. **C** An intact hypothalamus will remain if the human cerebral cortex is removed. Thus the capacity to adequately regulate body temperature will remain (p. 43).

A Although posture is at least partially controlled through the basal ganglia, postural control will be impaired by destruction of the cortex.

B,D Behaviors conditioned to patterned tones require association cortex, as does mediation of patterned vision.

E Decerebration would produce hypertonia (abnormal increase in muscle tension and a reduced ability of a muscle to stretch).

7. **B** Stimulation of the satiety center depresses food intake and produces anorexia, or loss of appetite (p. 44).

A,C Stimulation of the feeding center would override phasic satiety center inhibition and produce hyperphagia (overeating) with consequent obesity if stimulation were chronic.

D Such stimulation inhibits the feeding center.

E Bulimia (episodic binge eating) is a behavioral response of half of anorexics.

8. **B** Adaptation occurs when the sensitivity of a sensory receptor to a constant stimulus declines, decreasing the frequency of action potentials initiated by the receptor (pp. 47–48).

A A subthreshold stimulus will not generate any action potentials.

C Action potential amplitudes are constant.

D Accommodation in neurons is the increase in threshold due to inactivation of some Na^+ channels.

9. **B** To discriminate closely spaced stimuli, sensory nerves involved must have small receptive fields (p. 48).

A Large receptive fields for sensory nerves make discriminating closely spaced stimuli difficult.

C Low density of receptors implies larger areas, not smaller, for each receptive field.

D,E Neither receptor threshold nor afferent axon conduction velocity is relevant.

10. **B** The basis of referred pain is convergence of the afferent pain fibers from the heart and arm onto the same spinal neuron (p. 51).

A The heart attack limits blood flow to all parts of the body, but pain from this would not be localized to the left arm.

C Sensory input projects from the heart, but that does not explain pain in the arm.

D Afferent, not efferent, fibers are required to feel a sensation.

E There is no splinting pain associated with the heart.

11. **E** Endorphins act on opiate receptors in the midbrain to suppress descending pain transmission at the level of the spinal cord (p. 52).

A–C Prostaglandins, histamine, and serotonin are released as the result of tissue injury and lead to pain.

D Substance P is a common transmitter for nociceptive (pain) pathways.

12. **E** In myopia (nearsightedness), the near point is closer than normal. A concave, or negative lens, is needed to move the near point further from the eye (pp. 53–54).

A Cylindrical lenses are used to correct for astigmatism (refractive error that causes blurred vision).

B A negative, or concave, lens is needed to correct for myopia (nearsightedness), moving the near point further from the eye.

C In presbyopia (inability to focus on near objects), the near point gradually increases with age as the lens becomes less elastic.

D In hyperopia (farsightedness), the eyeball is too short or the lens is too flat. In either case, the closest focus is beyond the normal near point for a person's age. It is corrected with a convex, or positive, lens.

13. **A** The fovea centralis, located in the center of the macular region of the retina, is the area of most acute vision (p. 54).

B The fovea has the highest concentration of cones, not rods.

C–E Eye movements (saccades) normally tend to place targets on the fovea for the greatest visual acuity, away from the low threshold rods used under low levels of illumination.

14. **C** A pituitary tumor would put pressure on the optic chiasm where optic nerves cross. The pituitary is closer to nerves from the nasal fields of both retinas, which respond to light in the temporal visual fields (p. 57).

A Macular degeneration affects central vision in one or both eyes.

B Cataracts produce fuzzy vision.

D Damaged maculae affect central vision.

E Damage to the visual cortex produces generalized visual impairment rather than in specific fields.

15. **E** Because afferents from both ears relay on both sides of the brain, hearing from both ears is diminished (not answers C and D) (p. 61).

A,B Most auditory input is still transmitted to the auditory cortex because afferents from both ears relay on both sides of the brain; therefore, answers A and B are incorrect.

16. A The direction of nystagmus is determined by quick flick recovery after a slow movement in the opposite direction. Warm water in the left ear produces nystagmus to the left (not to the right, as in B) (p. 53).

C Warm water in the right ear does not produce nystagmus to the right.

D,E The slow movements from cold water are normally in the direction of the ear being tested. The mnemonic is COWS: <u>c</u>old, <u>o</u>ppo-site; <u>w</u>arm, <u>s</u>ame.

17. D The basal portion of the basilar membrane transduces high frequencies, therefore a lesion at this location would result in deafness to high tones (p. 61).

A Perforation of the tympanic membrane causes more loss of low frequencies, not high frequencies.

B Damage to the middle ear bones would cause general hearing deficits.

C The apical portion of the basilar membrane transduces low frequencies.

E Lesions of the medial geniculate body would produce hearing deficits dependent on the exact site of the lesion.

18. C Loss of cochlear hair cells is a common cause of sensorineural deafness (p. 60).

A,B,D Blocked eustachian tubes, punctured tympanic membranes (eardrums), or immobilized stapes would produce conduction deafness.

E Caloric nystagmus is a clinical test for vestibular pathology. It involves slow then fast eye movements induced by stimulation with warm or cool water in the ear.

19. A Sweet is associated with sugars and amino acids (p. 63).

B Most acids produce a sour, not a sweet, taste.

C, D Alkaloids taste bitter; no taste normally causes sneezes.

E Minty is not a basic taste.

20. B 1a afferents that arise from annulospiral endings signal the rate of change of muscle length (not A and C) (pp. 65–66).

D,E Golgi tendon organs, not 1a afferent fibers from the muscle spindle, signal muscle tension responses.

21. C The inverse myotatic reflex inhibits muscle contraction by inhibiting extensor motor neurons, allowing the limb to flex (pp. 67–68).

A,B The reflex is initiated by Golgi tendon organs and is at least disynaptic through an inhibitory interneuron to motoneurons.

D The body would collapse if supported by the affected limb.

E A crossed extensor, not flexor, reflex could support the body with the opposite limb.

22. A Voluntary movements involve coactivation, activation of both α and γ motoneurons (p. 66).

B Movements require activation of "lower" motoneurons that innervate skeletal muscle fibers.

C–E Alpha motoneurons may be activated from higher centers only a few milliseconds before γ motoneurons.

23. B Neurons of the lateral and medial reticulospinal tracts control axial and girdle muscles that are important for posture. The γ-motoneurons facilitate extensor tone for postural support (pp. 67, 70, 71).

A Purkinje cells inhibit cerebellar subcortical nuclei and do not affect muscle tone.

C Vestibulospinal neurons innervate neck muscles and are important for head movements.

D Rubrospinal neurons activate flexors and inhibit extensors, similar to corticospinal neurons.

E Corticospinal neurons initiate movements and usually decrease muscle tone.

24. D Lower motoneuron lesions sever the connection between the muscle and the CNS. This denervation results in muscle atrophy (p. 72).

A,E Upper motoneuron diseases release spinal lower motoneurons from descending inhibition, resulting in spasticity and exaggerated reflexes.

B Lower motoneuron lesions result in less activation of recurrent pathways.

C Pathologies of upper and lower motoneurons usually have different causes.

25. E The alpha rhythm is a normal response of the neocortex in the frequency range of 8 to 13 Hz, seen when the patient is relaxed with their eyes closed (not A-C) (p. 78).

D The predominant frequency in deep sleep is lower (2–4 Hz).

26. A REM sleep is characterized by an EEG pattern of low-amplitude desynchronized activity. This is paradoxical because this pattern is normally seen in an awake EEG (p. 80).

B Delta activity is characteristic of sleep stages 3 and 4, not REM sleep.

C There is a significant decrease in muscle tone during REM sleep.

D REM is cyclical with other sleep stages, occurring a few times during the night.

E The arousal threshold is high.

27. E Autonomic excitability is high during REM sleep (p. 80).

A–D Autonomic excitability decreases in deeper non-REM stages.

28. C Sleep apnea is characterized by airway blockage, typically in obese persons (pp. 79–80).

A Narcolepsy patients go quickly from awake to REM sleep.

B Hypersomnia patients have excessive slow-wave sleep.

D Somnambulism is sleepwalking.

E Enuresis is bedwetting.

29. D The blindness that can result from amblyopia (lazy eye) occurs in the cortex (p. 53).

A,B Patching his stronger, not weaker, eye is part of the treatment because the stronger eye suppresses the image from the weaker eye.

E The plasticity of the brain does not compensate for the loss of vision.

30. D A right-handed patient with a tumor in the right parietal lobe would neglect their left extremities because body image is maintained by the contralateral parietal cortex in parallel with contralateral somesthetic sensory projection (pp. 49, 69).

A,C,E Speech defects, recent memory loss, and psychomotor seizures are problems associated with the temporal, not parietal, lobes.

B Right homonymous hemianopsia is the lack of vision in the right half of the visual field due to a damaged left optic tract.

31. E Strychnine blocks transmission from spinal inhibitory interneurons postsynaptically (p. 67).

A This produces disinhibition of motoneurons, not excess excitation.

B–C Upper motoneurons are not affected.

D Afferent fibers are not affected.

8 Electrophysiology of the Heart

8.1 Conduction System of the Heart

The sinoatrial (SA) node is the primary pacemaker of the heart because it is able to spontaneously generate action potentials (inherent automaticity). These action potentials are conducted rapidly through the right atrial myocardium to the atrioventricular (AV) node, which delays the impulse before conducting it to the ventricles via the bundle of His and Purkinje fibers (**Fig. 8.1**). This provides an orderly contraction sequence from apex to base for efficient ejection of blood from the ventricles.

— The SA node spontaneously generates action potentials at a rate of ~80 to 100/min.
— The AV node also has inherent automaticity but at a slower rate than the SA node. If the SA node fails, the AV node will take over the pacemaker activity of the heart.
— The delay of the cardiac impulse at the AV node gives the contracting atria adequate time to empty their contents into the ventricles before ventricular contraction is initiated.

Note: The typical resting heart rate is 65 to 75 beats/min. This is due to vagal slowing of the heart below the intrinsic rate set by the SA node.

Fig. 8.1 ▶ **Cardiac conduction system.**
Contraction of cardiac muscle is modulated by the cardiac conduction system. This system of specialized myocardial cells generates excitatory impulses in the sinoatrial (SA) node that are conducted through the atria to the atrioventricular (AV) node, where the impulse is delayed before conducting it to the ventricles via the bundle of His and Purkinje fibers.
From *Atlas of Anatomy*, © Thieme 2008, Illustration by Markus Voll.

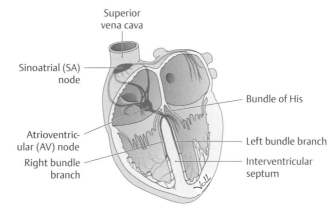

8.2 Events Causing Contraction of Cardiac Muscle

Contraction of cardiac muscle occurs when the excitation produced by an action potential is transmitted to cardiac myofibrils.

Cardiac Action Potentials

The general features of action potentials are discussed in Chapter 1.
The ionic conductances that are a feature of action potential phases at the SA node, cardiac muscle, and branches of the conduction system are discussed in this section.

▰ Arrhythmias

An arrhythmia is an abnormal heart rate or rhythm due to a fault in the normal, coordinated propagation of action potentials through the heart. There are many causes of arrhythmias, including coronary artery disease, hypertension, structural abnormalities, scarring of heart tissue (e.g., following a myocardial infarction), diabetes, hyperthyroidism, and drugs (e.g., caffeine and alcohol). Arrhythmias can be asymptomatic, or they may cause a fluttering sensation in the chest, a sensation of a racing/abnormally slow heartbeat, dyspnea (shortness of breath), dizziness, or syncope (fainting). Complications of arrhythmias include stroke and heart failure. There are four main classes of antiarrhythmic drugs, the selection of which will depend on the particular arrhythmia. Class I drugs are Na^+-channel blockers, class II drugs are β-blockers, class III drugs are K^+-channel blockers, and class IV drugs are Ca^{2+}-channel blockers. In addition to antiarrhythmic drugs, warfarin, an anticoagulant agent, is used in certain arrhythmias (e.g., atrial fibrillation) to prevent thromboembolic stroke.

Fig. 8.2 ▶ **Action potential phases at the sinoatrial node (A). Action potential phases at the atrial and ventricular myocardium, bundle of His, and Purkinje fibers (B).**

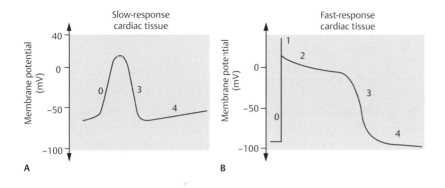

Action Potential Phases at the Sinoatrial Node

Refer to **Fig. 8.2.**

Phase 0

— This slow upstroke of the action potential shows membrane depolarization (it becomes less negative).

— It results from an increase in Ca^{2+} conductance, causing inward Ca^{2+} flow.

Phases 1 and 2

— These phases do not occur at the SA node.

Phase 3

— This is the repolarization phase.

— It results from an increase in K^+ conductance, which causes outward K^+ flow and repolarization of the membrane toward the K^+ equilibrium potential.

Phase 4

— The SA node does not have a stable resting membrane potential. During phase 4, it slowly depolarizes (which is responsible for its inherent automaticity).

— It results from an increase in Na^+ conductance, causing inward Na^+ flow and a decrease in K^+ conductance.

Note: The events at the AV node are similar to those at the SA node but slower. As a result, the AV node normally does not generate its own action potentials. This is because action potentials are conducted to the AV node more frequently than its own slow inherent rate of action potential generation.

Action Potential Phases at the Atrial and Ventricular Myocardium, Bundle of His, and Purkinje Fibers

Refer to **Fig. 8.2.**

Phase 0

— This rapid upstroke of the action potential shows membrane depolarization (it becomes less negative).

— It results from an increase in Na^+ conductance, causing rapid inward Na^+ flow.

— This inward Na^+ flow stops after a few milliseconds due to inactivation of Na^+ channels.

Phase 1

- This is an early slight membrane repolarization (membrane becomes more negative).
- It results from outward K^+ flow (due to a favorable electrochemical gradient) and a decrease in Na^+ conductance.

Phase 2

- This is the plateau of depolarization.
- In this phase, there is increased Ca^{2+} conductance. This creates an approximate balance between inward Ca^{2+} flow and outward K^+ flow.
- Phase 2 is responsible for the long duration of cardiac action potentials (150–250 msec) compared with those of nerve and skeletal muscle (1–2 msec), which have no such plateau phase.

Phase 3

- This is the repolarization phase.
- It results from a decrease in Ca^{2+} conductance and an increase in K^+ conductance. The net effect of this is a rapid outward K^+ flow, which repolarizes the membrane toward the K^+ equilibrium potential.

Phase 4

- This is the resting membrane potential.
- It is determined by high K^+ conductance, and its value is therefore close to the K^+ equilibrium potential (−96mV). At this potential, K^+ outflow and K^+ inflow are equal.

Table 8.1 compares the net ionic conductances of the atrial and ventricular myocardium, bundle branches, and Purkinje fibers and those at the SA node.

Table 8.1 ▶ Comparison Between the Net Ionic Conductances of the Atrial and Ventricular Myocardium, Bundle Branches, and Purkinje Fibers and Those at the Sinoatrial (SA) Node		
Phase	**Net Ionic Conductance at Atrial and Ventricular Myocardium, Bundle Branches, and Purkinje Fibers**	**Net Ionic Conductance at SA Node**
0	↑ inward Na^+ flow	↑ inward Ca^{2+} flow
1	↑ outward K^+ flow	-
2	Inward Ca^{2+} flow = outward K^+ flow	-
3	↑ outward K^+ flow	↑ outward K^+ flow
4	Resting membrane potential: outward K^+ flow = inward K^+ flow	Slow depolarization: ↑ inward Na^+ flow

Excitation–Contraction Coupling

- Depolarization of the cardiac muscle cell membrane triggers an action potential that passes through T tubules.
- During phase 2 (plateau) of the action potential, there is increased Ca^{2+} conductance, causing inward Ca^{2+} flow.

- This inward Ca^{2+} flow initiates the release of Ca^{2+} from the sarcoplasmic reticulum (Ca^{2+}-induced Ca^{2+} release). Such releases increase intracellular $[Ca^{2+}]$.
- Ca^{2+} binds to troponin C, and tropomysin moves out of its blocking position, allowing actin and myosin to form cross-bridges.
- The thick and thin filaments of actin and myosin slide past each other, resulting in cardiac muscle cell contraction.
- The contraction ends when Ca^{2+} ATPase facilitates the reuptake of Ca^{2+} into the sarcoplasmic reticulum, reducing the intracellular $[Ca^{2+}]$.

Note: The force of contraction of cardiac muscle cells is proportional to the amount of Ca^{2+} release, which varies depending on conditions.

8.3 Measuring the Electrical Activity of the Heart: The Electrocardiogram

Recording an ECG

The *electrocardiogram* (ECG) is a representation of the electrical activity in the heart (**Fig. 8.3**). It is measured by recording voltages through surface electrodes, which "look" at the heart from different positions. A wave of depolarization moving toward a lead causes an upward deflection of the ECG.

Interpretation of ECG Waves, Segments, and Intervals

Waves

- The P wave corresponds to atrial depolarization.
- The QRS complex corresponds to ventricular depolarization, which occurs from apex to base.

Note: Repolarization of the atria is masked by the QRS complex.

- The T wave corresponds to ventricular repolarization, which occurs from base to apex.

Fig. 8.3 ► **Electrocardiogram (ECG) curve.**
The ECG depicts electrical activity as waves, segments, and intervals.

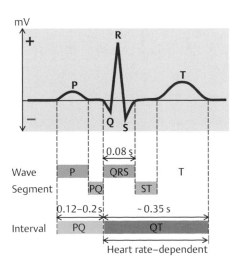

Cardiac axis

The cardiac axis is the direction of propagation through the ventricles when the QRS complex is at a maximum. The mean electrical axis is calculated by finding the QRS complex that is biphasic (deflections of the QRS complex are equally positive and negative), then finding the lead that is perpendicular to the biphasic lead that has a positive net deflection. This lead corresponds to the mean electrical axis. A normal axis is between −30 and +90 degrees. A value below −30° is left axis deviation and a value above +90° is right axis deviation. Many factors can alter the cardiac axis, including abnormal cardiac anatomy or position, infarction, ischemia, pulmonary embolism, cardiomyopathy, and conduction abnormalities.

Fig. 8.4 ► **Cardiac impulse spreading.**
The P wave corresponds to atrial depolarization. The PQ segment is an isoelectric period between atrial depolarization and the beginning of ventricular depolarization (impulse delayed at AV node). The Q wave of the QRS complex corresponds to depolarization of the septum from left to right. The R and S waves of the QRS complex corresponds to ventricular depolarization (in opposite directions). The ST segment is an isoelectric period between ventricular depolarization and repolarization. The T wave corresponds to ventricular repolarization.

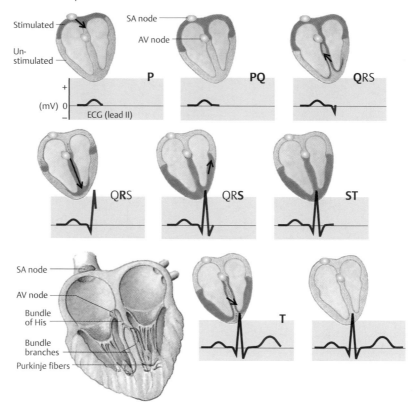

Figure 8.4 depicts cardiac impulse spreading and how this translates to waves of the ECG.

Segments

— The PQ segment is an isoelectric period between the P wave and the QRS complex. During this time, the wave of depolarization is traveling through the AV node into the bundle of His.
— The ST segment corresponds to the plateau phase of the cardiac action potential when all ventricular fibers are simultaneously depolarized.

Intervals

— The PQ interval corresponds to the conduction of the action potential through the AV node.
— The QT interval corresponds to ventricular depolarization and repolarization.

Information Provided by the ECG

Table 8.2 lists the information that may and may not be gathered by analysis of an ECG tracing.

Table 8.2 ▶ Electrocardiogram Information	
Information Provided	**Information Not Provided**
Heart rate and rhythm	Mechanical actions of the heart
Conduction pathways	Valve movements or pressures
Cardiac excitability	Contractile force of the myocardium
Cardiac refractoriness	
Relative size of the ventricles	
Some anatomical irregularities	
Pathological conditions	
(e.g., infarction and ischemia)	

9 The Heart as a Pump

9.1 Ventricular Performance

The ability of the heart to eject blood during a heartbeat is a function of three factors: the amount of cardiac filling (preload), the strength of contraction (contractility), and the pressure against which it ejects blood (afterload).
— *Systole* is the period when the ventricle is contracting, and ejection occurs.
— *Diastole* is the period when the ventricle is relaxing, and filling occurs.

Preload

Up to a point, the heart pumps more when it is filled more during diastole. This is often labeled Starling's law of the heart, or the *Frank–Starling mechanism.* The amount of filling is called *preload.* It is a reflection of forces in the vasculature acting to fill the ventricle. The amount of preload can be expressed in several ways, for example, end-diastolic volume, end-diastolic pressure, or stretch (sarcomere length).
— It will be low in cases of hypovolemia (low blood volume) or low systemic venous tone.
— It will be high in cases of fluid retention, many cases of heart failure, or excessive venous sympathetic stimulation.
— If the right ventricle is impaired but not the left, right ventricular preload will tend to be high, and left ventricular preload will tend to be low.

Contractility

Contractility (also called *inotropism*) expresses the ability of the heart to contract at a given preload. Greater contractility manifests as greater systolic pressure or greater systolic ejection. At the cellular level, contractility reflects the amount of Ca^{2+} released from the sarcoplasmic reticulum with each heartbeat. The supply of Ca^{2+} in the sarcoplasmic reticulum is increased (positive inotropy) by anything that stimulates more Ca^{2+} to enter the cell through Ca^{2+} channels, or that stimulates Ca^{2+} ATPase (sarco/endoplasmic reticulum Ca^{2+} ATPase [SERCA] pump) to take up Ca^{2+} from the cytosolic space. Both effects increase the amount of Ca^{2+} stored in the sarcoplasmic reticulum between beats.
— Positive inotropic effectors (i.e., those that cause an increase in contractility) include agents that cause a faster heart rate (more beats per minute allow more Ca^{2+} to enter per minute), for example, sympathetic stimulation via β_1-adrenergic receptors, drugs that are β_1-adrenergic receptor agonists, and cardiac glycosides (e.g., digoxin).
— Negative inotropic effectors (i.e., those that cause a decrease in contractility) include parasympathetic stimulation via muscarinic (M_2) receptors and β-adrenergic receptor antagonists (β-blockers).

Afterload

Afterload is the pressure that the ventricle works against to eject blood during systole. At rest, it is primarily a function of total peripheral resistance (TPR), i.e., it requires greater systolic pressure to eject blood in the face of high peripheral resistance. Afterload also depends on the output of the heart because pressure in the peripheral circulation is a function of the amount of blood ejected. For example, during exercise, peripheral resistance is low, which by itself would decrease afterload, but afterload is actually somewhat elevated because the heart is ejecting so much blood that mean arterial pressure is increased. High afterloads increase the work of the heart, and therapy for hypertension includes drugs that reduce TPR. A stenosed (narrowed)

Digoxin

Digoxin is a cardiac glycoside that was once one of the first-line agents used in the treatment of heart failure. Its use is now reserved for cases when symptoms are not fully treated by standard therapies or in cases of severe heart failure while standard therapies are initiated. The therapeutic and toxic effects of digoxin are attributable to inhibition of Na^+–K^+ ATPase (the digitalis receptor) located on the outside of the myocardial cell membrane. When the pump is inhibited, Na^+ accumulates intracellularly. The decreased Na^+ gradient that results from this affects Na^+–Ca^{2+} exchange, and Ca^{2+} accumulates intracellularly. Consequently, more Ca^{2+} (stored in the sarcoplasmic reticulum) is available for release and interaction with the contractile proteins in these cells during the excitation–contraction coupling process. At therapeutic doses of digoxin, there is an increase in contractile force. Toxicity to digoxin also relates to inhibition of Na^+–K^+ ATPase. Inhibition of the Na^+–K^+ pump affects the K^+ gradient; this may lead to a significant reduction of intracellular K^+, predisposing the heart to arrhythmias. Likewise, high levels of Ca^{2+} intracellularly may contribute to serious arrhythmias.

aortic valve can greatly increase afterload because the left ventricle has to generate a much higher pressure to eject through the restriction.

9.2 Cardiac Output

Cardiac output is the amount of blood pumped by either ventricle each minute. It is expressed as follows:

$$\text{Cardiac output} = \text{stroke volume} \times \text{heart rate,}$$

where *stroke volume* is the volume of blood ejected from the ventricles during each contraction. It is expressed as follows:

$$\text{Stroke volume} = \text{end-diastolic volume} - \text{end-systolic volume}$$

Ejection fraction is the percentage of ventricular end-diastolic volume that is ejected during a given contraction. It is expressed as follows:

$$\text{Ejection fraction} = \text{stroke volume/end-diastolic volume}$$

Stroke work is the work that the heart does during each heartbeat. *Cardiac work* is the work done by the heart each minute. For the left ventricle, these are expressed as follows:

$$\text{Stroke work} = \text{aortic pressure} \times \text{stroke volume}$$

$$\text{Cardiac work} = \text{stroke work} \times \text{heart rate}$$

Measuring of Cardiac Output

Cardiac output can be assessed by a variety of methods. Although now superseded by newer techniques, the simplest method conceptually is built around the Fick principle. In this method, the O_2 content of systemic arterial blood (from the pulmonary vein) and mixed venous blood (from the pulmonary artery) is measured. The difference between the two is the amount of O_2 taken up per unit volume of blood that passes through the lungs and heart. The rate of pulmonary O_2 uptake is also measured by spirometry. When the arteriovenous difference is divided into the rate of O_2 uptake, the result is cardiac output. The formal expression is

$$CO = V_{O_2}/(a_{O_2} - v_{O_2})$$

where CO is cardiac output (L/min), V_{O_2} is the rate of O_2 uptake (mL O_2/min), a_{O_2} is systemic arterial blood O_2 content (mL O_2/L), and v_{O_2} is mixed venous blood O_2 content. (mL O_2/L).

Example: A patient consumes O_2 at a rate of 230 mL/min. Systemic arterial blood contains 180 mL O_2/L, and mixed venous blood contains 130 mL O_2/L. What is his cardiac output?

$$CO = 230 \text{ mL } O_2/\text{min}/(180 \text{ mL } O_2/\text{L} - 130 \text{ mL } O_2/\text{L})$$
$$= 4.6 \text{ L/min}$$

Current methods for measuring cardiac output include the thermodilution technique, where a bolus of cold saline of known quantity and temperature is injected into the right atrium, where it mixes with warm blood and slightly cools it. Downstream, in the pulmonary artery, a transducer measures the amount of cooling. The less cooling, the greater the volume of warm

Heart failure

Heart failure is a pathophysiological state when cardiac output is insufficient to allow adequate perfusion and oxygenation of tissues. There are several etiological factors, including intrinsic disease of the heart muscle (e.g., ischemia, myocardial infraction [MI], cardiomyopathy, and myocarditis), chronic elevated preload (e.g., mitral regurgitation and fluid overload due to chronic renal disease), chronic elevated afterload (e.g., aortic stenosis and hypertension), disorders of cardiac filling (e.g., cardiac tamponade and pericarditis), congenital cardiac malformations, arrhythmias, diabetes, severe anemia, hyperthyroidism, and amyloidosis. Treatment is with the following drugs (in order of use): diuretics, angiotensin-converting enzyme (ACE) inhibitors, β-blockers, positive inotropic drugs (e.g., digoxin), and nitrates.

Cardiac muscle metabolism

Cardiac muscle requires an abundant supply of O_2-rich blood, as it depends almost exclusively on aerobic metabolism to supply the adenosine triphosphate (ATP) for its contractions. Other tissues can vary their extraction of O_2 from blood and survive with anaerobic metabolism. Cardiac tissue has a very high fractional extraction of O_2 and can only increase its uptake of O_2 by increasing coronary blood flow. At rest, the heart uses oxidation of fatty acids for its ATP, and only small quantities of glucose are used. When cardiac workload is increased, cardiac muscle removes lactic acid from coronary blood and oxidizes it directly.

blood mixed with the saline. Mathematical analysis then calculates cardiac output. A noninvasive method uses echocardiography to estimate chamber volumes and calculate cardiac output.

Autonomic Nervous System Regulation of Heart Rate, Conduction Velocity, and Contractility

The autonomic innervation of the heart is shown in **Fig. 9.1**.

Heart Rate and Conduction Velocity

— The parasympathetic nervous system decreases heart rate and conduction velocity.
  — It decreases heart rate by slightly hyperpolarizing pacemaker cells, decreasing the rate of diastolic depolarization in sinoatrial (SA) nodal cells and by increasing K^+ conductance in phase 4 of the action potential.
  — It decreases AV node conduction velocity by decreasing the inflow of Ca^{2+} and increasing the outflow of K^+.
— The sympathetic system increases heart rate and conduction velocity.
  — It increases heart rate by increasing the rate of diastolic depolarization in SA nodal cells by increasing Ca^{2+} or Na^+ conductance in phase 4 of the action potential.
  — It increases AV node conduction velocity by increasing the inflow of Ca^{2+}.

Contractility

— Parasympathetic action slows the heart rate, but it has little influence on contractility.
— Sympathetic (β-adrenergic) stimulation raises both heart rate and contractility. Key sites of sympathetic action include surface membrane Ca^{2+} channels and the SERCA pump.
— Phosphorylation of Ca^{2+} channels via cyclic adenosine monophosphate (cAMP) and protein kinase A allows more Ca^{2+} to enter the myocytes during phase 2 of the action potential.
— Phosphorylation of the protein phospholamban on the sarcoplasmic reticulum stimulates the SERCA pump to pump faster, thereby storing more Ca^{2+}, which is then released on subsequent beats. The increased pumping also decreases the period when Ca^{2+} is available to bind to troponin, thereby shortening the period of systole. This results in contractions that are more forceful and shorter in duration.

Fig. 9.1 ▶ **Autonomic innervation of the heart.**
Postganglionic parasympathetic fibers (*blue*) and sympathetic fibers (*red*) innervate the sinoatrial (SA) node and the atrioventricular (AV) node. This allows both divisions of the autonomic nervous system to modulate heart rate and conduction velocity. Postganglionic sympathetic fibers also innervate ventricular muscle, allowing it to modulate contractility.
From *Atlas of Anatomy*, © Thieme 2008, Illustration by Markus Voll.

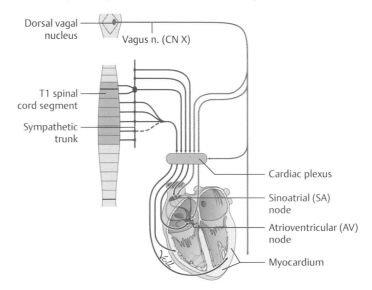

Table 9.1 summarizes the effects of the autonomic nervous system (ANS) on heart rate, conduction velocity, and contractility.

ANS Division	Neurotransmitter	Receptor	Effect on Heart Rate	Effect on Conduction Velocity	Effect on Contractility
Parasympathetic (vagus nerve)	Acetylcholine	M_2 receptors	↓	↓	–
Sympathetic (C7–T6)	Norepinephrine	β_1	↑ *	↑	↑

Table 9.1 ► Effects of the Autonomic Nervous System (ANS) on Heart Rate, Conduction Velocity, and Contractility

* Increased heart rate is caused by a simultaneous decrease in cardiac vagal activity (disinhibition) and an increase in cardiac sympathetic activity.

9.3 The Cardiac Cycle

The *cardiac cycle* is the set of repetitive events where the ventricles sequentially fill, contract, eject blood, and relax. The right and left ventricles do this in parallel and almost simultaneously. The cardiac cycle generates signals that can be detected and interpreted to determine the functional state of the heart. The most common of these signals used clinically are the electrocardiogram (ECG) and heart sounds. The ECG provides information about excitation of the myocardium, a process that is required to signal contraction. Heart sounds indicate the closure of valves and detect unusual blood flow patterns, such as turbulent flow (murmurs).

The sequence of events that comprise the cycle for the left ventricle is broadly divided into periods of systole and diastole. These are then further subdivided into phases. **Figure 9.2** illustrates the events that occur at each phase of the cardiac cycle.

Systole

Phase I

— Ventricular systole begins immediately after ventricular excitation (indicated by the QRS complex on the ECG).

— The rise in pressure causes closing of the mitral valve (and in the right ventricle, closing of the tricuspid valve). The sudden change in momentum of moving blood upon valve closure creates the first heart sound. Normally, the two valves close nearly simultaneously and are heard as one sound, but splitting into two may occur in some pathologies.

— As pressure rises before the aortic valve opens, there is a short period of *isovolumetric contraction*. The ventricle changes shape, but not volume, because both valves are closed.

Phase IIa

— When chamber pressure exceeds aortic pressure, the aortic valve opens, and a period of rapid ejection begins. Ventricular volume falls, but pressure continues to rise because active contractile force in the myocytes is still increasing and because the decreasing chamber radius allows wall tension to generate more pressure (as described by Laplace's law where pressure is inversely proportional to radius).

Phase IIb

— Pressure reaches a peak and starts to decline. Only a small amount of blood continues to be ejected. The ventricle does not empty completely but retains at least 30 to 50% of the end-diastolic volume.

— Usually two-thirds of the stroke volume is ejected during the first third of systole.

— The T wave of the ECG occurs, indicating ventricular repolarization.

"Cardioselective" β blockers

Blockage of β_1-adrenergic receptors in the heart reduces heart rate and contractility and delays atrioventricular (AV) conduction. Blockage of β_2 receptors causes arteriolar vasoconstriction (including the coronary arteries) and bronchoconstriction, so agents that act selectively on β_1 receptors (e.g., atenolol) are used to treat cardiovascular disease. These so-called cardioselective β_1 blockers are used to reduce myocardial oxygen (O_2) demand in angina and myocardial infarction (MI), to control hypertension, to depress AV nodal conduction in atrial fibrillation and atrial flutter, and to prevent ventricular fibrillation during the first 2 years following a myocardial Infraction (MI).

Murmurs

Murmurs are sounds produced by turbulent blood flow through heart valves. Stenosis is narrowing of a heart valve. It produces a murmur at the stage of the cardiac cycle when the valve is normally open. Regurgitation is when blood flows back across a valve when the valve is normally closed. It occurs at the corresponding part of the cardiac cycle. Murmurs may be more easily distinguished from one another by proper use of the stethoscope (bell vs diaphragm), by the position where the murmur is most audible, and how it is affected by inspiration, expiration, exercise, and the Valsalva maneuver.

Fig. 9.2 ► **Cardiac cycle**.
The cardiac cycle consists of systole (contraction) and diastole (relaxation). The electrocardiogram (ECG) measures the electrical activity of the heart. The pressures in the aorta, left ventricle and atrium, and right atrium (central venous pressure) vary according to the contraction stage. The left ventricular volume and aortic flow show the amount of blood pumped into systemic circulation. The four heart sounds are observed at specific stages in the cardiac cycle. The duration of systole is less variable than the duration of diastole, which varies considerably. (ESV, end-systolic volume)

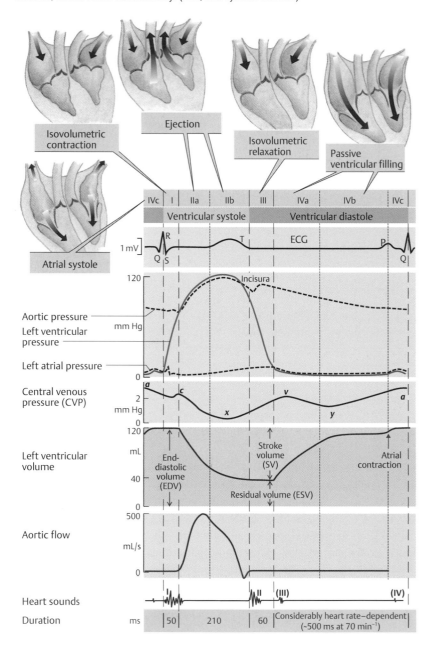

Diastole

Phase III

— Pressure falls rapidly and the aortic valve closes, beginning a period of *isovolumetric relaxation*. Normally, the pulmonary valve closes at almost the same time. The sudden change in momentum of moving blood creates the second heart sound.

Phase IV (a and b)

—   The mitral valve opens, and the left ventricle begins filling rapidly, then more slowly. Usually two-thirds of the filling occurs during the first third of diastole. This is the longest phase of the cardiac cycle, and the most variable. Its duration is greatly shortened at high heart rates and extended at very slow heart rates.

Phase IVc

—   The atria contract at the very end of ventricular diastole, producing the P wave of the ECG. Some additional blood flows into the ventricle. The additional volume is not significant at rest but takes on a more important role at high heart rates when the period of diastolic filling is much shorter.

Depicting the Cardiac Cycle: Graphing Pressure–Volume Loops

A useful way to display the events of the cardiac cycle is via a plot of chamber pressure versus chamber volume, known as a *pressure–volume loop* (**Fig. 9.3**).
—   The difference between end-diastolic and end-systolic volumes is the stroke volume.
—   The vertical portions of the loop (constant volume) represent the periods of isovolumetric contraction and isovolumetric relaxation.

Fig. 9.3 ▶ **Pressure–volume loop**.
A pressure–volume loop shows the relation between ventricular pressure and volume during a complete cardiac cycle. Valves open or close at the corners of the loop. As the stroke volume is ejected, ventricular pressure continues to rise to a peak, then falls until the aortic valve closes. The area within the loop is the stroke work.

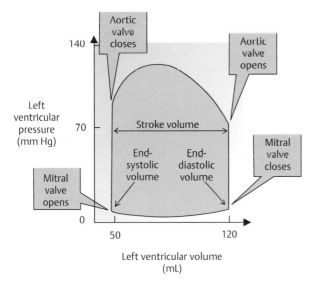

Atrial septal defect

Atrial septal defect (ASD) is a hole between the two atria of the heart that allows blood to shunt between these chambers. It is often not detected until adulthood, when it presents with cyanosis (bluish or purplish tinge to the skin and mucous membranes, caused by lack of oxygen to the blood), dyspnea (especially on exertion), fatigue, palpitations, and edema (a collection of excess fluid in the cavities or tissues of the body). Signs of ASD include arrhythmias (especially atrial fibrillation), a split second heart sound, and a pulmonary ejection systolic murmur that is appreciated best at the second intercostal space at the left sternal edge. Complications may include heart failure and shunt reversal. Small ASDs may heal spontaneously, but larger defects require surgical closure with a patch. Medications (e.g., β-blockers and digoxin) may be required for symptomatic control. Anticoagulant therapy (e.g., warfarin) may be required to reduce the risk of stroke.

Ventricular septal defect

Ventricular septal defect (VSD) is a hole between the two ventricles of the heart that allows blood to shunt between these chambers. This is the most common congenital heart defect and is often associated with Down syndrome, tetralogy of Fallot (see p. 101), and Turner syndrome. VSD may also occur due to septal rupture following a myocardial infarction (MI). Signs and symptoms include cyanosis, dyspnea (shortness of breath), fatigue, edema, tachycardia, and pansystolic murmur, which is appreciated best at the fourth intercostal space. Complications include Eisenmenger syndrome (pulmonary hypertension and shunt reversal), endocarditis (due to turbulent blood flow in a high-pressure system), stroke, and heart failure. Small VSDs may heal spontaneously, but larger defects require surgical closure with a patch. Medications (e.g., β-blockers and digoxin) may be required for symptomatic control. Diuretics may be used to reduce preload in heart failure.

Tetralogy of Fallot

Tetralogy of Fallot is a rare combination of four cardiac defects: pulmonary valve stenosis, overriding aorta, VSD, and right ventricular hypertrophy. Symptoms include cyanosis, dyspnea, fainting, finger clubbing, fatigue, and prolonged crying in infants. Signs include systolic murmur appreciated at the left sternal base, thrill (palpable murmur), and an abnormal second heart sound. Treatment includes corrective surgery in infancy. This usually involves intracardiac repair of the VSD with a patch and repair of the stenosed pulmonary valve. Complications following surgery, such as arrhythmias, can be treated with the appropriate antiarrhythmic agent or by implantation of a pacemaker or internal defibrillator.

10   The Circulation

10.1   Course of the Circulation

Figure 10.1 is an overview of the circulation. Its course is as follows:

— Blood from the superior and inferior vena cavae enters the right atrium.
— It then passes to the right ventricle through the tricuspid valve.
— From the right ventricle, it passes to the pulmonary artery through the pulmonic valve and into the lungs, where it is oxygenated.
— Oxygenated blood from the lungs enters the pulmonary vein and travels to the left atrium.
— It then passes to the left ventricle through the mitral valve.
— Blood is ejected from the left ventricle into the aorta through the aortic valve.
— It then travels through arteries to the tissues.
— Mixed venous blood from the tissues then drains into the vena cavae.

Fig. 10.1 ▶ **Circulation.**
Red: Oxygenated blood. Blue: Deoxygenated blood.
From *Atlas of Anatomy*, © Thieme 2008, Illustration by Markus Voll.

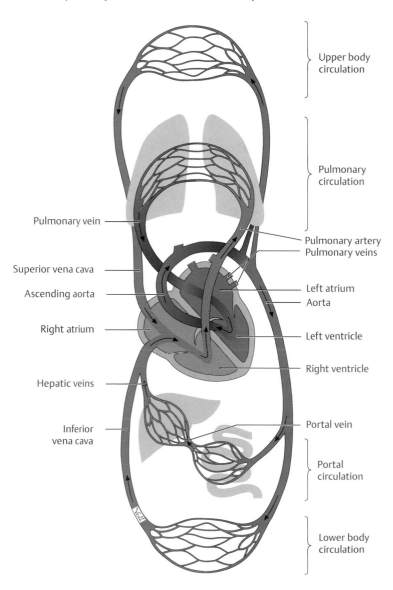

10.2 Vascular Components of the Circulatory System

Refer to **Fig 10.2.**

Arteries

- Arteries are high-pressure vessels that carry oxygenated blood from the heart to the rest of the body.
- They have thick walls that are composed of elastic tissue (elastin, collagen, and connective tissue) and smooth muscle.

Fig. 10.2 ▶ **Characteristics of vessels.**
The diameters of vessels leaving and entering the heart are large, whereas the diameters of capillaries are very small. The aggregated cross-sectional area of capillaries is large because they are an order of magnitude more numerous than arteries and veins. The holding capacity of veins is much greater than arteries or capillaries. Because they are extremely compliant, veins can expand to permit great increases in blood volume. (TPR, total peripheral resistance)

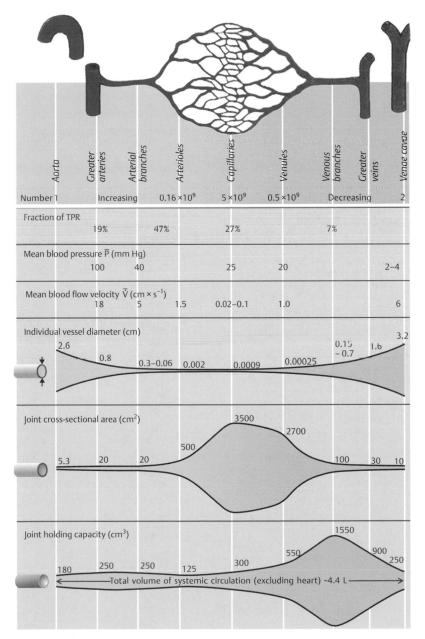

> **■ Aneurysms**
>
> Aneurysms are bulges in arteries, especially serious in the aortic arch, descending abdominal aorta, and the brain. Increasing dilatation causes increasing tension in the wall of the artery by the law of Laplace, which states that wall tension is proportional to pressure times radius. The increased tension contributes to increased diameter of a vessel and risk of rupture which is usually fatal. Aneurysms more than two inches wide are surgically repaired by inserting a synthetic graft into the vessel.

— The large arteries, particularly the aorta, expand elastically to contain the stroke volume and produce a continuous flow.

Arterioles

— Arterioles control blood flow and distribute blood into the capillary beds.
— Their walls are composed primarily of smooth muscle.
— They are the major sites of controllable resistance in the systemic circulation. Resistance is controlled by multiple factors, including local metabolism, vasoactive substances such as angiotensin II, and the sympathetic nervous system acting at α_1 receptors (producing vasoconstriction) or β_2 receptors (producing vasodilation).

Capillaries

— Capillaries facilitate the exchange of gases, fluid, and nutrients between the blood and the interstitial space.
— Their walls are thin, consisting of a single layer of endothelial cells surrounded by a basal lamina. In most tissues, the endothelial cells have small pores or gaps between cells that permit transcapillary exchange of small water-soluble substances by diffusion. Large water-soluble substances can cross capillary endothelial cell membranes by transcytosis in vesicles.
— The cumulative cross-sectional area of the capillaries is high.

Venules

— Venules are small veins that collect blood from the capillaries.

Veins

— Veins are low-pressure vessels that return blood back to the heart via the venae cavae. They also act as expandable reservoirs.
— They passively relax or actively constrict under sympathetic adrenergic stimulation of α_1 receptors.
— The contractile state of large veins is critically important in setting central venous pressure, which determines the preload of the right ventricle. In turn, this has a major effect on cardiac output, as the heart cannot pump more blood than is delivered to it by venous return.

10.3 Circulatory Hemodynamics

Blood Flow

Blood flow is defined as the volume of blood moving through a vessel per unit of time. It is expressed by

$$Q = v \times A$$

where Q is blood flow (mL/min), v is velocity (cm/sec), and A is cross-sectional area (cm^2).

— Blood flow is proportional to the product of velocity and cross-sectional area.
— The velocity of blood flow is faster through vessels with a low cross-sectional area; for example, the velocity of blood flow is faster in the aorta (small cross-sectional area) than in the capillaries (large cumulative cross-sectional area).

Relationship of Pressure and Resistance to Blood Flow

Blood flow through vessels is related to the driving pressure and the resistance to flow, as expressed by

$$Q = \Delta P/R$$

where Q is blood flow (mL/min), ΔP is pressure gradient (mm Hg), and R is resistance (mm Hg/mL/min).

— Blood flow is directly proportional to the driving pressure gradient and inversely proportional to resistance.
— The greater the driving pressure gradient, the greater the blood flow.
— The greater the resistance, the lower the blood flow.

Resistance is derived from Poiseuille's equation, as expressed by

$$R = 8\eta L/\pi r^4$$

where R is resistance, r is radius of the blood vessel, η is viscosity of the fluid, and L is length of the blood vessel.

— Resistance is directly proportional to the viscosity of blood, which is mainly determined by the concentration of cellular elements and proteins. If viscosity increases, then resistance increases, and blood flow will decrease. The converse is also true.
— Resistance is directly proportional to the length of the vessel.
— Because viscosity and the length of the vessels are relatively constant in the circulatory system, which is a closed system, the principal variable controlling resistance to blood flow is the radius of the vessel. Resistance is inversely proportional to the fourth power of the radius. Therefore, small changes in radius cause large changes in resistance. For example, a 19% increase in radius would halve the resistance.

Parallel and series resistance

Resistance to flow in a network of vessels depends on the characteristics of the individual vessels and how they are connected.

— When vessels are connected end to end (all the blood flows through each vessel in series), the resistance of a network is the sum of the individual vessel resistances and is greater than the resistance of any one vessel. This can be expressed as

$$R_{total} - R_1 + R_2 + R_3 + \ldots$$

— When vessels are connected in parallel (the blood divides among the different vessels), the resistance of a network varies in an inverse relationship. The total resistance is less than that of any one vessel. This can be expressed as

$$R_{total} = 1/(1/R_1 + 1/R_2 + 1/R_3 + \ldots)$$

Example: Two vessels each have a resistance value of 3 units. When connected in series, their combined resistance is 3 + 3 = 6 resistance units. When connected in parallel, their combined resistance is $1/(\frac{1}{3} + \frac{1}{3})$ = 1.5 resistance units.

Laminar and turbulent flow

— Laminar flow is streamlined and is the most efficient way to move blood. Maximum flow velocity occurs in the center of the bloodstream in nonturbulent vessels; minimum velocity occurs next to the wall of vessels.
— Turbulent flow, where the flow pattern is chaotic, causes *bruits* (audible vibrations). The tendency for turbulent flow increases with
   — ↑ velocity of blood flow (e.g., stenosed [narrowed] vessels or cardiac valves)
   — ↓ blood viscosity (e.g., severe anemia)

Regulation of Blood Flow

Blood flow through peripheral tissues is under multiple controls. Most controls can be broadly grouped into local control mechanisms and central control mechanisms that act via the sympathetic nervous system. In the majority of tissues, particularly the coronary and cerebral circulations, local control is the dominant influence. Central control is dominant in a few tissues (e.g., in the skin). There are other circulating regulatory mediators whose levels may increase sharply in pathological situations.

Local Control

Local control keeps blood flowing through a given vascular bed at a level appropriate for its metabolic activity, rising during high metabolic activity and falling during low metabolic activity. Local control reflects the interplay between numerous vasoactive substances produced by local tissues and the endothelium.

Autoregulation

Autoregulation is the process where tissues adjust the resistance of their arterioles to maintain blood flow constant in the face of changes in blood pressure (mean arterial pressure). Without autoregulation, flow would vary in parallel with mean arterial pressure, which varies throughout the day, and would often be too high or too low.

Active and reactive hyperemia

Active hyperemia is an increase in local blood flow in response to an increase in local metabolism, as occurs during muscle contraction or glandular secretion.

Reactive hyperemia is a transient increase in blood flow following release of an occlusion.

Mechanisms of Local Control

Myogenic mechanism

Arteriolar smooth muscle stretches in response to an increase in arterial pressure (**Fig. 10.3**). This opens stretch-gated channels, leading to depolarization, an influx of Ca^{2+}, and contraction that reduces blood flow back to its previous level. Conversely, a fall in arterial pressure relaxes arteriolar smooth muscle, allowing flow to rise to its previous level.

— This mechanism explains autoregulation but not active or reactive hyperemia.

Local tissue metabolic factors

Metabolizing tissue produces vasodilator metabolites (e.g., CO_2, H^+, and adenosine) and consumes O_2. Arteriolar smooth muscle dilates in response to an increase in vasodilator metabolites and fall in local O_2 concentration. These changes occur during a rise in local metabolism or a decrease in local blood flow.

— These factors are responsible for active and reactive hyperemia and for part of autoregulation.

Endothelium-produced factors

The endothelium that lines the arterioles is a source of substances that relax or constrict the surrounding smooth muscle. These substances are the following:

— *Nitric oxide (NO).* This is the most important endothelium-derived vasoactive factor. It is produced constitutively and increases in response to increased shear stress (faster blood velocity). NO acts by stimulating the production of cyclic guanosine monophosphate (cGMP), which activates a kinase whose actions inhibit smooth muscle contraction, causing vasodilation. NO is also an important inhibitor of platelet aggregation.

— *Prostacyclin (PGI$_2$).* PGI$_2$ is an eicosanoid that acts via cyclic adenosine monophosphate (cAMP) and protein kinase A to inhibit smooth muscle contraction, causing vasodilation.

— *Endothelium-derived hyperpolarizing factor (EDHF).* This factor relaxes vascular smooth muscle, causing vasodilation by opening K^+ channels. This hyperpolarizes vascular muscle cells which decreases their cystolic Ca^{2+} concentration.

— *Endothelin.* This is a peptide that causes potent vasoconstriction of vascular smooth muscle.

Sympathetic Control

— Vasoconstriction of blood vessels and subsequent decrease in blood flow occur as a result of norepinephrine binding to α_1-adrenergic receptors.

Collateral circulation

Collateral circulation is a type of long-term local blood flow regulation. When normal blood flow becomes partially or completely blocked, small collateral vessels enlarge and assume the major role in supplying blood to that region. At first, these vessels can supply only a small fraction of the normal blood supply, but with sufficient time, they can return flow to near-normal levels. Regular aerobic exercise may promote collateral vessel development in coronary arteries.

Angiogenesis

Angiogenesis is an increase in the number and size of the vessels supplying local vascular beds in response to a decrease in perfusion pressure. It is a mechanism for long-term regulation of blood flow. Cancer cells can trigger angiogenesis, causing nearby blood vessels to branch into them to provide nutrients and oxygen.

Nitrates

Nitrates are vasodilator drugs that break down or are metabolized in the body to produce nitric oxide (NO). Their sites of action depend on how and where this occurs. Some, such as nitroglycerin, act primarily in the venous system to reduce central venous pressure. This reduces cardiac preload and cardiac work and decreases the coronary demand for oxygen (O_2), which is one reason why they help relieve angina. Others have a more global action; for example, sodium nitroprusside spontaneously forms NO throughout the body and greatly reduces total peripheral resistance, thereby lowering blood pressure. The nitrates in general are short-acting drugs because they are metabolized quickly and because NO itself has a lifetime of only seconds.

Fig. 10.3 ▶ **Vasoconstriction and vasodilation.**
Neuronal regulation of blood flow is controlled by the sympathetic nervous system (**1a,b**).
Vasoconstriction is achieved by norepinephrine binding to α_1-adrenergic receptors on blood vessels.
Vasodilation is achieved by decreasing the tone of the sympathetic system. The salivary glands and the
genitals dilate in response to parasympathetic stimuli to which vasoactive substances (bradykinin and
nitric oxide [NO], respectively) act as the mediators. Local control of blood flow (**2a,b**) occurs in response
to myogenic effects and a change in oxygen concentration in tissues. Vasodilation may also occur in
response to an increase in metabolic products (e.g., CO_2, H^+, adenosine diphosphate [ADP], adenosine
monophosphate [AMP], adenosine, and K^+). Vasoactive hormones (**3a,b**) either have a direct effect on
the vascular musculature (e.g., epinephrine) or lead to the local release of vasoactive substances (e.g.,
NO and endothelin) that exert local paracrine effects. NO acts as a vasodilatory agent. It is released from
the endothelium when acetylcholine (via M receptors), endothelin-1 (via ET_B receptors), or histamine (via
H_1 receptors) binds with an endothelial cell. NO then diffuses to and relaxes vascular muscle cells in the
vicinity. Endothelin-1 causes vasodilation by causing the release of NO. It can also cause vasoconstriction
in the vascular musculature. When substances such as angiotensin II and antidiuretic hormone (ADH)
bind to the endothelial cells, they release endothelin-1, which diffuses to and constricts the adjacent
vascular muscles (via ET_A receptors). High concentrations of epinephrine have a vasoconstrictive effect
(via α_1-adrenergic receptors), whereas low concentrations exert vasodilatory effects (via β_2-adrenergic
receptors). Eicosanoids: prostaglandin $F_{2\alpha}$ ($PGF_{2\alpha}$) and thromboxane A_2 (released from platelets) have
vasoconstrictive effects, whereas prostacyclin (PGI_2), released from endothelium, and prostaglandin E_2
(PGE_2) have vasodilatory effects. Endothelium-derived hyperpolarizing factor (EDHF) causes vasodilation
by opening K^+ channels in vascular muscle cells, hyperpolarizing them, leading to a drop in cytosolic Ca^{2+}
concentration. Bradykinin and kallidin are vasodilatory agents cleaved from kininogens in plasma by the
enzyme kallikrein. (ATP, adenosine triphosphate)

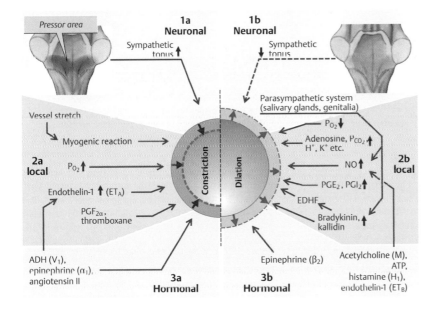

— Vasodilation of blood vessels and subsequent increase in blood flow occur due to a decrease
in sympathetic tone.

Vascular Compliance

Vascular compliance, or capacitance, describes the distensibility (ability to stretch) of blood
vessels. It is expressed by

$$C = \Delta V / \Delta P$$

where C is compliance, ΔV is change in the volume of a vessel (mL), and ΔP is change in pressure
(mm Hg).

— The large veins are highly compliant, which allows them to accommodate changes in blood
volume with relatively small changes in pressure.
— The order of compliance from high to low is veins > aorta > arteries > arterioles.

10.4 Arterial Pressure

Note: By convention, blood pressure, if not otherwise specified, refers to arterial pressure.

Systolic and Diastolic Pressure

Each cardiac ejection raises pressure in the aorta and major arteries, reaching a peak value called the *systolic pressure*. It begins to fall in late systole and continues to fall throughout diastole until the beginning of the next ejection. The value at that point is called the *diastolic pressure*. **Figure 10.4** illustrates how blood pressure is measured with a sphygmomanometer.

Pulse Pressure

Pulse pressure is the difference between the systolic and diastolic pressures.

— Stroke volume is the main determinant of pulse pressure. The capacitance of arteries is relatively low compared with veins; therefore, any increase in stroke volume during ventricular systole will increase systolic pressure. Diastolic pressure remains unchanged throughout systole, so the increase in systolic pressure will equal the increase in pulse pressure.

Mean Arterial Pressure

Mean arterial pressure (MAP) is the average pressure measured throughout the cardiac cycle. A typical value is ~95 mm Hg.

At rest, diastole occupies about two-thirds of the cardiac cycle, and systole occupies about one-third of the cycle; therefore, MAP is always less than their arithmetic average. It can be estimated from the following:

$$MAP = \text{diastolic pressure} + \tfrac{1}{3}\ \text{pulse pressure}$$

MAP is a function of cardiac output (CO) and total peripheral resistance (TPR):

$$MAP = CO \times TPR$$

— CO is determined by stroke volume and heart rate.
— TPR is determined by the series/parallel combination of the vessels in peripheral tissues.

▌ Hypertension

Hypertension is abnormally high blood pressure. *Essential hypertension* is the term used when the cause is unknown. It accounts for 90% of all cases. The other 10% of hypertension cases are secondary to diseases such as renal artery stenosis, polycystic kidneys, pyelonephritis, glomerulonephritis, diabetes mellitus, Cushing syndrome, Conn syndrome, pheochromocytoma, hyperparathyroidism, coarctation of the aorta, and preeclampsia. Pain can also be a cause of hypertension. Chronic hypertension can lead to end-organ damage, including left ventricular hypertrophy, renal failure, peripheral vascular disease, stroke, transient ischemic attacks (TIAs), MI, congestive heart failure, and cerebral encephalopathy. Drug treatment of hypertension includes diuretics (mainly thiazides), angiotensin antagonists (ACE inhibitors, angiotensin II inhibitors, and rennin inhibitors), β-blockers, Ca^{2+}-channel blockers, α_1-adrenergic receptor blockers, α_2-adrenergic receptor agonists, and direct vasodilators.

Fig. 10.4 ▶ Blood pressure (BP) measurement with a sphygmomanometer.
The BP is routinely measured at the level of the heart by a sphygmomanometer. An inflatable cuff is snugly wrapped around the arm, and a stethoscope is placed over the brachial artery. While reading the manometer, the cuff is inflated to a pressure higher than the expected systolic pressure (P_s; the radial pulse disappears). The air in the cuff is then slowly released (~2–4 mm Hg/s). The first sounds synchronous with the pulse (Korotkoff sounds) indicate that the pulse has fallen below the P_s. This value is read from the manometer. These sounds first become increasingly louder, then more quiet and muffled, and eventually disappear when the cuff pressure falls below the diastolic pressure (P_d; second reading).

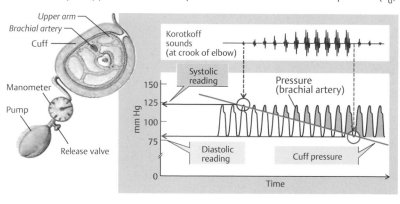

Neural Regulation of Arterial Pressure

Baroreceptor Reflex

The baroreceptor reflex allows the body to compensate rapidly for changes in arterial pressure. It is mediated by receptors sensitive to mechanical stretch that are located in the carotid sinuses and in the walls of the aortic arch.

The mechanism of the baroreceptor reflex is as follows:

— Decreased arterial pressure causes carotid sinus baroreceptors to undergo a reduced amount of stretch. This decreases the rate of action potential firing in the carotid sinus nerve, a branch of the glossopharyngeal nerve (cranial nerve [CN] IX). The aortic arch baroreceptors are innervated by branches of the vagus nerve (CN X) and acts in a similar manner. Impulses from these baroreceptors are relayed to the vasomotor center in the medulla oblongata, which increases sympathetic outflow to the heart and blood vessels and reduces parasympathetic outflow to the heart. This results in the following:

— ↑ heart rate
— ↑ contractility, stroke volume, and cardiac output
— ↑ venoconstriction (**Fig. 10.5**). This reduces the compliance of veins, resulting in an increase in venous return to the heart. According to the Frank–Starling mechanism, increased venous return increases filling pressures (preload) such that CO is increased.
— ↑ vasoconstriction of arterioles. This increases TPR and therefore arterial pressure.

There are additional baroreceptors in the walls of the heart, particularly the atria, and in the pulmonary vessels. These are called *cardiopulmonary* or *low-pressure baroreceptors* (because the pressures they detect are lower than arterial pressure). Their role is to detect fullness of the

Carotid sinus massage

Carotid sinus massage slows the heart rate via the baroreceptor reflex. It is a useful noninvasive procedure for termination of supraventricular tachycardia (SVT). It is also useful for differentiating SVT from ventricular tachycardia (VT), as VT will be unaffected by carotid sinus massage.

Jugular venous pressure

The internal jugular vein passes medial to the clavicular head of the sternocleidomastoid muscle up behind the angle of the mandible. It is a reliable manometer of right atrial pressure. It is not normally visible or palpable but may become distended in right ventricular failure.

Fig. 10.5 ▶ **Blood pressure regulation via vascular smooth muscle.**
Postsynaptic sympathetic fibers release norepinephrine, which acts at α_1-adrenergic receptors, causing contraction of vascular smooth muscle, thus increasing blood pressure. Conversely, circulating epinephrine acts at β_2 adrenergic receptors to cause dilation of vascular smooth muscle. Postsynaptic parasympathetic fibers do not innervate vascular smooth muscle.
From *Atlas of Anatomy*, © Thieme 2008, Illustration by Markus Voll.

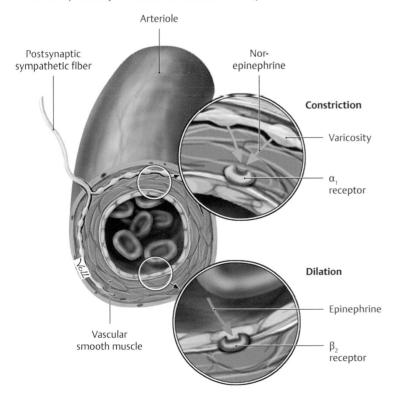

Arteriole

Postsynaptic sympathetic fiber

Norepinephrine

Constriction

Varicosity

α_1 receptor

Vascular smooth muscle

Dilation

Epinephrine

β_2 receptor

vascular system; that is, they serve as de facto blood volume detectors. Mechanistically, they work the same way as the arterial baroreceptors.

Table 10.1 summarizes the vasomotor response to the baroreceptor reflex.

Table 10.1 ► **Summary of Vasomotor Response to the Baroreceptor Reflex**		
Parameter Changed	**Mechanism**	**Overall Effect**
↑ Heart rate,* contractility, and stroke volume	Norepinephrine acting at β₁ receptors at the sinoatrial node and throughout the heart	↑ Cardiac output and ↑ blood pressure
↑ Venoconstriction	Norepinephrine acting at α₁ receptors on smooth muscles of veins	↑ Cardiac output and ↑ blood pressure
↑ Vasoconstriction of arterioles	Norepinephrine acting at α₁ receptors on smooth muscle of arterioles	↑ Total peripheral resistance and ↑ blood pressure
* ↑ Heart rate also occurs from reduced parasympathetic (vagal) tone at the sinoatrial node.		

Hormonal Control of Arterial Pressure

Renin–Angiotensin–Aldosterone System

The renin–angiotensin–aldosterone system (RAS) is involved in the longer-term regulation of arterial pressure by modulating blood volume (**Fig. 10.6**).

Fig. 10.6 ► **Renin–angiotensin–aldosterone system (RAS).**
If the mean renal blood pressure acutely drops below 100 mm Hg, renal baroreceptors will trigger the release of renin into the systemic circulation. Renin is a peptidase that catalyzes the cleavage of angiotensin I from angiotensinogen. Angiotensin-converting enzyme (ACE) cleaves two amino acids from angiotensin I to produce angiotensin II. Angiotensin II and aldosterone are the most important effectors of the RAS. Angiotensin II stimulates the release of aldosterone by the adrenal cortex. Both hormones directly and indirectly lead to normalization of plasma volume and arterial blood pressure. They also inhibit renin release via negative feedback. (GFR, glomerular filtration rate; RBF, renal blood flow)

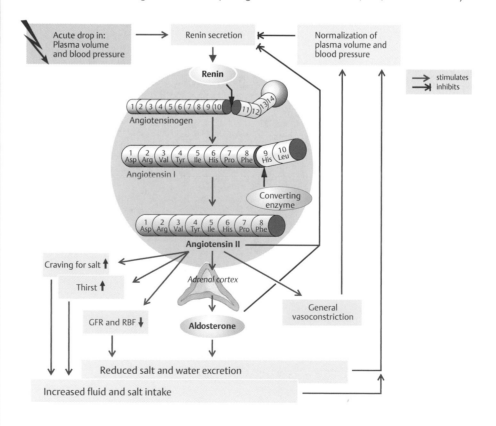

The response of the RAS to a decrease in arterial pressure is as follows:

— A decrease in arterial pressure causes a subsequent decrease in renal perfusion pressure and an increase in sympathetic stimulation of the kidney. Both influences stimulate juxtaglomerular cells of afferent arterioles to secrete the enzyme renin.
— Renin is secreted into the bloodstream, where it catalyzes the conversion of angiotensinogen to angiotensin I.
— Angiotensin I is transported throughout the peripheral vasculature, where it is converted to angiotensin II by the action of angiotensin-converting enzyme (ACE). Angiotensin II is physiologically active and causes
    — Vasoconstriction of arterioles, which increases arterial pressure by increasing TPR
    — Increased synthesis and secretion of aldosterone by the adrenal cortex. Aldosterone increases Na^+ (and water) reabsorption in the distal nephron. This increases arterial pressure by expanding blood volume, which increases CO.
    — Increased reabsorption of Na^+ and water by promoting Na^+–H^+ exchange in the proximal tubule.

Other Factors Involved in the Regulation of Arterial Pressure

Cerebral Ischemia

Whereas the changes in sympathetic and parasympathetic outflow that control blood pressure originate within vasomotor centers in the brain, the normal detectors of blood pressure are in the vasculature outside the brain. However, severe ischemia of the vasomotor centers (e.g., due to extremely low arterial pressure or to cerebral edema that compresses cerebral blood vessels) can excite the vasomotor centers directly. The result is a sharp rise in arterial pressure, known as the *Cushing reflex*.

Chemoreceptors: Carotid and Aortic Bodies

Carotid and aortic bodies contain sensory receptors that are sensitive to a reduction in the partial pressure of oxygen (PO_2).

— When PO_2 falls, impulses from these receptors are relayed to the vasomotor center, which increase sympathetic outflow. This leads to vasoconstriction of arterioles causing an increase in TPR and arterial pressure.

Antidiuretic Hormone

The secretion of antidiuretic hormone (ADH, or vasopressin) from the posterior pituitary is mainly controlled by plasma osmolality. However, the hypothalamic neurons that stimulate ADH secretion also receive input from baroreceptors. ADH is released by the posterior pituitary when arterial pressure or blood volume falls too low, such as with severe hemorrhage.

— It acts via V_1 receptors to cause direct arteriolar vasoconstriction, thus increasing TPR and arterial pressure.
— It also acts via V_2 receptors to promote the reabsorption of water in the collecting duct of the kidney. This increases blood volume, which, in turn, slowly increases CO and arterial pressure.

Atrial Natriuretic Peptide

Atrial natriuretic peptide (ANP) is released in response to the stretch of low-pressure mechanoreceptors located at the venoatrial junction of the heart. These atrial receptors are stretched by increased venous return.

— ANP causes vasodilation of arterioles, which decreases TPR and reduces arterial pressure.
— Vasodilation of renal arterioles by ANP increases renal blood flow and the glomerular filtration rate. This results in increased excretion of Na^+ and water, which reduces blood volume and CO and thus helps return arterial pressure to normal.

Note: Stretching of low-pressure mechanoreceptors also causes an increase in heart rate (*Bainbridge reflex*). This, in turn, increases CO and renal perfusion.

10.5 Microcirculation and Lymph

Capillary Exchange of Lipid and Water-soluble Substances

— Lipid-soluble substances (e.g., O_2 and carbon dioxide [CO_2]) cross the membranes of capillary endothelial cells by simple diffusion. They do not depend on the presence of pores.
— Small water-soluble substances (e.g., Na^+, Cl^-, glucose, and urea) move through the pores of capillary walls by diffusion.
— Large water-soluble substances (e.g., albumin and immunoglobulins) can cross capillary endothelial cell membranes by transcytosis in vesicles.

Capillary Exchange of Fluid

Fluid movement (filtration) across capillary membranes or through membrane pores is driven by differences in plasma and interstitial hydrostatic and osmotic pressures. These pressures are known as *Starling forces* (**Fig. 10.7**).

Capillary Hydrostatic Pressure
Capillary hydrostatic pressure (P_c) is the pressure of fluid within blood vessels. It is the most important factor in the capillary exchange of fluids.

— Filtration out of the capillary will increase with increasing P_c.
— P_c depends on changes in arterial blood pressure, capillary flow, and the ratio of the resistance in arterioles to that in venules.
— P_c is the only force that varies significantly between the proximal and distal ends of the capillary: 35 mm Hg at the arterial end, 15 mm Hg at the venous end.

Increases in venous pressure increase filtration into interstitial fluid much more than a comparable increase in arterial blood pressure.

Thin-walled capillaries are able to withstand a hydrostatic pressure of 35 mm Hg because of their small radius (the tension in their walls is also small by the law of Laplace).

Fig. 10.7 ▶ **Starling forces across a capillary wall.**
At the arterial end of a capillary hydrostatic pressure causes net filtration of fluids out of the capillary into interstitial fluid. Hydrostatic pressure reduces progressively throughout the capillary. At the venous end plasma oncotic pressure is the driving force for fluid reabsorption into the capillary.
P_c, capillary hydrostatic pressure; P_i, interstitial fluid hydrostatic pressure; π_c, plasma oncotic pressure; π_i, interstitial fluid oncotic pressure.

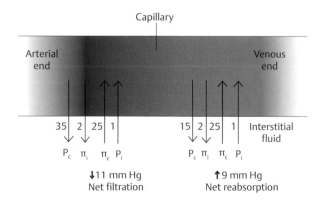

Interstitial Fluid Hydrostatic Pressure

Interstitial fluid hydrostatic pressure (P_i) is the pressure in the interstitial fluid.

— P_i is 1 to 2 mm Hg (or maybe slightly negative), so its contribution to capillary exchange is normally small.
— It is determined by the volume of interstitial fluid and by the distensibility of the interstitial space.
— P_i increases with lymphatic blockage and increased capillary permeability.

Plasma Oncotic Pressure

Plasma oncotic pressure (π_c) is the component of osmotic pressure contributed by macromolecules such as plasma proteins.

— It is ~25 mm Hg.
— It is the most important pressure opposing hydrostatic pressure.
— It decreases when plasma protein concentration is lower than normal.

Interstitial Fluid Oncotic Pressure

Interstitial fluid oncotic pressure (π_i) is proportional to the concentration of plasma proteins that are in the interstitial fluid.

— Interstitial fluid oncotic pressure is 1 to 2 mm Hg because the concentration of proteins in interstitial fluid is usually low.
— An increase in this oncotic pressure enhances the filtration of fluid out of the capillaries.
— If proteins leak out of the capillaries, then this pressure increases.

Net fluid movement (J_v) is predicted by the Starling hypothesis, expressed by the following equation:

$$\text{Net fluid movement } (J_v) = k\,[(P_c - P_i) - (\pi_c - \pi_i)]$$

where k is a filtration constant for the capillary membrane (a measure of the permeability of the capillary membrane to water) and is measured in mL/min/mm Hg. J_v is measured in mL/min.

— A positive value indicates that there is net fluid movement out of the capillary (filtration).
— A negative value indicates that there is net fluid movement into the capillary (reabsorption).

Note that for any one capillary the value of k is negligible and therefore in theoretical calculations of fluid movement it can be ignored. The net fluid flow therefore becomes a function of the theoretical net pressures acting across the capillary.

Example: P_c = 35 mm Hg; P_i = 5 mm Hg; π_c = 25 mm Hg; π_i = 5 mm Hg

J_v = [(35 mm Hg – 5 mm Hg) – (25 mm Hg – 5 mm Hg)] = +10 mm Hg (net filtration)

The two interstitial pressures are very low so the net filtration pressure is determined mainly by hydrostatic pressure and plasma oncotic pressure.

Note: Most capillaries have filtration at their arterial end and reabsorption at their venous end.

Lymph

Net filtration of fluid out of the capillaries usually exceeds net reabsorption of fluid into the capillaries by a slight margin because interstitial fluid hydrostatic pressure and interstitial fluid oncotic pressures are low. This leads to the formation of *lymph*.

— Lymph is a combination of excess filtered fluid, filtered proteins, and macromolecules (e.g., fats).
— The lymphatic system provides a route for lymph to flow from the interstitial space back into the circulatory system.

— Lymphatic vessels have valves similar to veins that establish unidirectional flow.

Edema

Edema is the accumulation of fluid in the interstitial fluid space, often seen as swelling. It is a sign that normal capillary/lymph exchange is disrupted.

— It can result from inadequate drainage of lymph due to obstruction of the lymphatic system or when capillary filtration exceeds capillary reabsorption.

    Table 10.2 summarizes many causes of edema.

Table 10.2 ▶ **Causes of Edema**		
Overall Cause of Edema	**Starling Parameter Affected**	**Underlying Cause**
Inadequate lymphatic drainage	↑P_i	Obstruction of the lymph nodes from parasites or cancer
Filtration exceeds reabsorption	↑P_c	↑Arteriolar dilation ↑Venous pressure Congestive heart failure Excessive fluid volume due to renal retention of salt and water Standing for prolonged periods
	↓π_c	↓Plasma proteins due to – Hypoalbuminemia – Liver disease – Nephrotic syndrome
	↑π_i	Leakage of proteins from the capillaries into interstitial spaces due to inflammation, or burns

Abbreviations: P_c, capillary hydrostatic pressure; P_i, interstitial fluid hydrostatic pressure; π_c, plasma oncotic pressure; π_i, interstitial fluid oncotic pressure.

10.6   Properties of Specific Circulations

The renal circulation is discussed in Chapter 15.

Coronary Circulation

Coronary Vascular Anatomy

The left and right coronary arteries, which supply blood to cardiac muscle, are the first two branches that emerge from the ascending aorta. The left coronary artery divides to form the left anterior descending branch and a circumflex branch.

— The left coronary artery supplies mainly the left ventricle.
— The right coronary artery supplies the right ventricle and also a major portion of the posterior wall of the left ventricle.

    Seventy-five percent of the coronary venous blood returns to the right atrium via the coronary sinus. Other coronary veins drain directly into the chambers of the heart.

Coronary Blood Flow

— Coronary blood flow exhibits autoregulation, active hyperemia, and reactive hyperemia.
— Coronary blood flow is predominantly under local metabolic control in response to changes in O_2 consumption and adenosine:
    — When cardiac activity increases, there is an increase in O_2 consumption. To compensate for this, coronary blood flow increases in proportion to the increase in O_2 consumption (active hyperemia).

■ Hypoalbuminemia

Lower than normal levels of albumin in the blood (hypoalbuminemia) and hence decreased plasma protein binding may occur in the following conditions: liver disease (e.g., hepatitis, cirrhosis, and hepatocellular necrosis), ascites, nephrotic syndrome, malabsorption syndromes (e.g., Crohn disease), extensive burns, and pregnancy. These all exhibit edema due to decreased plasma oncotic pressure.

■ Angina

Angina (angina pectoris) is sudden, crushing substernal chest pain that often radiates to the left shoulder and/or neck. It occurs as a result of myocardial ischemia. Classic, typical, stable angina is induced by exercise (typically), stress, cold weather, and heavy meals and is caused by atherosclerosis of the coronary arteries. Variant or unstable angina (angina that is worsening) occurs at rest or with minimal exertion and is due to coronary vasospasm. Other more rare causes of angina include aortic stenosis, hypertrophic obstructive cardiomyopathy, arrhythmias (causing hypoperfusion), and severe anemia. Electrocardiogram (ECG) analysis (performed at rest or during an exercise stress test) may show ST depression and flat or inverted T waves. Medical treatment of angina includes the use of nitrates, β-blockers, calcium channel blockers, and ACE inhibitors. Surgical treatment includes angioplasty with the placement of a stent or coronary artery bypass graft (CABG).

— Adenosine is a vasoactive substance that acts as a coupling agent between O_2 consumption and coronary blood flow and is therefore important in active hyperemia. It also plays a role in autoregulation.

— Most coronary flow to the left ventricle occurs during diastole when the muscular wall is relaxed, whereas the right ventricle receives about equal coronary flow in systole and diastole.

Note: Neural influences on vasodilation/vasoconstriction of the coronary vessels are overridden by the effects of the sympathetic system on heart rate and myocardial contractility (by stimulation of β_1 receptors), which leads to vasodilation by the metabolic mechanisms discussed above.

Cerebral Circulation

Cerebral Vascular Anatomy

The brain is supplied with blood from internal carotid and vertebral arteries. The basilar artery is formed by the convergence of two vertebral arteries. The two internal carotid arteries and the basilar artery enter the circle of Willis, which delivers blood to the brain by six large vessels.

The internal jugular veins provide the majority of the venous drainage via deep veins and dural sinuses.

Cerebral Blood Flow

— The brain exhibits autoregulation, active hyperemia, and reactive hyperemia.

— The distribution of flow to different areas of the brain varies according to their local immediate metabolic needs. The visual cortex, for instance, receives more blood flow during waking hours than during sleep.

— The main factor contributing to local metabolic control is the partial pressure of CO_2 (P_{CO_2}):
  — Increases in arterial P_{CO_2} produce marked vasodilation of cerebral arterioles and increased cerebral blood flow.
  — Decreases in arterial P_{CO_2} induce vasoconstriction.

— Vasoactive metabolites, such as adenosine, play a secondary role in the regulation of cerebral blood flow.

— Neural control of cerebral flow is minimal.

Pulmonary Circulation

The pulmonary circulation arises from the right ventricle in the pulmonary artery, which divides into the left and right pulmonary arteries that supply the left and right lungs. The arterial vessels subdivide many times to feed the pulmonary capillaries that surround the alveoli. Venules draining the capillaries recombine to form the pulmonary veins that carry blood to the left atrium.

— Pulmonary blood flow and CO from the left ventricle are always the same except for some of the relatively small amount of blood flowing in bronchial vessels.

— The total resistance of the pulmonary circuit is only ~15% of the TPR, and the corresponding mean pulmonary artery pressure is similarly ~15% of MAP.

— There is virtually no neural control of pulmonary vascular resistance. Resistance falls during elevated CO due to distention of small arterioles and recruitment of closed arterioles.
  — Local hypoxia causes vasoconstriction and a subsequent rise in pulmonary vascular resistance. This can occur in the whole lung, as often occurs in people who are not acclimatized to high altitude, or in localized lung regions that are inadequately ventilated. It functions to redistribute blood toward lung regions that can fully participate in gas exchange.

■ **Coronary artery bypass graft (CABG)**

CABG is a surgical procedure performed to bypass atherosclerotic narrowings of the coronary arteries that are the cause of anginal pain. These narrowings can eventually occlude if untreated, leading to MI. There are two main coronary arteries, left and right, and these have several branches. A CABG is denoted as single, double, triple, and so on, depending on the number of arteries that are to be bypassed. The internal thoracic artery, which supplies the anterior chest wall and breasts, is usually harvested to use as the bypass artery.

■ **Myocardial infarction (MI)**

MI is death of heart muscle. It is caused by complete occlusion of one or more coronary arteries by thrombosis. The pain of an MI is similar to that of angina, but it is more severe and of longer duration. It is also accompanied by nausea, vomiting, diaphoresis (sweating), dyspnea, and a feeling by the individual that he or she is going to die. Common complications of MI include arrhythmias, heart failure, hypertension, and emboli formation. ECG analysis shows ST elevation, T-wave inversion, and Q waves in the leads that "look at" the infarction. Cardiac enzymes are also measured and used as a basis for diagnosis of MI. Immediate treatment of an MI involves the use of thrombolytic drugs such as streptokinase (given as soon as possible after infarction) and aspirin, morphine, and nitrates. Longer-term treatment involves the use of β-blockers and ACE inhibitors. Surgical treatment is the same as for angina.

■ **Cardiac enzymes**

Several enzymes are released when cardiac muscle cells are damaged: troponin I, creatine kinase, myoglobin, and lactate dehydrogenase. However, troponin I is the only one that is specific for cardiac muscle damage and is routinely measured to help diagnose myocardial infarction (MI).

Fetal Circulation

Prenatal Fetal Blood Flow

Oxygenated blood from the placenta travels through the umbilical vein and ductus venosus to the inferior vena cava of the fetus (**Fig. 10.8**). Upon entering the right atrium, most of this blood travels through the foramen ovale to the left atrium and then into the left ventricle.

— Blood entering the right atrium from the superior vena cava travels preferentially to the right ventricle. Because the pulmonary circulation has very high resistance prior to birth (due to hypoxic vasoconstriction), blood pumped by the right ventricle moves from the pulmonary artery through the ductus arteriosus to the descending aorta.

— Some of the blood from the aorta flows through fetal tissues, but most flows back to the placenta via the umbilical arteries.

Changes in Fetal Blood Flow at Birth

At birth, the vascular resistance in the pulmonary vessels greatly decreases, with a corresponding fall in pulmonary artery pressure. At the same time, there is a large rise in systemic resistance due to the closure of the umbilical arteries. Aortic pressure rises considerably, and left atrial pressure rises slightly above right atrial pressure. The reversal of the pressure gradient in

Fig. 10.8 ▶ **Prenatal circulation.**
Red: Oxygenated blood. Blue: Deoxygenated blood.
From *Thieme Atlas of Anatomy*, *Neck and Internal Organs*, © Thieme 2006, Illustration by Markus Voll.

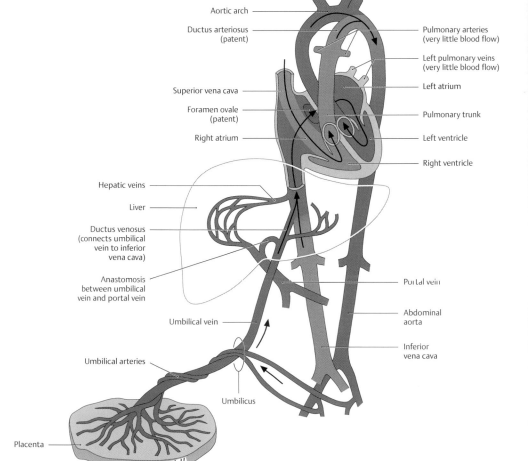

▋ Patent foramen ovale

A patent (open) foramen ovale is a small opening between the left and right atria that remains open (pathologically) after birth. It is often asymptomatic and does not require any treatment.

▋ Patent ductus arteriosus

A patent (open) ductus arteriosus (PDA) is an opening between the pulmonary artery and aorta that persists (pathologically) following birth. There may be no symptoms if the defect is small, but larger defects may cause failure to thrive, dyspnea (shortness of breath), fatigue, tachycardia, and cyanosis. Signs of PDA include systolic murmur, collapsing pulse, thrill, and a loud second heart sound. Complications include heart failure, pulmonary hypertension, endocarditis, pneumonia, and arrhythmias. In premature infants, spontaneous closure of a PDA may occur. If not, indomethicin (a nonsteroidal antiinflammatory drug [NSAID]) is used to block prostaglandin E_2, which keeps the ductus arteriosus open during fetal development. In full-term infants and in adults, surgical closure of the PDA is necessary.

the atria causes the flap covering the foramen ovale to close. Flow through the ductus arteriosus initially reverses, but increased oxygenation causes vasoconstriction, and it eventually closes permanently.

Splanchnic Circulation

The splanchnic circulation consists of blood flow through the intestines, spleen, pancreas, and liver. It accounts for ~25% of the cardiac output. Venous drainage from the intestines, spleen, and pancreas flows through the portal vein into the liver. Most of the blood entering the liver comes via the portal vein, and about one-quarter of it is supplied by the hepatic artery.

- At rest, splanchnic blood flow is mainly under local control. It increases during processing of a meal.
- Splanchnic blood flow is also under sympathetic control. During exercise, splanchnic blood flow decreases markedly, freeing up blood to go to working muscle. It also decreases severely in states of major hypovolemia. During hemodynamic shock, the intestines are at risk of damage from sympathetically mediated ischemia.

Cutaneous Circulation

The primary function of the cutaneous circulation is to regulate heat loss and help maintain body temperature.

Cutaneous blood flow is largely under neural control by sympathetic nerves. It is the most variable of the major vascular beds. It is not controlled by local metabolic factors, as the O_2 and nutrient requirements of skin tissue are small.

- When ambient temperature rises, there is vasodilation of cutaneous vessels, allowing heat loss to the environment.
- Trauma leading to the loss of blood (hemorrhage) activates sympathetic vasoconstriction throughout the body, including the skin. As a result, the skin becomes cold and sweaty.

Skeletal Muscle Circulation

- Skeletal muscle blood flow exhibits autoregulation, active hyperemia, and reactive hyperemia.
- Sympathetic control of skeletal blood flow predominates at rest.
  - Activation of α_1 receptors by norepinephrine causes vasoconstriction, and activation of β_2 receptors by epinephrine causes vasodilation.
- Local metabolic control of skeletal blood flow predominates during exercise.
- Exercise increases metabolism in muscles and the production of vasoactive substances, such as lactate and adenosine. These substances are responsible for vasodilation and increased skeletal blood flow. The local vasodilators override sympathetic vasoconstrictive signals.

11 Response of the Cardiovascular System to Gravity, Exercise, and Hemorrhage

11.1 Gravity

Air embolism

Surgery with the patient in a sitting position, as done in some neurosurgical procedures, poses the threat of air embolism, which is the introduction of air into the vascular system. If central venous pressure is not high enough to support the column of blood from the heart to a site of vascular penetration above the heart, pressure becomes negative at the site and will draw in air. Venous air emboli travel to the heart, where they may be trapped and greatly impede pulmonary blood flow, or travel to the pulmonary microcirculation, unleashing a host of damaging actions. Large air emboli are often fatal.

Effects of Gravity on the Cardiovascular System

Gravity has a major influence on vascular pressures and on the distribution of blood.

— When a person is sitting or standing, arterial pressures progressively decrease above the heart and increase below the heart, reaching ~90 mm Hg higher in the feet than at heart level.

— Similarly, when upright, venous pressures above the heart become negative, causing collapse of the jugular veins.

— The extent of gravity-induced venous pressure change below the heart depends on muscular activity.

  — Standing still for long periods allows pooling of blood in the lower extremities (due to the large capacitance of veins) and a rise in venous and capillary pressures (P_c); it also promotes edema. Edema causes a decrease in both blood volume and venous return to the heart. This, in turn, will cause a decrease in stroke volume and cardiac output (CO) via the Frank–Starling mechanism.

  — Walking and running force peripheral blood toward the heart. Venous valves inhibit backflow, thereby preventing a large buildup of pressure and preserving venous return to the heart.

— Moving from the supine to the standing position shifts at least 0.5 L of blood from the pulmonary vessels to the lower body and lowers central venous pressure.

Compensation

The baroreceptor reflex is usually able to compensate rapidly for the transient decrease in arterial pressure that occurs upon standing.

— Reduced afferent signals from the carotid sinus baroreceptors cause an increase in sympathetic outflow to the heart and blood vessels and a decrease in parasympathetic outflow to the heart. This results in

  — Increased heart rate, contractility, and CO

  — Constriction of arterioles, which increases total peripheral resistance (TPR)

  — Constriction of large veins, which increases venous return

— Orthostatic hypotension is syncope (fainting) that occurs upon standing. It is caused by a transient reduction in venous return that is not adequately compensated by the baroreceptor reflex. It can be exacerbated by drugs (e.g., β-blockers) that inhibit the sympathetic outflow induced by the baroreceptor reflex.

11.2 Exercise

Aerobic exercise causes cardiovascular changes that reflect a combination of central command effects and peripheral effects that are driven by the increase in metabolism in exercising muscle. These changes increase blood flow and oxygen delivery to skeletal muscle.

Central Command Effects

During exercise, central command receives input from the motor cortex or from mechanoreceptors in muscles and joints, leading to an increase in sympathetic outflow and a decrease in parasympathetic outflow. This, in turn, produces the following effects:

– Increased heart rate and contractility. This greatly increases CO with a modest increase in mean arterial pressure (MAP), as TPR is decreased by peripheral vasodilation.

– Increased venous return to match the increased CO. This is driven in part by the alternating contraction and relaxation of the exercising limbs and by venoconstriction.

– Vasoconstriction of the splanchnic vessels allows more of the CO to go to exercising muscle.

– Blood flow to the skin is initially reduced but soon rises to promote heat loss.

Peripheral Effects

– Local peripheral vasodilation occurs due to the release of vasodilator metabolites, such as adenosine and lactate. This causes a decrease in TPR.

11.3 Hemorrhage

Effects of Hemorrhage on the Cardiovascular System

Hemorrhage causes a decrease in blood volume, CO, and MAP.

Compensation

Hemorrhage invokes emergency responses (the sympathetic "fight-or-flight" response) that collectively act to keep MAP high enough to perfuse the brain and coronary vessels. This is accomplished by a combination of short-acting reflexes (baroreceptor reflex) and longer-term actions that promote retention of body fluid volume (**Fig. 11.1**).

Reduced afferent signals from both the arterial and cardiopulmonary baroreceptors cause an increase in sympathetic outflow to the heart and blood vessels and a decrease in parasympathetic outflow to the heart. This results in

– Increased heart rate and contractility

– Increased TPR caused by vasoconstriction of the splanchnic, skin, and renal vasculatures. This allows the limited CO to flow to the brain and coronary circulation.

– Constriction of large veins, which preserves venous return.

– The posterior pituitary secretes large amounts of vasopressin (antidiuretic hormone [ADH]), which raises TPR and causes the kidneys to reduce excretion of water.

– The renin–angiotensin–aldosterone system is activated, which provides additional peripheral vasoconstrictive actions via angiotensin II and reduced Na^+ excretion. Increased Na^+ reabsorption in the proximal tubule stimulated by angiotensin II and in the cortical collecting ducts stimulated by aldosterone are responsible for reduced Na^+ excretion.

– Thirst centers in the hypothalamus are stimulated, increasing the drive to drink water.

In extreme cases of peripheral hypoperfusion, reduced anaerobic metabolism leads to lactic acidosis, which activates chemoreceptors. These add to the afferent signals, causing increased sympathetic outflow.

Fig. 11.1 ► **Compensatory mechanisms when there is a risk of hypovolemic shock.**
Hypovolemic shock may occur due to acute heart failure, hormonal causes, or volume deficit
(e.g., hemorrhage). Compensation involves activation of the baroreceptor reflex and the
renin–angiotensin–aldosterone system, which cause physiological changes that increase blood pressure
and blood volume. (ADH, antidiuretic hormone; GFR, glomerular filtration rate)

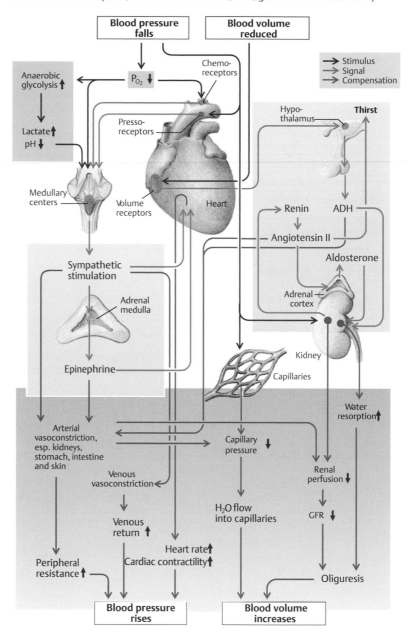

Review Questions

1. Which of the following observations shows that the heart rate at rest is predominantly under parasympathetic control?
   A. A totally denervated heart has a resting rate of 100 beats/min.
   B. Cutting the sympathetic efferent supply to the heart causes an increase in the heart rate.
   C. Administering a β-adrenergic antagonist, such as propranolol, causes the heart rate to increase.
   D. Infusion of norepinephrine increases heart rate.
   E. Cutting the parasympathetic efferent fibers to the heart causes a large increase in the heart rate.

2. Which of the following classes of drugs would cause an increased heart rate?
   A. Alpha-adrenergic blocking drug
   B. Beta-adrenergic blocking drug
   C. Cholinergic blocking drug
   D. Cholinergic stimulating drug
   E. Nitrovasodilator drug

3. What process is associated with phase 0 of the ventricular action potential?
   A. Rapid filling of the ventricle
   B. The membrane is at the resting potential between excitations.
   C. Propagation through the Purkinje network
   D. Calcium release from the sarcoplasmic reticulum
   E. Entry of sodium through voltage-gated channels

4. The relatively long plateau of ventricular action potentials coincides with which portion of the ECG?
   A. P wave
   B. PQ interval
   C. Q wave
   D. ST segment
   E. T wave

5. Cardiac glycosides, such as digitalis, enhance the contractile performance of cardiac muscle fibers by
   A. inhibiting the Na^+–K^+ ATPase of cell membranes.
   B. stimulating intracellular production of cyclic adenosine monophosphate (cAMP).
   C. decreasing the amount of Ca^{2+} available to myofibrils.
   D. stimulating cardiac β_1-adrenergic receptors.
   E. blocking cardiac muscarinic receptors.

6. Cardiac output (CO) can be defined by which three parameters?
   A. Heart rate; aortic pressure; end-diastolic volume
   B. Heart rate; end-systolic pressure; end-diastolic pressure
   C. R wave-to-R wave interval; end-diastolic volume; end-systolic volume
   D. Total peripheral resistance (TPR); pulse pressure; stroke volume
   E. TPR; central venous pressure; diastolic pressure

7. Which of the following is the best index of cardiac O_2 demand?
   A. Stroke volume multiplied by heart rate
   B. Systolic pressure minus diastolic pressure
   C. Diastolic pressure multiplied by heart rate
   D. Stroke work multiplied by heart rate
   E. Stroke volume multiplied by systolic pressure

8. A well-adjusted 45-year-old woman saw her family physician after experiencing episodes of headache, sweating, and heart "pounding" in her chest (heart rate 125–135/min; blood pressure 172/116 mm Hg). She reported feeling "antsy" and jittery. A urine sample showed elevated levels of metanephrines, which are metabolites of epinephrine. She is diagnosed with having a pheochromocytoma, a tumor of adrenal medullary tissue. It releases catecholamines in an unregulated manner, often episodically. The tumor also releases high levels of metanephrines. Why did the patient experience elevated heart rate and MAP?
   A. Actions of androgenic steroids
   B. Massive firing of the parasympathetic nervous system
   C. Actions of epinephrine and norepinephrine
   D. Actions of aldosterone
   E. Actions of corticosteroids

9. What is the effect of a constant intravenous (IV) infusion of norepinephrine?
   A. A decrease in TPR
   B. An increase in venous volume
   C. An increase in cardiac contractility
   D. A reflexively mediated increase in heart rate
   E. A decrease in afterload

10. A patient has an acute bout of reduced coronary blood flow, causing chest discomfort and weakness. Which treatment is not appropriate?
    A. Nitrovasodilators
    B. Beta-receptor antagonists
    C. IV norepinephrine
    D. Supplemental O_2
    E. Calcium channel blockers

11. Which mechanical event of the cardiac cycle is most closely associated in time with the P wave of the electrocardiogram (ECG)?
    A. Rapid diastolic filling
    B. Rapid systolic ejection
    C. Atrial contraction
    D. Atrial relaxation
    E. The start of ventricular contraction

12. Aortic pressure is greatest at which point during the cardiac cycle?
    A. When the aortic valve opens
    B. When the aortic valve closes
    C. The beginning of rapid ventricular filling
    D. The beginning of isovolumetric contraction
    E. The end of rapid ventricular ejection

13. In a normal cardiac cycle,
    A. the maximum and minimum left ventricular pressures are virtually the same as systolic and diastolic pressures measured in a peripheral artery.
    B. left ventricular pressure rises continuously during filling.
    C. > 90% of the end-diastolic volume is ejected.
    D. left ventricular pressure is greater than left atrial pressure at all stages of the cycle.
    E. the mitral valve is open longer than is the aortic valve.

14. In adults, what is the result of an uncorrected congenital ventricular septal defect?
    A. Low right ventricular pressure during diastole
    B. Right ventricular stroke volume is larger than left ventricular stroke volume.
    C. A low partial pressure of oxyen (PO_2) in systemic arterial blood
    D. A drop in pulmonary artery pressure
    E. A clear diastolic murmur

15. Which of the following might be expected to occur in a patient suffering from moderate right ventricular heart failure?
    A. Pulmonary edema
    B. Raised central venous pressure
    C. Increased systemic arterial blood pressure
    D. Lower right ventricular stroke volume compared with left ventricular stroke volume
    E. Enlarged left atrium

16. Most of the drop in pressure between the ascending aorta and the right atrium occurs across which vessel(s)?
    A. Peripheral arteries
    B. Arterioles
    C. Capillaries
    D. Peripheral veins
    E. Inferior vena cava

17. Why does an aging hypertensive patient often have a large pulse pressure?
    A. Because of low compliance of arterial walls
    B. Because typical drug therapy elevates heart rate
    C. Because of a very large stroke volume
    D. Because the heart is usually hypertrophic
    E. Because the TPR is decreased

18. Five-year-old children have a CO much smaller than normal adults. Both may have similar arterial blood pressures. As a result of normal growth, what change over time would you expect?
    A. The conductance to flow through individual arterioles increases.
    B. TPR decreases.
    C. The resistance to flow offered by the aorta increases.
    D. Heart rate increases.
    E. CO and TPR both increase.

19. What is the calculated vascular resistance to blood flow through the kidneys from the following data? Systolic arterial pressure = 130 mm Hg, diastolic arterial pressure = 70 mm Hg, mean renal venous pressure = 10 mm Hg, renal blood flow = 800 mL/min, renal arterial compliance = 4 mL/mm Hg.
    A. 0.1 mm Hg/mL/min
    B. 1.3 mm Hg/mL/min
    C. 8.0 mm Hg/mL/min
    D. 100 mm Hg/mL/min
    E. 200 mm Hg/mL/min

20. Therapy for the treatment of essential hypertension might include which of the following?
    A. Administration of a drug that causes the kidney to retain salt
    B. IV infusion of renin substrate (angiotensinogen)
    C. IV administration of acetylcholine
    D. Administration of an antagonist of adenosine receptor sites
    E. Administration of an α-adrenergic antagonist

21. Which of the following statements is true regarding the cardiovascular effects of a drug that causes selective dilation of peripheral arterioles?
    A. The activity in the carotid sinus nerves would likely increase.
    B. The heart rate would increase above normal.
    C. Arterial blood volume would be greater than before administration of the drug.
    D. Sympathetic adrenergic activity to the splanchnic region would be lower than normal.
    E. Venous pressure decreases.

22. An adult stranded in a desert has had no food or water for 2 days. Which statement is true concerning this condition?
    A. Below normal blood concentration of aldosterone
    B. High rate of firing of the carotid sinus nerves
    C. Renin secretion greater than normal
    D. Low circulating levels of vasopressin
    E. Low blood volume due to increased filtration pressure in the capillaries

23. Distention of the low-pressure mechanoreceptors located at the venoatrial junctions
    A. reduces the release of atrial natriuretic peptide (ANP).
    B. increases plasma renin.
    C. relaxes renal arterioles.
    D. stimulates smooth muscle contraction in the walls of large veins.
    E. causes vasoconstriction in the coronary circulation.

24. Filtration of fluid across capillary walls
    A. increases when arterioles vasoconstrict.
    B. is greater at the venule end than at the arteriolar end of the capillary.
    C. increases when large venules dilate.
    D. increases when plasma oncotic pressure is below normal.
    E. is promoted in cases of severe dehydration.

25. Edema can occur as a result of
    A. decreased mean capillary pressure.
    B. persistent, tonic contraction of precapillary sphincters.
    C. decreased albumin concentration of blood plasma.
    D. increased hydrostatic pressure in the interstitial fluid space.
    E. decreased venous pressure.

26. In a healthy adult, the pulmonary artery pressure cycles between 25 and 8 mm Hg, with a mean pressure of 15 mm Hg. What are the likely respective systolic and diastolic pressures of the right ventricle?
    A. 25/2 mm Hg
    B. 25/8 mm Hg
    C. 25/15 mm Hg
    D. 15/−2 mm Hg
    E. 15/8 mm Hg

27. In the splanchnic circulation,
    A. blood flow is regulated entirely by local factors independent of central control.
    B. venous drainage from the stomach flows directly into the ascending vena cava.
    C. more blood enters the liver from the venous drainage of other organs than from the hepatic artery.
    D. the hepatic portal vein carries blood from the liver to the pancreas, small intestine, and spleen.
    E. the total splanchnic blood flow accounts for < 10% of the CO.

28. When a person shifts from the supine to the standing position,
    A. arterial pressures in the legs and feet rise gradually over several minutes.
    B. vascular pressures in the rigid cranial vault are not affected.
    C. pressures in large vessels in the legs increase, but not in capillaries.
    D. over 1 or 2 minutes the volume of blood in pulmonary vessels decreases.
    E. valves in the veins prevent change in venous pressures.

29. If enough blood has been lost to cause a decrease in mean arterial pressure (MAP), which of the following would be true?
    A. Carotid sinus nerve activity would be increased.
    B. TPR would be decreased.
    C. Production of cAMP within cardiac myocytes would be increased.
    D. The arterioles of the intestine would be dilated.
    E. Heart rate would be decreased.

30. An untrained patient exercises on a treadmill at a moderate level. His control values are CO = 5 L/min, arterial pressure (systolic/diastolic) = 120/70 mm Hg, and heart rate = 70 beats/min. His exercise values are CO = 10 L/min, arterial pressure (systolic/diastolic) = 140/65 mm Hg, and heart rate = 120 beats/min. What is the main factor accounting for his increased CO?
    A. Increased heart rate
    B. Increased stroke volume
    C. Increased systolic pressure
    D. Increased pulse pressure
    E. Increased MAP

For Questions 31 and 32:
The following data are obtained during a catheterization procedure for a patient with diminished exercise tolerance:
Peak left ventricular pressure = 190 mm Hg
Mean left atrial pressure = 17 mm Hg
Peak aortic pressure = 110 mm Hg
Peak right ventricular pressure = 27 mm Hg
Peak pulmonary artery pressure = 23 mm Hg
Mean right atrial pressure = 6 mm Hg
CO = 3.2 L/min
Arterial O_2 content = 19 mL/dL
Mixed venous O_2 content = 15 mL/dL

31. What is the most likely cause of this patient's diminished exercise tolerance?
    A. A decrease in the number of cardiac β receptors
    B. Inadequate pulmonary ventilation
    C. An elevation in left ventricular end-diastolic pressure
    D. Stenosis of the aortic valve
    E. Poor cardiac muscle function after a myocardial infarction (MI)

32. Which of the following interventions would be most likely to improve the overall status of this patient?
    A. Administration of an α-blocking agent
    B. Administration of a rapidly acting diuretic
    C. Administration of a β-blocking agent
    D. Administration of supplemental oxygen (O_2)
    E. Replacement of the aortic valve

33. In severely anemic patients, a systolic ejection murmur can often be heard. Which of the following could account for this?
    A. Elevated heart rate
    B. Decreased velocity of blood flow
    C. Increased stroke volume
    D. Increased viscosity of blood
    E. Decreased blood density

34. Which of the following might be expected to occur in a patient suffering from moderate left ventricular heart failure?
    A. Increased cardiac ejection fraction
    B. Pulmonary edema
    C. Increased systemic arterial blood pressure
    D. Reduced central venous pressure
    E. Decreased end-diastolic volume

For Questions **35** and **36**:
On a warm summer day, a 64-year-old retiree experiences a dull aching in the calf muscle of his right leg each time he takes a walk. After standing quietly, he can continue to go the same distance before the pain reappears. He reports that a week ago he had been digging postholes for a new fence, using his right foot. Upon testing, only small amounts of radiopaque contrast material pass from the femoral to the right popliteal artery in an arteriogram. During rest and a treadmill exercise tolerance test, the calf muscle blood flow was measured; resting flow in right leg was 2 mL/100 g/min; exercise right calf flow was 8 mL/100 g/min. The respective flows for the left calf muscle were 3 and 40 mL/100 g/min.

35. What is the probable cause of this patient's problem?
    A. Myofascial pain due to overuse
    B. Spasm of small arteries
    C. Muscle cramps in the right calf
    D. Aneurysm in the right femoral artery
    E. Peripheral vascular disease of the right femoral artery

36. Why was the resting blood flow in the right calf muscle two-thirds of normal but the exercising flow only one-fifth of normal?
    A. Tissue injury limits exercise in the right leg.
    B. The obstruction becomes rate limiting when arterioles dilate with exercise.
    C. Local vasodilator metabolites are not produced in the right calf muscle.
    D. There is sympathetic vasoconstriction in the right leg.
    E. There is less activation of $β_2$-adrenergic receptors by epinephrine in the right leg.

37. During administration of a drug that directly dilates systemic arterioles, you would expect
    A. reduced plasma renin concentration.
    B. elevated 24-hour urinary Na^+.
    C. reduced plasma volume.
    D. reduced blood pressure.
    E. increased renal blood flow.

Answers and Explanations

1. **E.** Vagal tonic activity slows the heart at rest (p. 88). Therefore, cutting the vagus (parasympathetic efferent) nerve will cause a large increase in heart rate.
   A With the influence of both systems removed, one cannot tell which one is responsible for the rise in heart rate.
   B,C Cutting sympathetic supply to the heart or blocking adrenergic receptors has no effect or slightly slows the heart.
   D Norepinephrine increases heart rate, but this action is independent of parasympathetic control.

2. **C.** Tonic vagus nerve activity slows the SA nodal rate of depolarization via activation of muscarinic receptors. Blocking cholinergic muscarinic receptors would increase heart rate (p. 88).
   A Alpha-adrenergic blocking drugs would relax vascular smooth muscle but have little effect on heart rate.
   B Beta-blockers such as propranolol reduce heart rate and the force of contraction.
   D Cholinergic stimulating drugs such as carbachol decrease heart rate.
   E Nitrovasodilator drugs relax vascular smooth muscle.

3. **E.** Phase 0 is the rapid rise of the ventricular action potential due to sodium entry (p. 89).

A Rapid filling occurs before the beginning of the action potential.

B The membrane is at rest during phase 4.

C Propagation through the Purkinje network occurs before the beginning of the action potential.

D Calcium release occurs during phase 2.

4. **D.** The ST segment is the interval from ventricular depolarization to ventricular repolarization (p. 92).

A The P wave is produced by atrial depolarization and occurs prior to the ventricular action potential.

B The PQ interval indicates the time from atrial to ventricular depolarization and occurs prior to the plateau.

C The Q wave is part of the QRS complex and occurs prior to the plateau.

E The T wave indicates ventricular repolarization and occurs when the plateau ends.

5. **A.** Inhibiting the Na^+–K^+ ATPase slows the removal of Ca^{2+} via Na^+/Ca^{2+} antiport, thereby increasing the amount of Ca^{2+} available to signal contraction (p. 94).

B Catecholamines act by stimulation of cardiac β_1-adrenergic receptors, which increase intracellular cAMP.

C The amount of Ca^{2+} available for contraction is increased.

D,E Cardiac glycosides do not affect cardiac β_1-adrenergic receptors or muscarinic receptors.

6. **C.** CO is calculated either as the product of heart rate and stroke volume *or as mean arterial pressure (MAP) divided by TPR*. Heart rate can be calculated from the R wave-to-R wave interval, which is the time per beat, and stroke volume is the difference between end-systolic volume and end-diastolic volume (pp. 95, 106).

A,B The numbers do not yield stroke volume.

D,E The numbers do not yield MAP.

7. **D.** Demand for O_2 is proportional to cardiac work, which is the product of stroke work (systolic pressure times stroke volume) times heart rate (pp. 95, 99).

A Work cannot be calculated without knowing aortic, or systolic, pressure.

B To calculate cardiac work, pressures by themselves provide insufficient information.

C To calculate cardiac work, you need to know stroke volume and systolic, not diastolic, pressure.

E Without knowing how often the heart contracts, there is insufficient information.

8. **C.** The patient's tachycardia was due to the increased rate of depolarization of pacemaker cells in the SA node stimulated by the excess catecholamines epinephrine and norepinephrine. Epinephrine and norepinephrine also increase peripheral resistance by vasoconstriction and stimulate increased contractility, so blood pressure increases.

A Androgenic steroids build muscle mass with minimal cardiovascular effects.

B Parasympathetic activity slows the heart and may slightly decrease blood pressure.

D Aldosterone increases renal Na^+ and water reabsorption, which raises blood pressure but reflexly decreases heart rate.

E Corticosteroids increase contractility and vasoconstriction by enhancing the effects of catecholamines but do not affect heart rate.

9. **C.** Norepinephrine increases cardiac contractility via cardiac β receptors (p. 96).

A It increases total peripheral resistance via vascular smooth muscle α_1 receptors.

B Norepinephrine decreases venous compliance and therefore volume.

D The heart rate is directly increased by norepinephrine, not by reflex.

E Increased TPR and contractility increases afterload.

10. **C.** In the face of reduced coronary blood flow, the immediate goal is to reduce cardiac work, thereby reducing the demand for blood. Norepinephrine would stimulate cardiac contractility, increase cardiac work, and increase the demand for blood (p. 96).

A Nitrovasodilators dilate coronary arterioles.

B Beta-receptor antagonists reduce cardiac contractility.

D Supplemental O_2 may help the patient through the acute stage.

E Calcium channel blockers reduce peripheral resistance and cardiac work.

11. **C.** The P wave results from atrial depolarization, which occurs immediately before atrial contraction (p. 99).

A The P wave occurs after the phase of rapid diastolic filling.

B Rapid systolic ejection occurs after the QRS complex.

D Atrial relaxation follows later.

E Ventricular contraction follows later.

12. **E.** Aortic pressure is greatest when ventricular pressure is also greatest (p. 98).

A Pressure continues to rise after the aortic valve opens.

B Pressure starts falling before the aortic valve closes.

C Rapid ventricular filling occurs later, after the mitral valve opens, when aortic pressure is falling.

D At the beginning of isovolumetric contraction, aortic pressure is close to its lowest value.

13. **E.** Diastole, when the mitral valve is open, lasts longer than systole, when the aortic valve is open (p. 99).

A Once the aortic valve closes, left ventricular pressure falls well below arterial pressure.

B When the mitral valve opens, the ventricle is still relaxing, and pressure continues to fall during the initial phase of filling.

C Typically, ~30% to 50% of the end-diastolic volume remains at the end of systole.

D To move blood from the atrium to the ventricle, ventricular pressure must be slightly less than atrial pressure during phase IV.

14. **C.** A congenital septal defect initially causes a left-to-right shunt which elevates pulmonary artery pressure by allowing blood at the higher pressure of the left heart into the right heart. Over time, this leads to scarring of small pulmonary vessels, poor oxygenation, and greatly increased vascular resistance. Eventually, pulmonary hypertension causes reversal of the shunt, now being right to left, which further reduces oxygenation of systemic arterial blood (p. 100).

A The hypertrophied right ventricle requires a higher filling pressure.

B With a right-to-left shunt, the left ventricular stroke volume is larger.

D The high vascular resistance leads to high pulmonary artery pressure.

E The blood flowing through the defect causes a systolic murmur.

15. B. The right heart struggles to pump returning blood, so central venous pressure increases, and peripheral veins become engorged.

A,E Pulmonary edema and an enlarged left atrium occur with left heart failure.

C Systemic arterial pressure is unchanged.

D Because both ventricles beat at the same rate, their stroke volumes must be equal.

16. B. Arterioles are the sites of highest resistance and therefore greatest pressure drop (pp. 101–103).

A There is little drop in pressure until the arterioles are reached.

C Capillaries are individually very resistant, but their huge number in parallel reduces their combined resistance.

D,E Neither peripheral veins nor the inferior vena cava offer significant resistance to blood flow, so there is little pressure drop.

17. A. Stiff (low-compliance) arterial walls cause a large pulse pressure even with normal stroke volume (C) (p. 105).

B Common drug therapy such as β-blockers lowers heart rate.

D Cardiac hypertrophy results from the heart working against high afterloads.

E TPR is increased, but that has little effect on pulse pressure.

18. B. CO is inversely related to TPR so if CO increases, TPR decreases (p. 106).

A Most of the TPR is accounted for by arteriolar resistance (inverse of conductance). As the body grows, the CO is divided through more arterioles in parallel, which accounts for their overall conductance increase and not an increase in individual arteriolar conductance.

C The large-diameter aorta offers little resistance to flow.

D Heart rate declines as children grow, from over 120 beats per minute in infancy to a normal 70 in adulthood.

E Because CO must increase with body size, TPR must decrease.

19. A. Resistance is the pressure difference driving flow (MAP minus mean venous pressure) divided by flow (p. 102). MAP equals diastolic pressure plus one-third of pulse pressure, which is $70 + (130 – 70)/3 = 90$ mm Hg (p. 106). Thus, the pressure difference is $90 – 10 = 80$ mm Hg and resistance is 80 mm Hg $\div 800$ mL/min $= 0.1$ mm Hg/mL/min.

20. E. Alpha-adrenergic antagonists decrease sympathetic stimulation of arterioles and thereby reduce blood pressure (p. 106).

A If the kidneys retain salt, then the extracellular compartment expands and raises blood pressure.

B It is renin, rather than angiotensinogen availability, that determines the production of angiotensin II. However, angiotensin II raises blood pressure.

C Acetylcholine slows the heart, but it is not given IV, as it is hydrolyzed by circulating esterases.

D Blocking adenosine receptors decreases coronary blood flow, which compromises cardiac function.

21. B. Dilation causes decreased peripheral resistance and decreased blood pressure, causing the baroreceptor reflex to increase heart rate (p. 107).

A The drop in arterial pressure decreases carotid sinus nerve activity.

C Arterial blood volume does not change.

D Sympathetic activity to the splanchnic region also increases.

E Venous pressure increases via reflex vasoconstriction.

22. C. In the face of decreased extracellular fluid volume, the renin concentration increases (p. 109).

A Aldosterone also increases.

B Carotid sinus nerves fire more slowly because blood pressure is reduced.

D Vasopressin (ADH) is also increased.

E Peripheral vasoconstriction would reduce the net filtration pressure.

23. C. Distention of the low-pressure mechanoreceptors located at the venoatrial junctions reduces sympathetic stimulation of the kidneys and causes the release of ANP. Both actions relax renal arterioles and permit increased glomerular filtration (p. 109).

A Atrial natriuretic peptide (ANP) release would be increased

B ANP and reduced sympathetic drive inhibit renin secretion.

D The afferent neurons from the baroreceptors inhibit central stimulation of large veins.

E There would be no direct influence on coronary vessels.

24. D. Low plasma oncotic pressure increases the net filtration pressure by reducing the force opposing capillary hydrostatic pressure (p. 111).

A Arteriolar vasoconstriction reduces capillary pressure and reduces filtration.

B Filtration is greater at the arteriolar end of the capillary, where pressure is higher.

C Venodilation reduces mean capillary hydrostatic pressure.

E Dehydration tends to concentrate plasma proteins, raising oncotic pressure, and leads to reflex arteriolar vasoconstriction. Both effects reduce or even reverse net filtration pressure.

25. C. Decreased albumin concentration decreases plasma oncotic pressure, resulting in less reabsorption and more fluid (edema) in the interstitial space (p. 112).

A Decreased capillary pressure produces less filtration and less interstitial fluid.

B Contraction of precapillary sphincters decreases capillary pressure.

D Increased interstitial hydrostatic pressure can result from edema, but it is not a cause of edema.

E Decreased venous pressure decreases capillary pressure.

26. A. The systolic pressure in the right ventricle is virtually the same as the systolic pulmonary arterial pressure.

B–E Right ventricular diastolic pressure is virtually the same as central venous pressure, which is typically about 2 mm Hg.

27. **C.** The majority of hepatic blood is drainage from the stomach, spleen, pancreas, and intestines (p. 115).

A Splanchnic blood flow is severely decreased by central control during exercise and dehydration.

B Venous blood from the stomach flows via the portal vein to the liver.

D Portal blood flows into the liver.

E At rest, splanchnic blood flow is ~25% of the CO.

28. **D.** Over the course of a minute or so, up to 500 mL of blood shifts from the pulmonary circuit to the abdomen and lower extremities (p. 116).

A Pressures in the arterial system are affected immediately.

B The rigidity of the cranium does not prevent a fall in both arterial and venous pressures.

C Capillary pressures in the legs gradually rise, as they can never be less than venous pressure.

E Valves prevent backflow, but as blood accumulates in the lower extremities, venous pressure rises if the person remains stationary.

29. **C.** The fall in arterial pressure would trigger reflex adrenergic stimulation of the heart via β_1 receptors, which would lead to the production of cAMP (p. 117).

A Low blood pressure causes decreased carotid sinus nerve activity.

B,D Reflex sympathetic activation would cause peripheral vasoconstriction, including vessels in the gastrointestinal (GI) tract.

E Sympathetic stimulation of the heart would increase heart rate.

30. **A.** The CO doubled and heart rate almost doubled; thus, increased heart rate accounts for most of the change in CO (p. 117).

B A small increase in stroke volume accounts for the rest.

C–E Increased systolic and pulse pressure with little change in diastolic pressure is typical of exercise, but pressure changes by themselves do not change CO.

31. **D.** The large difference between peak left ventricular pressure and peak aortic pressure indicates stenosis of the aortic valve. This prevents adequate amounts of blood from being pumped out of the left ventricle into the aorta.

A,E A decrease in the number of β receptors or poor cardiac muscle function would decrease cardiac contractility, and the left ventricle would be unable to generate pressures of 190 mm Hg.

B Pulmonary ventilation is adequate, as arterial blood is well oxygenated.

C The diminished ejection causes a backup of blood, elevating left ventricular diastolic and left atrial pressures. High diastolic pressures are a result, not a cause, of the inability of the left ventricle to eject blood during systole.

32. **E.** The stenosed aortic valve needs replacement.

A,B Administration of an α-blocking agent or a rapidly acting diuretic would reduce the systemic arterial pressure, which might further impair the cardiovascular status of the patient.

C A β-receptor blocker would reduce cardiac contractility and reduce the CO.

D Blood oxygenation is adequate in this patient and does not need to be supplemented.

33. **C.** To maintain adequate delivery of O_2 in severe anemia, the body responds with an increased heart rate and high stroke volume. The high blood velocity caused by the high stroke volume leads to turbulence, which is the source of the murmur (B).

A The high heart rate by itself does not cause increased ejection velocity.

D Decreased hematocrit results in decreased viscosity, which is another cause of turbulence.

E Low density decreases the likelihood of turbulence.

34. **B.** Left heart failure, because of a high end-systolic volume, increases left atrial pressure. This raises pulmonary capillary pressure and produces pulmonary edema.

A The cardiac ejection fraction is reduced in the failing heart.

C With moderate heart failure, there is no consistent change in arterial blood pressure.

D Heart failure is characterized by fluid retention, which increases central venous pressure.

E End-diastolic volume increases due to the large volume remaining after systole and the high left atrial pressure.

35. **E.** Atherosclerosis in the right femoral artery restricts blood flow.

A The test results indicate a blood flow problem rather than a muscle tissue problem.

B Spasm of small arteries of the fingers and toes may occur in Raynaud disease in response to exposure to cold. Anything that stimulates the sympathetic nervous system can cause arterial spasms.

C Muscle cramps can limit blood flow, but they are relieved by stretching the muscle and do not come and go with exercise and rest.

D An aneurysm (a bulge in the artery wall) would not limit blood flow.

36. **B.** At rest, the arterioles are mostly constricted and present greater overall resistance than the affected femoral artery. As the arterioles dilate from locally-produced metabolites with exercise, the obstruction somewhat limits flow in the right leg, while flow in the unobstructed leg increases greatly.

A The right leg still exercises until pain from the hypoxia in muscles stops it.

C Vasodilator metabolites are still produced with exercise, causing a 4-fold increase in blood flow.

D,E In exercising muscle, local control overrides neurohumoral control, and the neural innervation of the right calf is normal.

37. **D.** Arteriolar vasodilation reduces systemic blood pressure.

A,B The lower pressure reduces filtered Na^+, and the kidney reflexly retains Na^+, so urine Na^+ is lower, and renin rises.

C Plasma volume rises from Na^+ retention.

E Lower blood pressure slightly reduces renal blood flow in the range of autoregulation.

12   Ventilation and Pulmonary Blood Flow

Anatomy

The respiratory system is composed of the conducting airways, the respiratory airways, and alveoli (**Fig. 12.1**).

Conducting and Respiratory Airways

The conducting airways include the nose, mouth, pharynx, larynx, trachea, bronchi, bronchioles, and terminal bronchioles. As their name suggests, these airways merely conduct air to the respiratory airways; they do not participate in gas exchange.

— The bronchi are > 1 mm in diameter and have cartilaginous rings that protect them from collapsing during expiration. They are not embedded in the lung parenchyma, so their diameter is not dependent on lung volume.

— The bronchi branch to form bronchioles that are smaller in diameter and have no supporting cartilage. They are embedded within lung parenchyma, and their diameter expands and contracts with lung volume.

The respiratory airways include the respiratory bronchioles (i.e., bronchioles with alveoli in their walls; **Fig. 12.2**) and alveolar ducts.

Alveoli.   There are ~300 million alveoli in adult lungs, each being ~250 μm in diameter. Their walls are composed of a simple squamous epithelium, primarily type I pneumocytes. Each alveolus is encased by pulmonary capillaries, which are sandwiched between the lumens of adjacent alveoli. The total surface area available for gas exchange is ~150 m^2.

Pressures in the Lungs

To understand the mechanics of ventilation and airflow during breathing, it is necessary to review the pressure in the lungs.

— *Intrapleural pressure* is the pressure that is generated between the lungs and chest wall by the opposing forces created by the elastic recoil of the lungs and the elastic recoil of the chest wall.

— *Alveolar pressure* is the pressure within the alveoli.

— *Transpulmonary pressure* is alveolar pressure minus intrapleural pressure.

Intrapleural pressure is always less than alveolar pressure; therefore, transpulmonary pressure is always positive. It is the positive transpulmonary pressure that keeps the lungs inflated (like a balloon) against the chest wall.

Clinical implications of the anatomy of the bronchi

Improperly placed endotracheal (ET) tubes or aspirated foreign bodies are more likely to become lodged in the right main bronchus than the left. This is because the left bronchus diverges sharply at the tracheal bifurcation, whereas the right bronchus is relatively straight.

Drug delivery via endotracheal tubes

Endotracheal (ET) tubes are used to maintain airway patency and facilitate ventilation in an emergency situation. When intravenous (IV) access cannot be obtained, drugs (e.g., epinephrine, nalaxone, atropine, and lidocaine) are sometimes given via ET tubes. They can exert local or systemic effects.

Fig. 12.1 ▶ Conducting and respiratory parts of the bronchial tree.

The bronchial tree branches into successively finer divisions. The bronchi are reinforced by cartilage rings or plates and are lined by pseudostratified columnar, ciliated epithelium that contains goblet cells. The bronchioles do not have cartilage.

From *Thieme Atlas of Anatomy, Neck and Internal Organs*, © Thieme 2006, Illustration by Markus Voll.

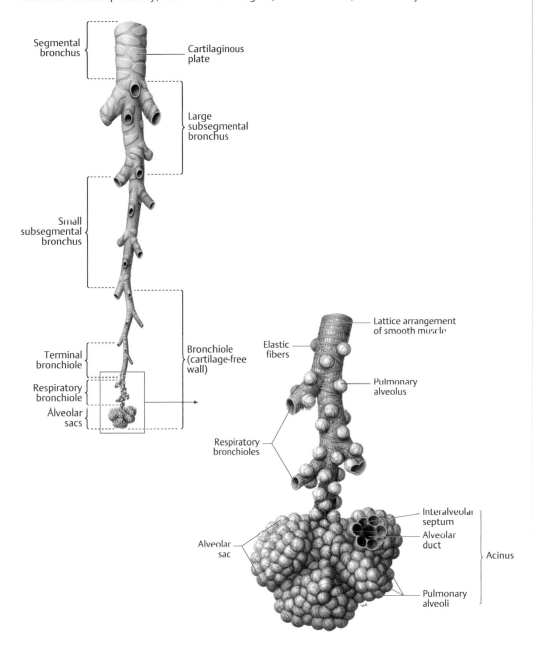

Fig. 12.2 ▶ Structure of a respiratory bronchiole.

Alveoli first begin to appear on the respiratory bronchioles, marking the start of the respiratory portion of the lung. These alveoli are isolated initially, then become more numerous and are collected into sacs. Each sac has a central open space, or alveolar duct, that is continuous with the lumen of its respiratory bronchiole. The alveolar walls are composed of squamous epithelium and are in direct contact with the pulmonary capillaries for gas exchange to occur. Connective tissue with abundant elastic fibers is found throughout the branches of the bronchial tree and the alveoli. These contribute substantially to the elastic recoil of the lungs during expiration.

From *Thieme Atlas of Anatomy, Neck and Internal Organs*, © Thieme 2006, Illustration by Markus Voll.

12.1 Ventilation

Ventilation (breathing) is the process by which air enters and exits the lungs.
Respiration is the overall term for ventilation, gas exchange, and utilization in cells.

Mechanics of Ventilation

Ventilation occurs in a cyclical manner with alternating inspiratory and expiratory phases.

Inspiration

Inspiration is an active process and is principally mediated by the diaphragm during quiet breathing.

— Contraction of the diaphragm enlarges the chest cavity, reducing intrapleural pressure. This increases the transpulmonary pressure and expands the lungs (**Fig. 12.3**). Minimal movement of the diaphragm (a few centimeters) is sufficient to move several liters of gas.
— The external intercostal and accessory muscles are not necessary for resting respiration, but they contribute substantially to deep respiration during exercise and respiratory distress.

Expiration

Expiration is a passive process during quiet breathing. When the diaphragm relaxes, air is expelled from the lungs due to the elastic recoil of the lung–chest wall system. Active expiration (using muscles of expiration) occurs during exercise or in obstructive lung disease.

Fig. 12.3 ▶ **Mechanics of ventilation.**
When the diaphragm moves to the inspiratory position (*red*), the ribs are elevated by the intercostal muscles (chiefly the external intercostals) and scalene muscles. Because the ribs are curved and directed obliquely downward, elevation of the ribs expands the chest transversely (toward the flanks) and anteriorly. Meanwhile, the diaphragm leaflets are lowered by muscle contraction causing the chest to expand inferiorly. These processes result in overall expansion of the thoracic volume. When the diaphragm moves to the expiratory position (*blue*), the chest becomes smaller in all dimensions, and the thoracic volume is decreased. This process does not require additional muscular energy. The muscles that are active during inspiration are relaxed, and the lung contracts as the elastic fibers in the lung tissue that were stretched on inspiration release their stored energy, causing elastic recoil. For forcible expiration, however, the muscles that assist expiration (mainly the internal intercostal muscles) can actively lower the rib cage more rapidly and to a greater extent than is possible by passive recoil alone.
From *Thieme Atlas of Anatomy*, *Neck and Internal Organs*, © Thieme 2006, Illustration by Markus Voll.

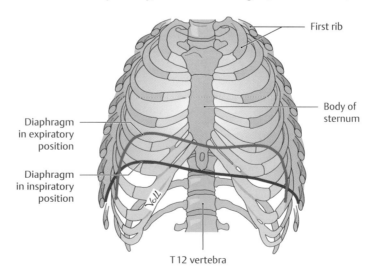

- The rectus abdominus, external and internal obliques, and transverse abdominals contribute to active expiration by forcing abdominal contents upward against the diaphragm. This causes an increase in intrapleural pressure, which compresses the alveoli, allowing the expulsion of air.
- The internal intercostal muscles stiffen the chest during expiration, preventing bulging of the chest and minimizing changes in chest volume.

Compliance of the Respiratory System

Lung Compliance

Lung compliance expresses the distensibility of the lungs, that is, how easily the lungs expand when transpulmonary pressure increases. It is expressed by the following equation:

$$C = \Delta V/\Delta P$$

where

C = lung compliance
ΔV = increase in lung volume (mL)
ΔP = increase in transpulmonary pressure (mm Hg).

- Compliance is inversely related to stiffness.
- Compliance is inversely related to the elastic recoil, or *elastance,* of the lung. Recoil causes the lungs to return to their previous volume when stretching ceases following an increase in transpulmonary pressure. It is mediated by surface tension in the alveoli and by elastic fibers in the lung connective tissue.

Compliance of the Lung–Chest Wall Combination

Because the lungs and chest wall expand and contract together, the overall compliance of the respiratory system is that of the lung–chest wall combination. The compliance of the lung–chest wall combination is lower than the compliance of the lungs alone or chest wall alone.

- The compliance of the lung–chest wall combination varies with lung volume. Compliance is highest at the normal resting volume (functional residual capacity [FRC]) and decreases at both very low and very high volumes.
- At low volumes, compression of the chest wall reduces compliance.
- At high volumes, the increased stretch of elastic tissues in the lung parenchyma causes the lungs to get stiffer (less compliant). High transpulmonary pressure is required to drive this increase in volume, but it is not responsible for the decrease in compliance.

Changes in Lung Compliance in Disease States

- Lung compliance is decreased in pulmonary fibrosis because the interstitium surrounding the alveoli becomes infiltrated with inelastic collagen.
- Lung compliance is increased in emphysema because many small alveoli are replaced by fewer but larger coalesced air spaces that have less elastic recoil.

Surface Tension in the Alveoli

Surface tension is due to the cohesive forces between water molecules at the air–water interface in the alveoli of lungs. It acts to contract the alveoli and is a major contributor to the force of elastic recoil of the lung.

▪ Pulmonary fibrosis

Pulmonary fibrosis is a restrictive lung disease in which repeated lung injury leads to the inflammation of alveoli and interstitium and eventual scarring. In restrictive lung disease, the ability of the lungs to expand is reduced (↓ compliance of the lung–chest wall). There is also an impairment of diffusion due to a decrease in diffusion surface area and an increase in diffusion distance. Pulmonary fibrosis may be idiopathic, or it can be caused by several factors, including connective tissue disease (e.g., rheumatoid arthritis and systemic lupus erythematosus), drugs (e.g., bleomycin and amiodarone), hepatitis, ulcerative colitis, radiation, and environmental pollutants. Symptoms include exertional shortness of breath (dyspnea), dry cough, fatigue, weight loss, and joint pain (arthralgia). Complications are hypoxemia (low oxygen content of blood), pulmonary hypertension, and right-sided heart failure secondary to pulmonary hypertension (cor pulmonale). Respiratory failure may occur at the end stage. Lung function tests reflect decreased lung volumes and gas exchange. Treatment involves the use of immunosuppressant drugs (e.g., prednisone and methotrexate). Lung transplantation may be considered in younger patients when fibrosis is severe.

If there were no opposing force, surface tension would cause the alveoli to collapse (atelectasis). However, the collapsing force is opposed by transpulmonary pressure, which is always positive, allowing the alveoli to remain open.

According to the law of Laplace, the transpulmonary pressure P (in dynes/cm^2) required to prevent collapse of an alveolus is directly proportional to surface tension T (in dynes/cm), and inversely proportional to alveolar radius r (in cm), as expressed by

$$P = 2T/r$$

— All alveoli in a given region of the lungs have about the same transpulmonary pressure. If they all had the same surface tension, the Laplace relationship predicts that the smaller alveoli would collapse and force their volume into larger alveoli. However, surface tension is reduced by pulmonary surfactant, and the reduction is greater in small alveoli than in larger ones because small alveoli concentrate the surfactant. Thus, the increased tendency to collapse because of small radius is just balanced by a greater reduction in surface tension.

Surfactant

Surfactant is a complex substance, consisting of proteins and phospholipids (mainly dipalmitoyl lecithin), that is produced in type II pneumocytes. It lines alveoli and lowers surface tension by the same mechanism as detergents and soaps (i.e., it coats the water surface and reduces cohesive interactions between water molecules).

As an extension of its role in lowering surface tension, surfactant also produces the following effects:

— It increases compliance at all lung volumes, which allows for easier lung inflation and greatly decreases the work of breathing.

— It reduces the otherwise highly negative pressure in the interstitial space, which reduces the rate of filtration from pulmonary capillaries. This assists in maintaining lungs without excessive water.

Failure of surfactant production and/or excessive surfactant breakdown occurs in neonatal respiratory distress syndrome (RDS).

Airflow through the Bronchial Tree

Airflow through the bronchial tree obeys the same principles as blood flow through blood vessels except that the viscosity of air is much lower than that of blood. Airflow is related to the driving pressure and the resistance to flow by

$$Q = \Delta P/R$$

where Q is airflow (mL/min), ΔP is pressure gradient between the mouth/nose and alveoli (cm H$_2$O), and R is airway resistance (cm H$_2$O /mL/min).

— Airflow is directly proportional to the pressure difference between the mouth/nose and the alveoli and inversely proportional to airway resistance.

Airway Resistance

Resistance is derived from Poiseuille's equation as expressed by

$$R = 8\eta L/ \pi r^4$$

where R is airway resistance, r is radius of the airway (cm), η is viscosity of air, and L is length of the airway.

— Like the circulatory system, the length of the bronchial tree is relatively constant, as is the viscosity of inspired air. Therefore, any changes in resistance to airflow are mainly due to changes

---

Neonatal respiratory distress syndrome

Neonatal RDS is a hyaline membrane disease that is caused when a deficiency in surfactant leads to alveolar collapse (atelectasis). It normally occurs in babies born before 28 weeks. Symptoms usually start within minutes of birth and may include absence of breathing (apnea), shortness of breath (dyspnea), increased rate of breathing (tachypnea), grunting, nasal flaring, chest wall retractions, and cyanosis. Complications include pneumothorax (air in the pleural space) and intracranial hemorrhage. Treatment involves the use of continuous positive airway pressure (CPAP) to help maintain airway patency and the administration of beractant (a natural bovine-derived surfactant) or colfosceril (a synthetic surfactant). These agents are given by tracheal instillation. Neonatal RDS may be prevented by the administration of corticosteroids to the mother 24 hours prior to delivery of a premature baby to hasten surfactant production.

Test for fetal lung maturity

Fetal lung maturity can be tested by extracting a sample of amniotic fluid and measuring the lecithin:sphingomyelin ratio (S/L ratio). An S/L ratio < 2:1 indicates surfactant deficiency and therefore lung immaturity.

in the radius of the airways. Because resistance is inversely proportional to the airway radius to the fourth power, small changes in diameter cause large changes in resistance.

— The large airways offer little resistance to airflow. The small airways individually have high resistance, but their enormous number in parallel reduces their combined resistance to a small value. Therefore, the sites of highest resistance in the bronchial tree are normally in the medium airways.

Regulation of Airway Resistance. Airway resistance is primarily regulated by modulation of airway radius by the parasympathetic and sympathetic nervous systems.

— *Parasympathetic nervous system:* Vagal stimulation releases acetylcholine that acts on muscarinic (M_3) receptors in the lungs, leading to bronchoconstriction. This increases the resistance to airflow.
— *Sympathetic nervous system:* Postganglionic sympathetic nerves release norepinephrine that act on β_2 receptors, leading to bronchodilation. This decreases the resistance to airflow.

Relationship of Pressures and Airflow during the Breathing Cycle

— Prior to inspiration, pressures within the airways and alveoli are zero (i.e., equal to external barometric pressure), and intrapleural pressure is about –5 mm Hg, yielding a transpulmonary pressure of +5 mm Hg. This positive pressure just balances the elastic recoil of the lung–chest wall combination. There is no airflow.
— During inspiration, the thoracic cavity expands, making intrapleural pressure more negative by several mm Hg and making transpulmonary pressure more positive by the same amount (**Fig. 12.4**). The increase in transpulmonary pressure causes expansion of the lungs. The increasing volume of the lungs lowers alveolar pressure by ~1 mm Hg, thereby creating a positive driving pressure between the trachea and alveolar space. This causes air to flow through the bronchial tree into the alveoli.
— During expiration, the diaphragm relaxes. The lung–chest wall, now expanded at the end of inspiration, has greater elastic recoil than at rest and contracts. The compression of the lungs increases alveolar pressure by ~1 mm Hg, driving air out of the lungs through the trachea. As the lungs get smaller during exhalation, their elastic recoil diminishes, transpulmonary pressure falls, and intrapleural pressure rises (becomes less negative) back to its resting value.

Fig. 12.4 ► **Alveolar pressure (P_A) and intrapleural pressure (P_{pl}) during respiration.**
During inspiration, intrapleural pressure becomes more negative, and the transpulmonary pressure (P_{TP}) becomes more positive. This causes alveolar pressure (P_A) to fall. During expiration, P_A increases, P_{TP} falls, and P_{pl} rises back to its original value. V_{pulm}, respiratory volume.

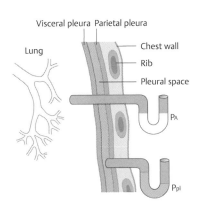

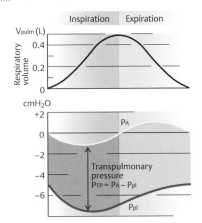

Pneumothorax

Pneumothorax occurs when air is admitted into the pleural space between the lung and the chest wall. This causes an increase in intrapleural pressure and a decrease in transpulmonary pressure, which may lead to partial or complete lung collapse. A primary pneumothorax can occur spontaneously due to rupture of a pleural bleb. Pneumothoraces may also occur secondarily to lung disease, for example, chronic obstructive pulmonary disease (COPD), asthma, tuberculosis (TB), pneumonia, cystic fibrosis, and lung cancer. They may also occur due to penetrating chest trauma. Symptoms include shortness of breath (dyspnea), the severity of which depends on the degree of lung collapse; sudden, sharp chest pain that is exacerbated by taking a deep breath or coughing; chest tightness; and increased heart rate (tachycardia). Signs may be subtle but include diminished breath sounds and chest expansion on the affected side and hyperresonance on percussion. No treatment may be required for small pneumothoraces that heal spontaneously. Larger pneumothoraces require placement of a chest tube or surgical repair (rarely).

Tension pneumothorax

Tension pneumothorax is a life-threatening condition in which air accumulates in the pleural space and becomes trapped when the injured tissue acts as a one-way valve. This causes complete collapse of the lung on the affected side and the heart to shift to the opposite side, thus compromising venous return and cardiac output. The ventilatory capacity of the other lung is also impaired due to compression. Signs include respiratory distress, increased heart rate (tachycardia), distended neck veins, hypotension, and tracheal deviation (away from the side of tension pneumothorax). Complications include hypoxemia (low blood oxygen), respiratory failure, shock, and cardiac arrest. Treatment for this condition is emergency needle decompression of the affected side (the side opposite that to which the trachea is deviated) followed by placement of a chest tube.

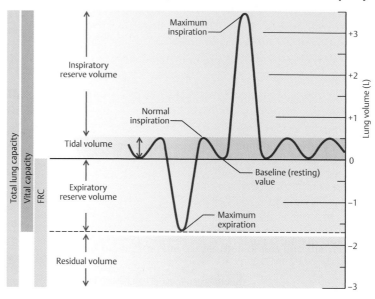

Fig. 12.5 ▶ **Lung volumes and capacities. FRC, functional residual capacity.**

Lung Volumes and Capacities

— *Lung volumes* are a way to functionally divide volumes of air that occur during different phases of the breathing cycle (**Fig. 12.5**). They are all measured by spirometry, except for residual volume. They vary with height, sex, and age.
— *Lung capacities* are the sums of two or more lung volumes.
— Tidal, inspiratory, and expiratory reserve volumes and inspirational and vital capacities are used in basic pulmonary function tests.

Lung Volumes

— *Tidal volume* (TV) is the volume of air that moves in or out of the lungs during one normal, resting inspiration or expiration.
— *Inspiratory reserve volume* (IRV) is the volume of air that can be inspired beyond a normal inspiration.
— *Expiratory reserve volume* (ERV) is the volume of air that can be expired beyond a normal expiration.
— *Residual volume* (RV) is the volume of air left in the lungs and airways after maximal expiration.

Table 12.1 contains the normal approximate lung volumes and expresses them as a percentage of total lung capacity (TLC).

Table 12.1 ▶ Normal Lung Volumes and Their Relationship to Total Lung Capacity		
Lung Volume	**Normal Volume (mL)**	**Percentage of Total Lung Capacity**
Tidal volume	500 mL	9%
Inspiratory reserve volume	3000 mL	52%
Expiratory reserve volume	1300 mL	22%
Residual volume	1000 mL	17%

Lung Capacities

— *Inspirational capacity* (IC) is the maximum volume of air that can be inspired with a deep breath following a normal expiration. It is the sum of TV and IRV.

Chronic obstructive pulmonary disease

COPD is a term used to describe chronic obstructive bronchitis and emphysema, which always coexist, to varying degrees. With chronic bronchitis, inflammation (most commonly caused by cigarette smoke) causes the bronchial tubes to thicken and scar and produce excess mucus. This causes narrowing of the airway lumen and obstruction. Emphysema occurs when the walls of the alveoli are progressively destroyed. This decreases the surface area of the alveoli for gas exchange with pulmonary capillary blood and causes the small airways to collapse during expiration, trapping air in the lungs. This may be caused by cigarette smoke or α_1-antitrypsin deficiency. The characteristic symptoms of COPD are persistent cough, sputum, dyspnea, and wheeze. Signs of COPD include hyperinflation of the lungs, causing a barrel chest appearance, hypertrophy of accessory muscles of respiration, descended trachea, respiratory distress, crepitations, and wheeze. Lung function tests show that the forced expiratory volume/forced vital capacity is < 70%, and residual volume and total lung capacity are high. Drugs used to treat COPD include bronchodilators (β_2-agonist drugs), inhaled corticosteroids, and antibiotics (when necessary).

Pink puffers and blue bloaters

Some patients with COPD increase their respiratory rate (hyperventilate) to try to cope with their shortness of breath (dyspnea). In this way, they manage to achieve relatively normal oxygenation of arterial blood, and their blood CO_2 concentration can be either normal or low. They are termed "pink puffers" because they are breathless and pink from the exertion. Other patients with COPD do not have the muscle or lung capacity to increase their respiratory rate. They have low oxygenation of arterial blood and high blood CO_2 concentrations, and so appear blue. Right-sided heart failure may develop secondary to pulmonary hypertension (cor pulmonale), resulting in edema and "bloating."

— *Functional residual capacity* (FRC) is the volume of the lungs after passive expiration with relaxed respiratory muscles. It is the sum of ERV and RV.
— *Vital capacity* (VC), or *forced vital capacity* (FVC), is the maximum volume of air that can be expired in one breath after deep inspiration. It is the sum of TV, IRV, and ERV.
— TLC is the total volume of air that can be contained in the lungs and airways after a deep inspiration. It is the sum of all four lung volumes: TV, IRV, ERV, and RV.

Note: TLC and FRC cannot be measured by spirometry because residual volume is needed for their calculation.

Table 12.2 contains the normal lung capacity volumes.

Table 12.2 ▶ Normal Lung Capacities	
Lung Capacity	**Normal Volume (mL)**
Inspirational capacity	3500 mL
Functional residual capacity	2300 mL
Vital capacity or forced vital capacity	4900 mL
Total lung capacity	5800 mL

Forced Expiratory Volume

FEV_1 is the volume of air that can be forcibly expired in the first second following a deep breath (**Fig. 12.6**). It is usually > 70% of the FVC ($FEV_1/FVC > 70\%$).

— In obstructive lung disease (e.g., asthma and COPD), FEV_1 is reduced proportionally more than FVC; therefore, $FEV_1/FVC < 70\%$.
— In restrictive lung disease (e.g., fibrosis), both FEV_1 and FVC are reduced. This means that FEV_1/FVC is normal or increased.

Dead Space

Dead space is volume within the bronchial tree that is ventilated but does not participate in gas exchange.

— *Anatomical dead space* is the volume of the conducting airways (pharynx, trachea, and bronchi) that do not contain alveoli and therefore cannot participate in gas exchange. It is ~150 to 200 mL.
— *Physiological dead space* is the total volume of the bronchial tree that is ventilated but does not participate in gas exchange.

Fig. 12.6 ▶ **Volume exhaled versus time during a forced exhalation.**
The total volume exhaled is the forced vital capacity (FVC), and the volume exhaled in the first second is the forced expiratory volume (FEV_1).

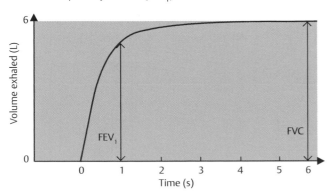

Asthma

Asthma is predominantly an inflammatory disease with associated bronchospasm (sudden constriction of the smooth muscle in the walls of the bronchioles), mucosal swelling, and increased mucus production. There is episodic bronchial obstruction causing wheezing, shortness of breath (dyspnea), cough, and mucosal edema. The inflammatory response may be triggered by allergens such as animal hair, dust mites, feathers, pollen, and mold. In nonallergic asthma, bronchial hyperreactivity may be caused by the inhalation of chemicals, cigarette smoke, viral infections, cold air, exercise, and stress. Drugs (e.g., aspirin and or other nonsteroidal antiinflammatory drugs [NSAIDS]) can also precipitate bronchospasm. This can be serious and sometimes fatal; therefore, these drugs are contraindicated in patients with asthma who have a history of hypersensitivity reactions and should be used with caution in all asthmatics. Acetaminophen can be used by asthmatics to treat mild to moderate pain.

— In healthy lungs, physiological dead space is approximately equal to anatomical dead space. However, physiological dead space may be increased in lung diseases where there are mismatches between ventilation (V) and perfusion (pulmonary blood flow [Q]).

— Physiological dead space can be calculated using Bohr's equation. This calculation assumes that the partial pressure of CO_2 (Pa_{CO_2}) in the alveoli is the same as that in systemic arterial blood.

$$V_D/V_T = (Pa_{CO_2} - Pe_{CO_2})/Pa_{CO_2}$$

where V_D is physiological dead space (mL), V_T is tidal volume (mL), Pa_{CO_2} is arterial P_{CO_2} (mm Hg), and Pe_{CO_2} is mixed expired P_{CO_2} (mm Hg).

— This equation, which expresses physiological dead space as a fraction of the exhaled TV, accounts for the fact that mixed expired P_{CO_2} is lower than alveolar P_{CO_2} due to the dilution of alveolar P_{CO_2} with dead space air (containing no CO_2).

Example: Pa_{CO_2} = 41 mm Hg; Pe_{CO_2} = 28.7 mm Hg

$$V_D/V_T = (41 - 28.7)/41 = 0.3$$

So, dead space is 30% of tidal volume, or V_D = 150 mL when V_T = 500 mL.

Ventilation Rate

Minute ventilation refers to the total ventilation per minute. It is expressed as

$$\text{Minute ventilation} = \text{TV} \times \text{breaths/min}$$

Alveolar ventilation refers to ventilation of alveoli that participate in gas exchange per minute. It is expressed as

$$\text{Alveolar ventilation} = (\text{TV} - \text{physiological dead space}) \times \text{breaths/min}$$

12.2 Pulmonary Blood Flow

The pulmonary circulation is discussed in Chapter 10.

Distribution of Pulmonary Blood Flow

When a person is upright, the force of gravity affects the distribution of pulmonary blood flow within the lungs (but not the total amount of blood flow) because vascular pressures progressively fall at locations above the heart. This distribution of blood flow is described in terms of "zones" of the lung.

Zone 1: Lung Apex.   If pulmonary artery pressure is not high enough to support the column of blood from the right ventricle all the way to the apices of the lungs, the uppermost blood vessels collapse, and there is no flow in this region. This does not normally occur in healthy lungs but may occur if right ventricular pressure is extremely low (e.g., due to hemorrhage). Also, if alveolar pressure is increased to the point where it exceeds vascular pressure, blood vessels collapse (e.g., due to positive pressure ventilation).

Zone 2: Middle of the Lung.   In zone 2, blood flow is intermittent. Pulmonary artery pressure drives blood flow at its peak during systole, but not during the rest of the cardiac cycle.

Zone 3: Lung Base.   Zone 3 has no gravitational impediment to blood flow because regions located below the heart always have vascular pressures greater than alveolar pressure. Therefore, blood flow is continuous.

Shunts

Right-to-Left Shunts

— Right-to-left shunts allow blood to move directly from the pulmonary circulation to the systemic circulation.
— Intracardiac right-to-left shunts occur in cases where right ventricular pressure is greater than left ventricular pressure due to right ventricular hypertrophy (increased size of the muscle of the right ventricle).
— Right-to-left shunts can also occur through regions of the lung that are perfused but not ventilated due to infection (e.g., pneumonia) or injury (e.g., chemical inhalation).
— The result of a right-to-left shunt is decreased oxygenation of the blood (hypoxemia).

Left-to-Right Shunts

— Left-to-right shunts allow blood to recirculate from the systemic circulation back to the pulmonary circulation.
— Left-to-right shunts usually occur through atrial or ventricular septal defects and result in the right ventricle pumping more blood than the left ventricle. This increases the work of the right ventricle and may lead to right ventricular hypertrophy.
— There is no hypoxemia.

Capillary Exchange of Fluid in the Lungs

Fluid outside the alveoli in the interstitial space is governed by Starling forces as it is in all other capillary beds (see Chapter 10). Although pulmonary capillary hydrostatic pressure (P_c) is normally low, the negative interstitial hydrostatic pressure (P_i) and the finite interstitial fluid oncotic pressure (π_i) favor net filtration of fluid out of pulmonary capillaries into the interstitium at a modest rate. The filtered fluid is removed via the pulmonary lymphatic system. If the rate of filtration exceeds the capacity of the lymphatic system to remove it, the result is pulmonary edema.

■ **Pulmonary edema**

Pulmonary edema is the accumulation of fluid in the lungs. It begins as fluid in the interstitial spaces and if severe enough, spreads to the alveoli. There are two types: cardiogenic and noncardiogenic. The cardiogenic type occurs in heart failure when left atrial pressure, and hence pulmonary venous pressure, rises too high. The noncardiogenic type occurs in lung disease or injury that allows protein to leak from the capillaries, thereby raising the oncotic pressure in the interstitium. The main consequences of pulmonary edema are reduced lung compliance, which increases the work of breathing, and poor oxygenation due to a ventilation/perfusion mismatch. Signs and symptoms begin with shortness of breath (dyspnea), rapid breathing (tachypnea), anxiety, and hypoxemia (low blood oxygen). The patient may cough up blood (hemoptysis) and produce pink, frothy sputum. Treatment is directed at the underlying cause, and supplemental oxygen is administered.

13   Gas Exchange and Transport

13.1   Partial Pressures

In a gas mixture, each gas species exerts a pressure, the partial pressure of that gas. The sum of the partial pressures of the gases in a mixture equals the total gas pressure.

Partial pressure for an individual gas = the fraction of that gas in the gas mixture × total gas pressure

Calculation of Partial Pressure of Oxygen (Po$_2$) in Dry Inspired Air

O_2 comprises 21% of air; total gas pressure = 760 mm Hg (at sea level)

$$Po_2 = 0.21 \times 760 \text{ mm Hg}$$

$$= 160 \text{ mm Hg}$$

At high altitude, the Po$_2$ is reduced because barometric pressure is lower.

Correction of Po$_2$ for the Presence of Water Vapor

Dry air entering the lungs becomes completely saturated with water as air passes through moist airways. This displaces some of the other gases and slightly reduces their partial pressures.

Partial pressure of water vapor (P$_{H_2O}$) is 47 mm Hg at body temperature.

$$\text{Total pressure of gases other than water} = 760 \text{ mm Hg} - 47 \text{ mm Hg}$$

$$= 713 \text{ mm Hg}$$

Therefore, the Po$_2$ in warm, humidified inspired air is

$$Po_2 = 713 \text{ mm Hg} \times 0.21$$

$$= 150 \text{ mm Hg}$$

13.2   Gas Exchange

Diffusion of Gases

> **▮ Dry lungs and gas exchange**
>
> The epithelial cells that make up most of the alveolar wall actively reabsorb Na+ and water, thereby preventing accumulation of water in the alveoli, a situation usually referred to as "dry" lungs, although they are not literally dry. The lack of excess water prevents it from acting as an impediment to gas exchange.

O_2 and carbon dioxide (CO_2) diffuse between alveolar gas and pulmonary capillary blood according to standard physical principles (Fick's law; see Chapter 1).

— The total amount moved per unit of time is proportional to the area available for diffusion and to the difference in partial pressure between alveolar gas and pulmonary capillary blood, and inversely proportional to the thickness of the diffusion barrier.
— Gas will diffuse from the alveoli (higher partial pressures) to the pulmonary capillaries (lower partial pressures) until they equilibrate and no partial pressure gradient exists. As a result, blood entering the pulmonary veins from the pulmonary capillaries has virtually the same partial pressures as gases in the alveoli (**Fig. 13.1**).
— The diffusion barrier, composed of alveolar epithelial cells (type I pneumocytes) and capillary endothelial cells, is very thin, which ensures that the diffusion distance between alveolar gas and pulmonary capillary blood is very short. This allows blood in the pulmonary

Fig. 13.1 ▶ **Alveolar gas exchange.**
The capillaries surrounding an alveolus contain less gas than does the alveolus. During transit through a pulmonary capillary, blood equilibrates with alveolar gas and acquires the same partial pressures. (P_v, partial pressure of mixed venous blood; P_a, partial pressure of arterial blood; P_{ACO_2}, alveolar partial pressure of CO_2; P_{AO_2}, alveolar partial pressure of O_2)

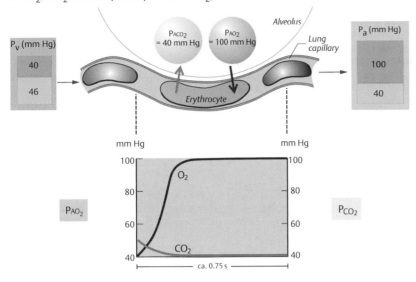

capillaries to equilibrate with alveolar gas during the short time (< 1 sec) that the blood is in the capillaries.

Limitations to the Exchange of Gases

There is an upper limit to the rate at which O_2 or any other gas can be exchanged by the lungs. These limits can be divided into categories of diffusion- and perfusion-limited exchange.

Diffusion-limited Exchange

Gas exchange may be limited by the rate at which it can diffuse across the alveolar membrane, either because of thickening of the alveolar membrane (e.g., due to fibrosis [rare]) or because the surface area available for gas exchange is reduced (e.g., in emphysema). If the surface area is reduced, pulmonary blood is constrained to flow through fewer capillaries. Consequently the blood spends less time in the capillaries, and complete equilibration of gas between alveolar air and pulmonary capillary blood will not occur.

Diffusion-limited uptake can be illustrated using carbon monoxide (CO) as an example. Because hemoglobin in red blood cells so avidly binds CO, the partial pressure of CO (P_{CO}) in the blood remains effectively zero as CO diffuses into pulmonary capillaries. At a given alveolar P_{CO}, the gradient remains constant during uptake, and the amount taken up is therefore purely a function of the area available for diffusion. Variations in blood flow have little influence on the diffusion of CO because all of the CO that diffuses into the blood is immediately bound to hemoglobin.

Perfusion-limited Exchange

Gas exchange of O_2 or other gases (except CO) can also be limited by the rate at which the pulmonary capillary blood removes the gas. A greater rate of perfusion would lead to greater diffusion of gas.

Perfusion-limited uptake can be illustrated using nitrous oxide (N_2O) as an example. N_2O diffuses rapidly from alveolar gas into pulmonary capillaries and does not bind to hemoglobin; thus, the partial pressure of N_2O (P_{N_2O}) in the blood immediately rises to match its value in the

■ **Cystic fibrosis**

Cystic fibrosis is an autosomal recessive disease in which there is a defect in the epithelial transport protein CFTR (cystic fibrosis transmembrane conduction regulator) found in the lungs, pancreas, liver, genital tract, intestines, nasal mucosa, and sweat glands. This alters Cl^- transport in and out of cells and inhibits some Na^+ channels. In the lungs, Na^+ and water are absorbed from secretions that then become thick and sticky. In the pancreas, secretions are thick and sticky because duct cells cannot secrete Cl^- via the CFTR, and water normally follows this ion movement. Sweat is salty because Cl^- is not being absorbed via the CFTR, so Na^+ also remains in the duct lumen. Symptoms include cough, wheezing, repeated lung and sinus infections, salty taste to the skin, steatorrhea (foul-smelling, greasy stools), poor weight gain and growth, meconium ileus (in newborns), and infertility in men. Complications of this disease include bronchiectasis (abnormal dilation of the large airways), deficiency of fat-soluble vitamins (A, D, E, and K), diabetes, cirrhosis, gallstones, rectal prolapse, pancreatitis, osteoporosis, pneumothorax, cor pulmonale, and respiratory failure. Treatment involves daily physical therapy to help expectorate secretions from the lungs, antibiotics to treat lung infections, mucolytics (e.g., acetylcysteine) to decrease the viscosity of mucus, and bronchodilators.

■ **Diffusion hypoxia with nitrous oxide**

N_2O is a common inhalational anesthetic. It is administered in high concentration along with supplemental oxygen. When anesthesia is terminated, the N_2O that accumulated in the body now diffuses into the alveoli. This partially displaces other gases that are there, including O_2 and CO_2. If the patient is breathing room air, this displacement lowers the alveolar P_{O_2} and P_{CO_2} and leads to a transient hypoxemia. To prevent this from occurring, the patient is temporarily given 100% O_2.

alveolar gas. Diffusion in a given capillary stops before blood completes its transit through the capillary because there is no longer a gradient. Diffusion can only be increased if blood flow increases.

Partial Pressure Changes of Oxygen and Carbon Dioxide Following Gas Exchange

Partial Pressure Changes of Oxygen

- The P_{O_2} of humidified inspired air is 150 mm Hg.
- The P_{O_2} of alveolar air is 100 mm Hg. This is due to the diffusion of O_2 from alveolar air into pulmonary capillary blood.
- The P_{O_2} of systemic arterial blood is 95 mm Hg. It is almost the same as the P_{O_2} of alveolar air because the partial pressure of pulmonary capillary blood equilibrates with alveolar air. However, ~2% of the cardiac output bypasses the pulmonary circulation, which accounts for the slight discrepancy in partial pressures.
- The P_{O_2} of venous blood is 40 mm Hg because O_2 has diffused from arterial blood into the tissues.

Partial Pressure Changes of Carbon Dioxide

- The P_{CO_2} of humidified inspired air is almost zero.
- The P_{CO_2} of alveolar air is 40 mm Hg because CO_2 from venous blood entering the pulmonary capillaries diffuses into alveolar air.
- The P_{CO_2} of systemic arterial blood is 40 mm Hg because pulmonary capillary blood equilibrates with alveolar air.
- The P_{CO_2} of venous blood is 46 mm Hg. It is higher than systemic arterial blood due to the diffusion of CO_2 from the tissues into venous blood following cellular respiration.

Table 13.1 summarizes the partial pressure changes of O_2 and CO_2 following gas exchange.

Table 13.1 ▶ Partial Pressures of Oxygen and Carbon Dioxide				
	Humidified Inspired Air	Alveolar Air	Arterial Blood	Venous Blood
P_{O_2}	150	100	95	40
P_{CO_2}	0.3	40	40	46

13.3 Ventilation and Perfusion Ratios for Optimum Gas Exchange

Ventilation/perfusion ratio is the ratio of alveolar ventilation V to perfusion (pulmonary blood flow) Q.

- In healthy lungs, the V/Q ratio is close to 1:1, resulting in optimum gas pressures and oxygenation in systemic arterial blood.

Distribution of V/Q Ratios

There are regional differences in alveolar ventilation and blood flow in the upright individual.

- Alveolar ventilation is higher at the base of the lungs than the apices because the base is more compliant and changes more in volume during each breathing cycle.
- Blood flow is very low at the apex of the lung and very high at the base due to the effects of gravity.

The differences in regional blood flow are greater than the differences in regional ventilation. This creates different V/Q ratios at various levels of the lung. Typical values are as follows:

— Apex V/Q is ~3:1.
— Middle of lungs (heart level) V/Q is ~1:1.
— Base of lungs V/Q is ~1:2.

Despite these regional differences, which are mainly attributed to gravity, the overall V/Q ratio of the lung is matched 1:1.

V/Q Mismatching

A V/Q mismatch (inequality), beyond the modest regional differences in V/Q ratio that is seen in healthy lungs due to gravity, occurs because of a lung disorder that affects normal ventilation and/or perfusion. This can be illustrated using airway obstruction and pulmonary embolism as examples.

— V/Q ratio with airway obstruction: In complete airway obstruction, perfusion is normal, but there is no ventilation.

  — V/Q = 0, which is termed a *shunt*.
  — Gas exchange does not occur; therefore, alveolar air has a composition closer to that of systemic venous blood: low Po_2 and high Pco_2.

— V/Q ratio with pulmonary embolism (PE): In PE causing complete blockage of the pulmonary artery, ventilation is normal, but there is no blood flow.

  — V/Q = ∞, which is termed *dead space*.
  — Gas exchange does not occur; therefore, alveolar air has a composition closer to that of inspired air: high Po_2 and low Pco_2.

V/Q mismatches ultimately cause reduced oxygenation of systemic arterial blood (low Po_2). The lungs are able to compensate somewhat for V/Q mismatches by hypoxic vasoconstriction.

Hypoxic Vasoconstriction

If an area of the lung become hypoxic (has a low Po_2), for example, due to a blocked bronchus, there is reflex vasoconstriction of pulmonary arteriolar smooth muscle. This is the opposite of the response seen in the peripheral vasculature, where hypoxia stimulates vasodilation. Hypoxic vasoconstriction increases pulmonary vascular resistance in the hypoxic region so that blood flow decreases and is diverted to other areas of the lung with higher Po_2. This prevents perfusion of poorly ventilated areas of the lungs and thus optimizes the V/Q ratio and gas exchange in the rest of the lung. This mechanism is limited, and in cases of severe lung malfunction, it cannot prevent pathological V/Q mismatch. Furthermore, a low Po_2 in the whole lung, as occurs at high altitude, can lead to high resistance and pulmonary hypertension in susceptible individuals.

Table 13.2 summarizes the possible causes of impairment of gas exchange.

◼ Pulmonary embolism

Pulmonary embolism (PE) is an obstruction in the pulmonary arterial system, usually caused by blood clots from the periphery, particularly the deep veins of the legs, which are transported to the lung. Symptoms include shortness of breath (dyspnea), chest pain exacerbated by taking a deep breath or coughing, and cough +/ blood (hemoptysis). PE decreases the area available for diffusion of gases (increases dead space) and therefore causes a V/Q mismatch. In severe cases, PE can cause death due to hypoxia and cor pulmonale (right heart failure due to chronic pulmonary hypertension). Treatment involves the use of the anticoagulants (e.g., heparin and warfarin) or thrombolytics (e.g., streptokinase; not normally required). Surgical clot removal may be necessary for large pulmonary emboli.

Table 13.2 ► Impairment of Gas Exchange	
Impairment	**Possible Causes**
Abnormal ventilation	Obstructive lung diseases (e.g., asthma, COPD, and cystic fibrosis) Foreign body obstruction
Abnormal diffusion	Pulmonary edema Restrictive lung diseases (e.g., pulmonary fibrosis, neonatal respiratory distress syndrome, and pneumothorax) Emphysema Lung resection
Abnormal perfusion	Pulmonary embolism
Abbreviation: COPD, chronic obstructive pulmonary disease.	

13.4 Oxygen Transport in the Blood

Oxygen is mainly transported in the blood bound to hemoglobin. However, some is always transported as dissolved O_2, which determines the Po_2 of the blood.

Transport of Oxygen Bound to Hemoglobin

Hemoglobin is a metalloprotein consisting of four subunits. Each subunit contains a globular protein chain bound to a heme group. Heme is a porphyrin containing an iron atom in the ferrous state (Fe^{2+}), which can bind an O_2 molecule. Thus, each hemoglobin molecule has four binding sites for O_2. The fraction of the binding sites that are occupied (the percentage saturation of O_2, Sao_2) is mainly a function of the existing Po_2, but it also depends on the structure of the protein chains and other factors.

Hemoglobin–Oxygen Dissociation Curve (Fig. 13.2)

— The hemoglobin–O_2 dissociation curve is a plot of hemoglobin–oxygen saturation as a function of Po_2. When one hemoglobin subunit binds an O_2 molecule, this increases the affinity of the other subunits for O_2, resulting in a sigmoid–shaped relation between Po_2 and Sao_2.

— Normal hemoglobin is 50% saturated at a Po_2 of 26 mm Hg, 75% saturated at a Po_2 of 40 mm Hg, (the Po_2 of mixed venous blood), 90% saturated at a Po_2 of 60 mm Hg, and 100% saturated at 95 mm Hg (the Po_2 of systemic arterial blood).

— Because hemoglobin is > 90% saturated at all values of arterial Po_2 > 60 mm Hg, people with healthy lungs can tolerate a wide range of inspired Po_2 values without difficulty. Furthermore, raising arterial Po_2 by breathing air enriched with O_2 adds little additional O_2 content to the blood. However, the situation is quite different for people with lung pathology, for whom supplemental O_2 is of great benefit.

Note: Fetal hemoglobin (Hb F) has a slightly higher affinity for O_2 than does the adult form (Hb A) because it binds 2,3-diphosphoglycerate (2,3-DPG) less avidly. Hb F is 50% saturated at a Po_2 of 19 mm Hg. This facilitates the uptake of O_2 by the fetus at the relatively low values of Po_2 that exist in the placenta.

Shifts in the Hemoglobin–Oxygen Dissociation Curve

The affinity of hemoglobin for O_2 is somewhat dependent on conditions. A decrease in affinity (meaning it requires a higher Po_2 to reach a given level of saturation) is called a "shift to the right" of the hemoglobin–O_2 dissociation curve, whereas an increase in affinity is a "shift to the left" (**Fig. 13.3**).

— A right shift facilitates the unloading of O_2 in peripheral tissues. Conditions favoring a shift to the right (high temperature, low pH, and high Pco_2) exist in exercising muscle. Unloading more O_2 raises local Po_2 and drives O_2 diffusion into nearby tissues. In muscle, this benefits

Fig. 13.2 ▶ **Hemoglobin–O_2 dissociation curve.**

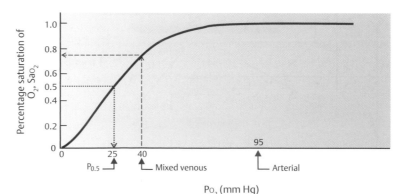

Fig. 13.3 ► **Shifts in the hemoglobin–O₂ dissociation curve**.
The O_2 dissociation curve is shifted to the right (*green line*) indicates a decrease in affinity of hemoglobin for O_2 and therefore an increase in O_2 unloading in peripheral tissues. The curve is shifted to the left (*blue line*) when there is an increase in the affinity of hemoglobin for O_2 and reduced O_2 unloading into peripheral tissues. (DPG, 2,3-diphosphoglycerate.)

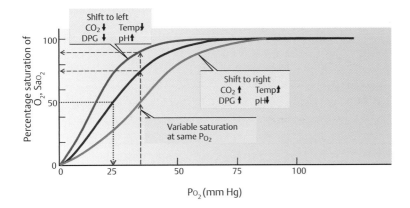

Carbon monoxide poisoning

CO is an odorless, tasteless gas that can poison an individual without the person being aware. CO binds to the same sites on hemoglobin as O_2, but ~200 times more avidly. When CO binds to hemoglobin, the binding sites are no longer available to load O_2 even if alveolar P_{O_2} is normal. Thus, delivery of O_2 to the tissues is greatly reduced. CO poisoning can be treated with hyperbaric O_2. By greatly raising arterial P_{O_2}, the dissolved O_2 level increases so much that the dissolved component, normally trivial, is high enough to help oxygenate body tissues. The high P_{O_2} also raises the rate at which the residual CO dissociates from hemoglobin.

oxidative metabolism. A shift to the right also occurs when there is increased levels of 2,3-DPG in the blood at high altitude.

— A left shift reduces the unloading of O_2 in peripheral tissues. It occurs with decreased temperatures, high pH, low high P_{CO_2}, and decreased 2,3-DPG levels.

Hypoxia and Hypoxemia

— *Hypoxia* is a general term meaning low delivery of O_2 to tissues. It may occur in local tissue sites as a result of vascular occlusion or in the whole body at high altitude or in conditions of impaired ventilation or low cardiac output.

Hypoxemia is a more specific term meaning low P_{O_2} in arterial blood. It may occur due to hypoventilation, low inspired P_{O_2}, diffusion impairment, V/Q mismatch, or a right-to-left shunt.

Comparing Causes of Hypoxemia: A-a Gradient

An A-a gradient refers to the difference between average alveolar P_{O_2} (P_{AO_2}) and arterial P_{O_2} (P_{aO_2}), as expressed by

$$\text{A-a gradient} = P_{AO_2} - P_{aO_2}$$

— It can be used as a tool to analyze causes of hypoxemia.

— Alveolar P_{O_2} can be assessed by sampling exhaled gas at the end of expiration (after dead space gas has been washed out).

— There is always a small A-a gradient of ~5 mm Hg, due to the shunting of pulmonary blood flow and to the small gravitational V/Q mismatch in the lungs.

— An increased A-a gradient (> 5 mm Hg) indicates that inspired air is reaching alveoli that are not transferring O_2 to the blood. This may occur due to an impairment of diffusion (e.g., fibrosis or emphysema), V/Q mismatch, or right-to-left shunt.

Table 13.3 compares the causes of hypoxemia.

Table 13.3 ► Comparison of the Causes of Hypoxemia		
Cause of Hypoxemia	A-a Gradient	Mechanism
Hypoventilation	Normal	If ventilation is depressed due to drugs, chest injury, or very high work of breathing, not enough O_2 enters the lungs. Both alveolar and arterial Po_2 are low; therefore, the A-a gradient is normal.
Low inspired Po_2	Normal	If the inspired air has a low Po_2, as at high altitude, then alveolar and arterial Po_2 will also be low, even if there is hyperventilation in response; therefore, the A-a gradient is normal.
Diffusion impairment	↑	A diffusion impairment means that oxygen in the alveoli cannot fully equilibrate with pulmonary capillary blood. It occurs when damage to the lungs reduces the surface area for gas exchange. Blood flows through the now-limited capillaries too rapidly to permit full equilibration, so arterial Po_2 is decreased, and the A-a gradient is increased.
V/Q mismatch	↑	The presence of ventilated (V) but not perfused (Q) lung regions leads to poorly oxygenated blood (decreased arterial Po_2), so the A-a gradient is increased.
Right-to-left shunt	↑	A right-to-left shunt, which might occur as a result of congenital cardiac defects, mixes nonoxygenated blood with oxygenated blood, thus lowering arterial Po_2. The A-a gradient is increased.

13.5 Carbon Dioxide Transport in Blood

CO_2 is mainly transported in blood in the form of bicarbonate (HCO_3^-). It is also transported as dissolved CO_2 (5%) or carbaminohemoglobin (5%). The content of CO_2 (the sum of its concentration in all forms) is ~2.5 times the content of O_2.

Transport of Carbon Dioxide as Bicarbonate

CO_2 is produced by tissues as a result of aerobic respiration. It freely diffuses into red blood cells and is then transported in blood by the following steps (**Fig. 13.4**):

— CO_2 combines with H_2O within red blood cells to form carbonic acid (H_2CO_3). This is catalyzed by carbonic anhydrase. H_2CO_3 dissociates into H^+ and HCO_3^-.

— Much of the HCO_3^- moves into the plasma via an antiporter in exchange for Cl^- (the "chloride shift"). Plasma, as well as the red blood cells, is therefore a vehicle for the transportation of CO_2 to the lungs.

— The H^+ produced is largely buffered by hemoglobin. Removal of O_2 from the blood by uptake into tissues increases the amount of CO_2 that can be converted to HCO_3^- because deoxygenated hemoglobin is a better buffer than oxygenated hemoglobin. In addition, deoxygenated hemoglobin binds more carbamino-CO_2 than oxygenated hemoglobin.

— In the lungs, the reverse of the process described above occurs: HCO_3^- enters red blood cells in exchange for Cl^-, and HCO_3^- binds to H^+ to form H_2CO_3. H_2CO_3 then breaks down to form CO_2 and H_2O, and CO_2 is expired.

Fig. 13.4 ▶ **CO$_2$ transport in blood**.

CO$_2$ is an end product of energy metabolism (**1**). It enters red blood cells, where it combines with H$_2$O to form H$^+$ and HCO$_3^-$ (**2**). This is catalyzed by carbonic anhydrase. HCO$_3^-$ leaves red blood cells via an HCO$_3^-$/Cl$^-$ antiporter (**3**). CO$_2$ also forms carbaminohemoglobin within red blood cells (**4**). H$^+$ ions liberated in the formation of HCO$_3^-$ and carbaminohemoglobin are buffered by hemoglobin (Hb) (**5**). In the lung, these reactions proceed in the opposite direction, and CO$_2$ diffuses from red blood cells into the alveoli.

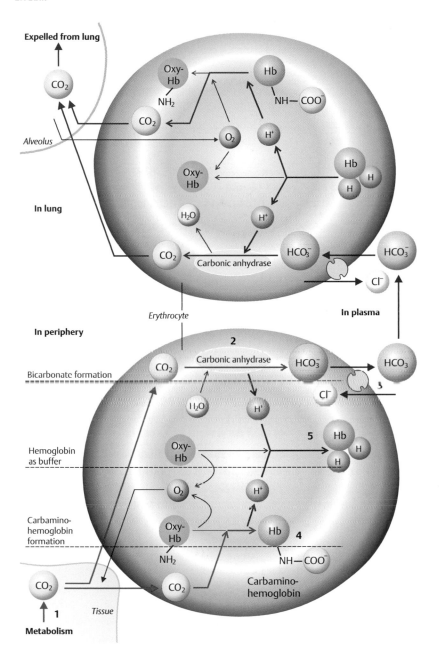

The Carbon Dioxide Dissociation Curve

The CO_2 dissociation curve is the relationship between CO_2 concentration in the blood and P_{CO_2} (**Fig. 13.5**).

— It is nearly linear over P_{CO_2} values between normal venous blood (~46 mm Hg) and normal arterial blood (~40 mm Hg), as compared with the highly nonlinear hemoglobin–O_2 dissociation curve in the normal physiological range.

Fig. 13.5 ▸ **CO_2 dissociation curve**.
Increasing P_{CO_2} increases total CO_2 concentration. Because O_2 binding to hemoglobin decreases its capacity to bind CO_2, the total CO_2 concentration at any given P_{CO_2} is somewhat lower at high O_2 saturation (*red curve*) than at a lower O_2 saturation (*purple curve*). The CO_2 concentration values in normal arterial and mixed venous blood are indicated at points **a** and **v**, respectively. The normal range of CO_2 dissociation is determined by connecting these two points by a line called "physiologic CO_2 dissociation curve".

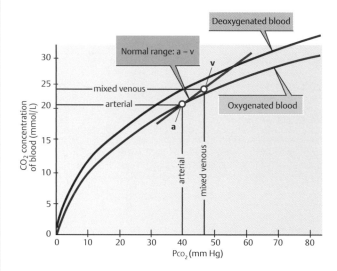

14 Control of Respiration

Respiration is controlled to maintain arterial gas pressures at appropriate levels, to minimize the work of breathing, and to alter respiration transiently for conscious activities (e.g., speech), reflexes (e.g., sneezing), and voluntary acts (e.g., blowing and breath holding).

14.1 Central Control of Respiration

Medullary Respiratory Center

Regular quiet breathing is directed by interacting groups of neurons in the medulla (**Fig. 14.1**). Neurons located in diffuse groups in the dorsal and ventral regions of the medulla act as pattern generators whose activity waxes and wanes during each breathing cycle. Efferent neurons rhythmically stimulate motoneurons in the spinal cord, whose peripheral axons exit in spinal nerves and travel in the phrenic nerve to the diaphragm and in other nerves to reach the accessory muscles of ventilation.

The activity of the medullary pattern generators is modulated by many inputs.

Cerebral Cortex

Parallel pathways originating in the cerebral cortex take over control of the spinal motoneurons during voluntary control of breathing.

Inputs that Modulate Respiration

Ventilation is normally adjusted to maintain arterial P_{CO_2} close to a set point of 40 mm Hg. Excursions from the set point occur under many conditions that then excite or inhibit respiratory neurons to increase or decrease ventilation (**Fig. 14.2**).

Chemoreceptors

Central chemoreceptors

The primary detection of arterial P_{CO_2} occurs via chemoreceptors in the floor of the fourth ventricle in the brainstem.

− CO_2 from the blood equilibrates with the cerebrospinal fluid (CSF) so that any change in arterial P_{CO_2} causes a change in CSF P_{CO_2}. This also causes a change in CSF pH, for example, a rise in arterial P_{CO_2} causes a drop in CSF pH ($CO_2 + H_2O \rightleftarrows H^+ + HCO_3^-$). In response, the chemoreceptive neurons, which are sensitive to pH, stimulate neurons in the medullary respiratory center to increase ventilation. A lowering of arterial P_{CO_2} (causing ↑ pH) results in inhibition of the medullary respiratory center.

Peripheral chemoreceptors

Chemoreceptive neurons in the carotid bodies (located at the bifurcation of the internal and external carotid arteries) and aortic bodies (on the arch of the aorta) respond to changes in arterial P_{CO_2}, P_{O_2}, and pH and send afferent impulses to the medullary respiratory centers via the glossopharnygeal and vagus nerves, respectively (**Fig. 14.1**). The carotid body chemoreceptors are the most important peripheral chemoreceptors.

− The carotid body chemoreceptors are not sensitive to small changes in arterial P_{O_2} in well-oxygenated blood but markedly increase their stimulation of the respiratory centers as arterial P_{O_2} falls below 60 mm Hg, a situation called *hypoxic drive*.

■ **Drugs causing respiratory depression**

Barbiturates, benzodiazepines, and opioids are all known to cause respiratory depression. Barbiturates and benzodiazepines act by facilitating the effects of gamma-aminobutyric acid (GABA), the main inhibitory neurotransmitter in the central nervous system (CNS), at the α subunit of the $GABA_A$ receptor. Opioids act at μ receptors throughout the body, the effects of which can be both excitatory and inhibitory. These drugs depress the response of the medullary respiratory center to hypercapnia (↑ CO_2), leading to respiratory depression.

■ **Cheyne-Stokes and Kussmaul breathing**

Cheyne-Stokes breathing is characterized by the absence of breathing (apnea) followed by an increased rate of ventilation (hyperventilation). It is caused by a delayed response of respiratory neurons to changes in the partial pressures of oxygen (P_{O_2}) and carbon dioxide (P_{CO_2}), which may occur when there is hypoperfusion of the brain or when hypoxic drive is regulating respiration.

Kussmaul breathing is characterized by regular deep breathing that occurs (appropriately) in response to metabolic acidosis (e.g., diabetic ketoacidosis; see box page 282).

Fig. 14.1 ► **Respiratory control and stimulation.**

The medullary rhythm generator(s) repetitively excite spinal motoneurons involved in ventilation. Their activity is modulated by feedback from central chemoreceptors, which respond to changes in the P_{CO_2} and pH of cerebrospinal fluid (CSF); peripheral chemoreceptors, which respond to changes in P_{CO_2}, pH, and P_{O_2} in the blood; mechanoreceptors, which respond to stretching of intercostal muscles to modulate the depth of breathing; and higher centers, which modulate the basic rhythm of respiration during times of emotion, during reflexes (e.g., coughing or sneezing), and during voluntary control of respiration (e.g., while speaking or singing). During physical work, the total ventilation increases due to coinnervation of the respiratory centers by collaterals of cortical efferent motor fibers and through impulses transmitted by proprioceptive fibers from the muscles.

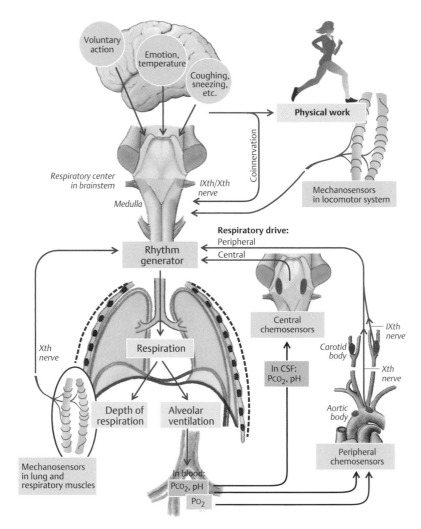

Fig. 14.2 ► **Modulators of respiratory neurons.**

Many factors have excitatory and inhibitory effects on respiratory neurons that then increase or decrease ventilation accordingly.

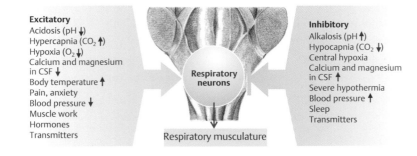

- An increase in arterial P_{CO_2} and a fall in pH will activate peripheral chemoreceptors, resulting in an increase in respiratory rate. Likewise, when arterial P_{CO_2} falls and arterial pH rises, their stimulation decreases.
- Peripheral chemoreceptors respond more quickly than central chemoreceptors to sudden changes in P_{CO_2}, but are not required for the long-term maintenance of arterial P_{CO_2}. Thus peripheral chemoreceptors serve as rapid detectors of changes in P_{CO_2}, while central chemoreceptors play a major role in keeping arterial P_{CO_2} at a normal steady-state value.

Other Receptors that Modulate Respiration

Irritant receptors
- There are receptors in the walls of the bronchi that respond to inhaled irritants (e.g., dust, pollen, and chemicals) and trigger reflexes such as coughing and sneezing.

Pulmonary stretch receptors
- There are a variety of stretch receptors in the smooth muscle of the bronchial tree that influence the medullary respiratory center. These receptors are responsible for the *Hering–Breuer reflex*, which exerts an inhibitory influence as the lungs inflate, thereby limiting the depth of respiration.

Muscle and joint receptors
- Muscle and joint receptors are activated during exercise and trigger an increase in ventilation.

14.2 Response of the Respiratory System to Exercise and High Altitude

Exercise

- The onset of exercise causes a rapid initial increase in depth and frequency of breathing, followed by a slower secondary rise. The precise triggers for increased ventilation are unknown but are thought to involve receptors in activated muscles and joints, an increase in body temperature, conscious awareness of exercise, and other cerebral cortical activation.
- During prolonged exercise, a ventilatory plateau is reached, representing a steady state. Ventilation is matched to increased metabolic demands of exercise ($\uparrow O_2$ consumption and $\uparrow CO_2$ production).
- During exercise, there is an increase in pulmonary blood flow due to an increase in cardiac output. There is adequate time for gas exchange despite this increased pulmonary blood flow.
- The ventilation/perfusion (V/Q) ratio progressively increases during exercise because ventilation increases more than cardiac output (and therefore pulmonary blood flow). The ratio may reach 4:1.
- P_{O_2}, P_{CO_2}, and pH are maintained at levels of a resting person during light to moderate cardiovascular exercise. At peak cardiovascular exercise, anaerobic respiration causes lactic acid buildup in blood. This causes arterial pH to drop and stimulation of central chemoreceptors, leading to an increase in respiratory rate and subsequent decrease in P_{CO_2}.
- After cessation of exercise, respiratory rate only gradually returns to resting values while lactic acid is metabolized (repayment of the "oxygen debt").

High Altitude

— At high altitudes, barometric pressure and therefore alveolar P_{O_2} are decreased, resulting in decreased arterial P_{O_2} (hypoxemia) and hypoxia. The severity of hypoxia is proportional to the altitude.

— Immediate hypoxemia stimulates peripheral chemoreceptors, producing an acute increase in the ventilation rate (hyperventilation) and an increase in the V/Q ratio. This hyperventilation then causes a respiratory alkalosis that is corrected by renal compensatory mechanisms after several days (see Chapter 18).

— Hypoxemia also stimulates hypoxic vasoconstriction in the lungs. This increases pulmonary vascular resistance and may lead to right heart failure secondary to pulmonary hypertension (cor pulmonale).

— Acclimatization occurs with prolonged exposure to high-altitude conditions.

  — Hypoxemia stimulates the release of renal erythropoietin, which stimulates bone marrow to increase production of red blood cells (polycythemia). This results in an increase in hemoglobin concentration and a larger O_2 transport capacity. However, polycythemia leads to an increase in the hematocrit (percentage of blood volume that is red blood cells) after 1 to 2 weeks and increased blood viscosity.

  — Increased production of 2,3-diphosphoglycerate (2,-3-DPG) causes a shift of the hemoglobin–O_2 dissociation curve to the right, which facilitates the unloading of O_2 in the tissues at a given P_{O_2}.

■ **Acute mountain sickness**

Acute mountain sickness is often seen with rapid changes from sea level to 10,000 ft (3000 m). If severe, hypoxic stimulation of the entire lung causes hypoxic vasoconstriction throughout pulmonary vessels and a large increase in pulmonary artery pressure, leading to high-altitude pulmonary edema (HAPE). Cerebral edema may also occur. Symptoms include headache, malaise, nausea, vomiting, dizziness, shortness of breath (dyspnea), and increased heart rate (tachycardia). Treatment involves descending to a lower altitude as soon as possible and O_2 administration. HAPE is treated with nifedipine (a calcium-channel blocker) or sildenafil (a phosphodiesterase inhibitor). Dexamethasone is used for cerebral edema.

Review Questions

1. During expiration,
   A. airway compression may occur if pressure within the airways is less than the pressure in the pleural space.
   B. airway compression will most likely occur at high lung volumes.
   C. pressures due to recoil properties of the lungs are greatest at low lung volumes.
   D. airway resistance is highest at high lung volumes.
   E. muscular effort is required for airflow.

2. In a premature infant without pulmonary surfactant,
   A. surface tension of alveoli is less than normal.
   B. a greater than normal pressure difference between the inside and outside of the lungs is required for lung inflation.
   C. large alveoli collapse more easily than small alveoli.
   D. pressures due to surface tension in large and small alveoli are equal.
   E. the lungs can be inflated with normal muscular effort.

3. What is the cause of a decrease in arterial P_{O_2} with chronic bronchitis?
   A. Increased V/Q for all regions of the lungs
   B. Abnormally uneven V/Q
   C. Decreased shunting of blood from the right to the left side of circulation
   D. A higher P_{O_2} in arterial blood than in alveolar gas
   E. Decreased functional residual capacity

4. A patient with a normal vital capacity expires only 40% of his forced expiratory volume/forced vital capacity in 1 second (FEV_1/FVC). Which problem does this suggest?
   A. Abnormally low lung compliance
   B. Obstructive disease
   C. Restrictive disease
   D. Weak inspiratory muscles
   E. Airflow limitation due to airway compression is less than normal.

5. Physiological dead space volume is unchanged by
   A. increases in volume of the alveolar dead space.
   B. widening of the airways above the respiratory zone.
   C. increases in breathing frequency.
   D. increases in volume of the anatomical dead space.
   E. breathing through a tube (e.g., a snorkel).

For Question **6** and **7**, the following data were obtained from a man during a pulmonary screening test: tidal volume = 447 mL, respiratory rate = 10 breaths/min, dead space volume =147 mL, Pa_{CO_2} = 44 mm Hg.

6. What is his minute ventilation?
   A. 40 mL/min
   B. 44.7 mL/min
   C. 4000 mL/min
   D. 4200 mL/min
   E. 4470 mL/min

7. What is his alveolar ventilation?
   A. 1470 mL/min
   B. 2940 mL/min
   C. 3000 mL/min
   D. 4323 mL/min
   E. 5940 mL/min

8. When the ambient pressure in an aircraft is suddenly reduced to 190 mm Hg (¼ atmosphere, altitude 34,000 ft [10,363 m]), the pilot discovers that his O_2 equipment fails to deliver any O_2. What is the O_2 tension in the cabin of the aircraft?
   A. 20 mm Hg
   B. 40 mm Hg
   C. 60 mm Hg
   D. 80 mm Hg
   E. 100 mm Hg

9. Which of the following is true in pulmonary embolism?
   A. The surface area available for gas exchange decreases.
   B. The dead space is decreased.
   C. Loose emboli travel to the brain.
   D. Loose emboli cause strokes.
   E. Pulmonary embolism is never fatal.

10. An elderly person is found unconscious after her gas house heater malfunctions. One-half the hemoglobin in her blood is bound to carbon monoxide. Which of the following is true?
    A. P_{50} for O_2 will be decreased from normal.
    B. O_2 content of blood at a P_{O_2} of 100 mm Hg will be normal.
    C. Arterial P_{CO_2} will be increased by 50%.
    D. Arterial P_{CO_2} will be decreased by 50%.
    E. O_2 carried in physically dissolved form at a P_{O_2} of 100 mm Hg will be less than normal.

11. An alveolar-to-arterial difference (A-a gradient) in the partial pressure of oxygen (PO_2) will increase when
    A. alveolar ventilation is reduced by a drug overdose.
    B. shunting of blood from the right to the left side of the circulation is increased when a lobe of the lung fills with fluid (pneumonia).
    C. diffusing capacity decreases because one lobe of a lung is removed.
    D. breathing through a tracheostomy tube.
    E. ventilation and perfusion in all parts of the lung are doubled.

12. Which of the following stimulates activity in central chemoreceptors?
    A. Decreased partial pressure of oxygen (PO_2)
    B. Increased PaO_2
    C. Normal A-a PO_2
    D. Increased partial pressure of carbon dioxide (PcO_2)
    E. Increased pH

13. During light to moderate exercise, i.e., before anaerobic metabolism becomes significant (and lactic acid production is increased), then you would expect that
    A. $PaCO_2$ increases linearly with the severity of exercise.
    B. PaO_2 decreases linearly with the severity of exercise.
    C. alveolar ventilation increases in proportion to the increased CO_2 production.
    D. minute ventilation increases in proportion to the increased $PaCO_2$.
    E. the ratio of the dead space volume to tidal volume (V_D/V_T) increases.

Answers and Explanations

1. **A.** If pressure in the pleural space is greater than that within the airways, the pressure difference can compress the airways (p. 126).
   B,D Airway compression is least likely to occur at high lung volumes because elastic recoil of the lungs is maximal, and airway resistance is minimal.
   C Pressures due to recoil properties of the lungs are greatest at high lung volumes.
   E Expiratory airflow is normally produced by elastic recoil, the potential energy stored in the lungs from the previous inspiration, and not by muscular effort.

2. **B.** According to the law of Laplace, the transpulmonary pressure (measured in dynes/cm^2) required to prevent collapse of an alveolus (P) is directly proportional to surface tension (T) and inversely proportional to alveolar radius (r), as expressed by P = 2T/r. Therefore, a greater pressure difference is required to inflate alveoli to overcome surface forces (p. 130).
   A Lack of surfactant will increase surface tension.
   C Small alveoli collapse more easily than large alveoli.
   D Lack of surfactant makes pressures different in large and small alveoli.
   E More effort is needed to inspire.

3. **B.** Chronic bronchitis is associated with uneven ventilation to perfusion ratio (p. 139).
   A V/Q does not increase.
   C Blood flow does not change.
   D PO_2 in blood can never be higher than alveolar O_2.
   E The functional residual capacity is increased in chronic obstructive pulmonary disease (COPD).

4. **B.** Constricted bronchioles and bronchi in obstructive lung disease slow the rate of expiration (and inspiration) (p. 133).
   A,D Abnormally low lung compliance and weak inspiratory muscles are typical of restrictive lung disease.
   C In restrictive lung disease, the ratio FEV_1/FVC is not reduced.
   E Airway compression produces decreased vital capacity.

5. **C.** Increasing the frequency of breathing increases the ventilation of the dead space per minute, although it does not significantly affect the volume of the dead space for each breath (p. 134).
   A,B,D Increases in alveolar or anatomical dead space, or widening the airways above the respiratory zone change physiological dead space.
   E A snorkel also increases dead space, changes airway volume, and increases anatomical dead space.

6. **E.** His minute ventilation is tidal volume multiplied by the respiratory rate, or 447 mL × 10 breaths/min = 4470 mL/min (p. 134).

7. **C.** His alveolar ventilation is alveolar volume multiplied by the respiratory rate. Alveolar volume is tidal volume minus dead space, or 447 mL − 147 mL = 300 mL. 300 mL × 10 breaths/min = 3000 mL/min (p. 134).

8. **B.** The partial pressure of O_2 is still 21% of the total pressure, so 0.21 × 190 mm Hg = 40 mm Hg (p. 136).

9. **A.** Emboli block the pulmonary arterial system, decreasing the area for gas exchange (p. 139)
   B Pulmonary embolism increases dead space.
   C,D The pulmonary capillary circulation filters emboli, so they do not travel to the brain and cause stroke.
   E Severe emboli cause death due to hypoxia.

10. A. Poisoning of hemoglobin with CO causes the O_2 dissociation curve to shift to the left, so the partial pressure where the available hemoglobin is 50% saturated with O_2 is reduced (p. 141).

B Because CO competes with O_2 for sites on the hemoglobin molecule, less O_2 will be carried at a given P_{O_2} level.

C, D P_{CO_2} is determined by alveolar ventilation and virtually unaffected by the presence of CO.

E The presence of CO has no effect on O_2 solubility in blood.

11. B. The A-a gradient increases when arterial P_{O_2} is decreased, such as when blood is shunted from right-to-left. In this latter situation, the arterial P_{O_2} is lowered when nonoxygenated blood mixes with oxygenated blood (p. 142).

A,D Hypoventilation by itself or a change in dead space is not associated with an increased alveolar-to-arterial P_{O_2} difference.

C While removal of one lobe of a lung reduces diffusing capacity, an increase in the A-a gradient will only occur under extreme conditions (e.g., strenuous exercise).

E Uniformly doubling the ventilation (V) and perfusion (Q) in all parts of the lungs will not change the ratio, V/Q.

12. D. Central chemoreceptors are highly sensitive to increased P_{CO_2} or [H$^+$] (p. 145).

A A low arterial P_{O_2} stimulates peripheral chemoreceptors.

B, C An increased alveolar P_{O_2} or normal A-a P_{O_2} difference has no effect on chemoreceptors.

E Increasing pH means decreasing [H$^+$], which decreases chemoreceptor activity.

13. C. Ventilation matches CO_2 production during light to moderate exercise (p. 147).

A,B Pa_{CO_2} and Pa_{O_2} change little during light to moderate exercise.

D Increased minute ventilation keeps Pa_{CO_2} from increasing.

E The ratio V_D/V_T decreases because light to moderate exercise causes the dead space volume (V_D) to decrease, and the tidal volume (V_T) to increase.

15 Renal Anatomy, Body Fluids, Glomerular Filtration, and Renal Clearance

15.1 Functions of the Kidney and Functional Anatomy

Functions of the Kidney

The kidneys perform three basic processes:

— *Ultrafiltration* (filtration across a semipermeable membrane that is driven by hydrostatic pressure) of large volumes of water and solutes from the blood into the renal tubular system
— *Reabsorption* of filtered substances that are needed by the body from the renal tubules into the bloodstream. These include water, Na^+, glucose, and bicarbonate (HCO_3^-).
— *Secretion* of substances from the bloodstream into the renal tubules

By extension of these three processes, the kidneys are able to

— Adjust salt and water excretion to maintain a constant extracellular fluid (ECF) volume and osmolality. Blood pressure homeostasis is maintained in part via modulation of ECF volume.
— Maintain acid–base balance
— Remove waste substances and ingested toxins from the blood and excrete them as urine. Waste substances are produced from metabolism and include urea and ammonia from protein catabolism and uric acid from nucleic acid metabolism.

In addition to homeostatic and excretory functions, the kidneys produce several hormones, including erythropoietin, which stimulates the production of red blood cells in response to hypoxia (low partial pressure of oxygen [P_{O_2}]); calcitriol, which increases serum [Ca^{2+}] and [$P_{O_4}^{3-}$] for the mineralization of new bone; and renin, which forms part of a system that helps to regulate Na^+, ECF volume, and blood pressure.

Functional Anatomy of the Kidney

Renal Cortex, Renal Medulla, and Renal Pelvis

The kidney is divided into three main anatomical areas: the cortex, the medulla, and the renal pelvis (**Fig. 15.1**).

— The *renal cortex* is the area where ultrafiltration occurs.
— The *renal medulla* is the area that serves to concentrate the urine.
— The *renal pelvis* collects urine and drains into the ureter and bladder.

Nephron

The functional unit of the kidney is the nephron (**Fig. 15.2**), where the three basic processes of ultrafiltration, reabsorption, and secretion occur. Substances that are filtered and secreted but not reabsorbed are excreted as final urine.

There are five parts of the nephron: the glomerulus, the proximal convoluted tubule, the loop of Henle, the distal convoluted tubule, and the collecting ducts.

Glomerulus

— The glomerulus is the site of plasma ultrafiltration, or *glomerular filtration* (**Fig. 15.3**). Arterial blood is delivered to the glomerular capillaries via afferent arterioles. Plasma then undergoes ultrafiltration across the glomerular barrier, and the ultrafiltrate passes into Bowman's

---

■ Erythropoietin

Erythropoietin is a renal hormone that regulates the production of red blood cells in the bone marrow. Patients with chronic renal failure develop anemia secondary to inadequate levels of erythropoietin. Human recombinant erythropoietin has been shown to be effective in treatment of anemia associated with uremia (increased blood [urea]). There are no direct adverse effects of replacement therapy, although ~25% of patients experience hypertension during treatment (mechanism not understood). Patients on renal dialysis require erythropoietin.

■ Renal osteodystrophy

In chronic renal failure, the failing kidneys are unable to perform the necessary α_1-hydroxylation reactions to produce calcitriol (the active form of vitamin D), and they have a reduced capacity to excrete phosphate. This leads to hyperparathyroidism due to hypocalcemia and hyperphosphatemia. Derangement of bone remodeling occurs, which is referred to as *renal osteodystrophy*. The symptoms of renal osteodystrophy include bone and joint pain, bone deformation, and increased likelihood of bone fractures. Chronic renal failure requires hemodialysis several times per week until renal transplantation can occur. Renal osteodystrophy is treated with calcium and calcitriol, restricting the dietary intake of phosphate, and by the administration of medications that bind phosphate, such as calcium carbonate and calcium acetate.

Fig. 15.1 ► Midlongitudinal section through a right kidney, posterior view.
From *Thieme Atlas of Anatomy, Neck and Internal Organs*, © Thieme 2006, Illustration by Markus Voll.

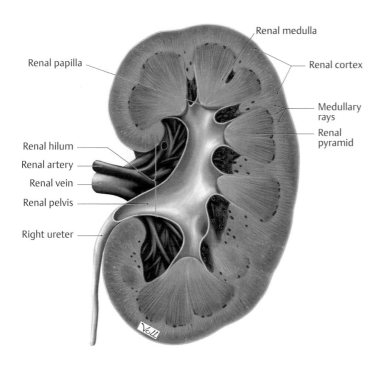

capsule and the renal tubular system. The fraction of plasma and substances that do not undergo filtration leave the glomerulus via efferent arterioles and flow into peritubular capillaries surrounding the nephron.

— The glomerular barrier consists of capillary endothelial cells, the endothelial basement membrane, and the epithelial cells of Bowman's capsule.
  — Capillary endothelial cells are fenestrated (have holes), thus increasing capillary permeability.
  — The epithelial cells of Bowman's capsule have podocytes with slitlike pores between them that are closed by a slit membrane.
  — Both the endothelial basement membrane and the epithelial cells of Bowman's capsule are lined by negatively charged glycoproteins. This makes the glomerular barrier relatively impermeable to negatively charged plasma proteins (e.g., albumin). In glomerular disease, these negatively charged proteins may be destroyed, allowing proteins to enter the urine (proteinuria).
  — The glomerular barrier is most permeable to small, neutral or positively charged molecules.

Renal Tubular System

— The proximal convoluted tubule is the site of reabsorption of most of the filtered load from the tubular lumen into pericapillary blood.
— The loop of Henle that extends into the medulla is involved in the concentration or dilution of urine by changing the osmolality of the tissue surrounding it.
— The distal convoluted tubule and collecting ducts reabsorb Na^+ and water as necessary to maintain ECF and electrolyte balance. The collecting duct also collects final urine and drains into the renal pelvis for excretion.

Urinary tract infections

Urinary tract infections (UTIs) are common, especially in women due to the proximity of the urethra to the vagina (allowing easier spread of sexually transmitted infections and diseases) and due to the short length of the urethra compared with men. UTIs present with any of the following symptoms: frequency of urination, urgency, strangury (frequent, painful expulsion of small amount of urine despite urgency), hematuria (blood in the urine), cloudy urine, incontinence, fever with diarrhea and vomiting, and pain (usually suprapubic pain in women and anal pain in men). Trimethoprim with sulfamethoxazole is given to treat uncomplicated UTIs caused by susceptible bacteria (*Escherichia coli*, *Staphylococcus* spp., *Streptococcus* spp., *Pseudomonas*, and *Proteus*). In addition, patients are advised to drink plenty of fluids and urinate often.

Nephrotic syndrome

Nephrotic syndrome results in severe proteinuria (loss of proteins into the urine), hypoalbuminemia (see box page 112), and edema due to the decrease in capillary oncotic pressure. Causes of nephrotic syndrome include glomerulonephritis (inflammation of the glomerulus), diabetes, neoplasia, and drugs. Signs include peripheral edema, ascites (accumulation of fluid in the peritoneal space), and swelling of the eyelids. Venous thrombosis and emboli may occur due to excretion of certain clotting factors and antithrombin III in the urine. Treatment involves the administration of a loop diuretic with a K^+-sparing agent, plasma protein replacement, and anticoagulation (if necessary) to prevent thrombosis or emboli. The underlying cause should also be sought and treated appropriately.

Renal oxygen consumption and metabolism

Per unit of tissue weight, the kidneys are perfused by more blood, and they consume more oxygen (O_2) than does any other organ except the heart. Yet the renal arteriovenous O_2 content difference is lower than that of other organs. This unique feature reflects the high filtering capacity of the kidneys; consequently, RBF is far in excess of the kidneys' basal O_2 requirements. The high renal O_2 consumption reflects the amount of energy required for the reabsorption of filtered Na^+. Energy is derived from oxidative metabolism (mostly of fatty acids) in the renal cortex. In contrast, the renal medullary structures derive energy from anaerobic metabolism of glucose.

Fig. 15.2 ► Anatomy of the nephron.

The smallest functional unit of the kidney is the nephron, which consists of the glomerulus, renal tubules, and collecting ducts.

From *Thieme Atlas of Anatomy*, *Neck and Internal Organs*, © Thieme 2006, Illustration by Markus Voll.

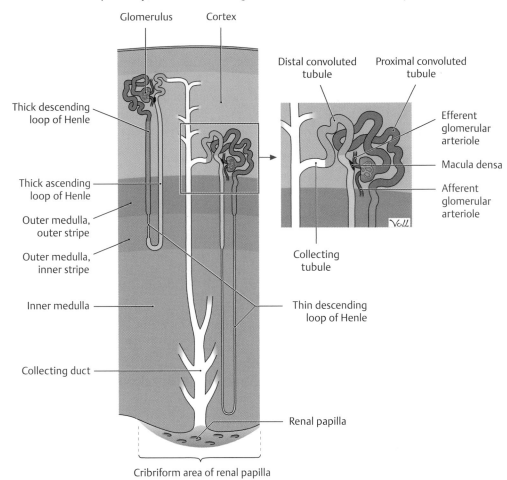

Fig. 15.3 ► Glomerulus and Bowman's capsule.

Blood enters the glomerulus by an afferent arteriole and exits via an efferent arteriole from which the peritubular capillary network arises. The glomerular barrier separates the blood side from the Bowman capsular space. The glomerular barrier comprises the fenestrated endothelium of the glomerular capillaries, followed by the basement membrane as the second layer, and the visceral epithelial cells of Bowman's capsule on the urine side. The latter is formed by podocytes with numerous interdigitating footlike processes (pedicels). The slitlike spaces between them are closed by the slit membrane, with pores of ~5 nm in diameter.

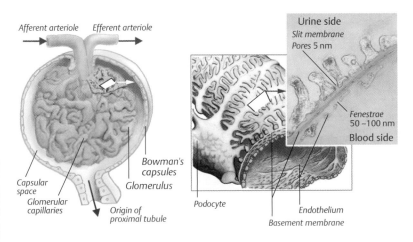

15.2 Body Fluid Compartments and Their Composition

Total Body Water

Total body water (TBW) accounts for ~60% of body weight. The percentage of TBW declines with increased age and increased amounts of body fat.

TBW is distributed between two major compartments within the body: intra- and extracellular fluid (ICF and ECF).

Intracellular Fluid

ICF is the fluid within the body's cells (cytoplasm).

— It accounts for ~40% of body weight (two-thirds of TBW).
— K^+ and Mg^{2+} are the major cations.
— Proteins and organic phosphates (e.g., adenosine triphosphate [ATP], adenosine diphosphate [ADP], and adenosine monophosphate [AMP]) are the major anions.

Extracellular Fluid

ECF is plasma and interstitial fluid (including lymph).
— ECF accounts for ~20% of body weight (one third of TBW), of which ~5% is attributed to plasma and ~15% of which is interstitial fluid.
— Plasma is the fluid portion of the blood that remains after blood cells are removed. It contains important proteins (e.g., albumin and globulins).
— Interstitial fluid is the fluid that occupies the spaces between cells. Its composition is the same as plasma except that it is relatively free of proteins.
— Na^+ is the major cation.
— Cl^- and HCO_3^- are the major anions.

Measuring the Volume of Body Fluid Compartments

Indicator Dilution Method

The indicator dilution method involves the administration of a known amount of a marker substance that will distribute within the body compartment under investigation.

— TBW can be measured using tritiated water or deuterium oxide (D_2O), as its distribution is the same as water.
— ECF volume is measured using substances that will not enter cells (e.g., mannitol, inulin, sucrose, and thiosulfate).
— Plasma volume is measured using substances that neither leave the vasculature nor enter red blood cells (e.g., Evans blue dye or radioactive serum albumin).

The volume of the compartment is then found by determining the final concentration of the marker substance that has been added to the compartment using the following equation:

$$V = Q/C,$$

where V is the volume of the body fluid compartment or volume of distribution (L), Q is the quantity of the marker substance added to the compartment minus the amount lost from the compartment by excretion or metabolism during the measurement (mg), and C is the measured final concentration of marker substance (mg/L).

Note:
— Interstitial fluid volume is calculated as the difference between ECF volume and plasma volume.
— Intracellular volume is calculated as the difference between TBW and ECF volume.

Movement of Fluid between Compartments

Under normal circumstances, the *osmolality* (i.e., the concentration of osmotically active solutes) of the ECF compartment and ICF compartment are equal. The osmotic content of ICF is determined by the concentration of K+ and charged proteins, and the osmolality of ECF is primarily determined by its NaCl content.

Any alteration in the osmolality of the ECF will cause water movement between the ECF and ICF toward the compartment with the higher osmolality (i.e., higher solute concentration) until equilibrium is reached between the two compartments.

Table 15.1 summarizes circumstances/disease states that may cause fluid movement between the ECF and ICF, as well as the effects on hematocrit (percentage of blood volume composed of red blood cells [RBCs]), plasma protein concentration, and blood pressure (BP).

Hematocrit

The hematocrit is the percentage of blood volume that is RBCs. It is normally ~48% for men and ~38% for women. The hematocrit is elevated in polycythemia, a disorder in which the bone marrow produces excessive RBCs. This is driven by the increased secretion of erythropoietin by the kidneys in response to hypoxia (low Po_2, for example, due to chronic obstructive pulmonary disease (COPD). It may also be elevated in dehydration due to loss of ECF volume, which concentrates RBCs. The hematocrit is lowered in hemorrhage (due to loss of RBCs) and iron-deficiency anemia (due to defective RBC synthesis).

Dextrose as an intravenous fluid

Hypovolemia (decreased blood volume) may require the administration of intravenous (IV) fluids to restore ECF volume. Distilled water given IV would lyse RBCs due to osmotic forces. This is prevented by giving a 5% dextrose solution with normal osmolality. The body metabolizes the glucose quickly, leaving an increase in water without ions.

Syndrome of inappropriate antidiuretic hormone secretion

Syndrome of inappropriate antidiuretic hormone secretion (SIADH) occurs when excessive amounts of antidiuretic hormone (ADH) are secreted from the posterior pituitary gland. This leads to hyponatremia (low blood sodium levels) and fluid overload. Causes include head injury, meningitis, cancer, and infections (e.g., brain abscess and pneumonia). Treatment involves management of the underlying cause and the use of demeclocycline or lithium carbonate (vasopressin [ADH] antagonists) for symptomatic control.

Table 15.1 ▶ Fluid and Electrolyte Imbalances and Their Effects			
Circumstance/ Disease State	**Term Given**	**Effects on ECF and ICF**	**Effects on Hematocrit, Plasma Protein Concentration, and BP**
Infusion of isotonic NaCl, heart failure, renal failure	Isosmotic volume expansion	↑ ECF volume Osmolarity of ECF and ICF is unchanged; thus, there is no water shift between ECF and ICF	↓ hematocrit and plasma protein concentration due to increase in ECF volume ↑ BP due to increase in ECF volume
Diarrhea, vomiting, diuretics, blood loss, burns	Isosmotic volume contraction	↓ ECF volume Osmolarity of ECF and ICF is unchanged; thus, there is no water shift between ECF and ICF	↑ hematocrit and plasma protein concentration due to decrease in ECF volume ↓ BP due to decrease in ECF volume
Excess intake of NaCl, increased aldosterone (causing retention of NaCl)	Hyperosmotic volume expansion	↑ ECF volume ↑ osmolarity of ECF as NaCl (osmotically active) is added to ECF Shift of water from ICF to ECF, thus increasing ICF osmolarity and decreasing ICF volume	↓ hematocrit and plasma protein concentration due to increase in ECF volume ↑ BP due to increase in ECF volume
Aldosterone deficiency (causing loss of NaCl)	Hyposmotic volume contraction	↓ ECF volume ↓ osmolarity of ECF (aldosterone deficiency causes the kidneys to excrete more NaCl than water) ↑ ICF volume and ↓ ICF osmolarity (until equilibrium with ECF is reached) as water moves into cells	↑ hematocrit and plasma protein concentration due to decrease in ECF volume ↓ BP due to decrease in ECF volume
Sweating	Hyperosmotic volume contraction	↓ ECF volume ↑ osmolarity of ECF, as sweat is hyposmotic (water loss is greater than salt loss) ↓ ICF volume and ↑ ICF osmolarity (until equilibrium with ECF is reached), as water shifts from ICF to ECF	↑ plasma protein concentration due to loss of ECF Hematocrit unchanged as water shifts out of red blood cells
Syndrome of inappropriate antidiuretic hormone secretion (SIADH)	Hyposmotic volume expansion	↑ ECF volume ↓ osmolarity of ECF, as excess water is retained (due to inappropriate ADH secretion) ↑ ICF volume and ↓ ICF osmolarity (until equilibrium with ECF is reached), as water shifts into cells	↓ plasma protein concentration due to increase in ECF volume Hematocrit unchanged as water shifts into red blood cells

Abbreviations: ADH, antidiuretic fluid; BP, blood pressure; ECF, extracellular fluid; ICF, intracellular fluid.

15.3   Renal Blood Flow, Renal Plasma Flow, and Glomerular Filtration Rate

— *Renal blood flow (RBF)* is the volume of blood entering the kidneys per minute. It is ~1 L/min (20% of cardiac output).
— *Renal plasma flow (RPF)* is the volume of plasma entering the kidneys per minute. The average RPF is ~600 mL/min.
— RBF is calculated as follows:

$$RBF = RPF/(1-hematocrit)$$

where (1-hematocrit) is the fraction of total blood volume that is plasma and *glomerular filtration rate (GFR)* is the volume of plasma filtered per minute by all glomeruli in the kidneys. The average GFR for a healthy 70 kg man is 125 mL/min (~20% of RPF) and declines with age.

— The magnitude of the GFR is an index of general kidney function.
— *Filtration fraction* is the fraction of total plasma volume that is filtered across the glomerulus. It is expressed by the following equation:

$$Filtration\ fraction = GFR/RPF$$

— It is normally one-fifth (20%) of RPF. The other 80% flows into the peritubular capillaries from the efferent arterioles.

Determinants of GFR

Ultrafiltration of plasma occurs as plasma moves from glomerular capillaries into Bowman's capsule under the influence of Starling forces (**Fig. 15.4**). Glomerular filtration is the same mechanism as systemic capillary filtration, i.e., the balance between hydrostatic and oncotic forces across the glomerular membrane determines the direction of fluid movement.

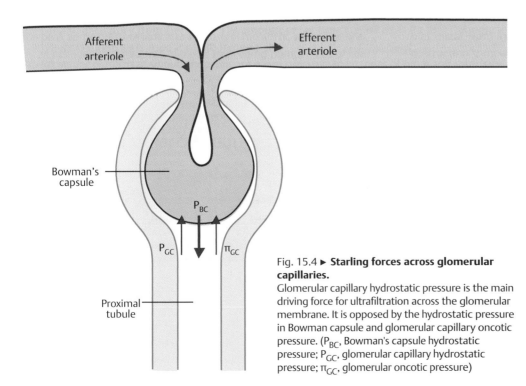

Fig. 15.4 ▶ **Starling forces across glomerular capillaries.**
Glomerular capillary hydrostatic pressure is the main driving force for ultrafiltration across the glomerular membrane. It is opposed by the hydrostatic pressure in Bowman capsule and glomerular capillary oncotic pressure. (P_{BC}, Bowman's capsule hydrostatic pressure; P_{GC}, glomerular capillary hydrostatic pressure; π_{GC}, glomerular oncotic pressure)

▮ Rhabdomyolysis

Rhabdomyolysis is the rapid breakdown of skeletal muscle due to injury to muscle tissue. The muscle breakdown product, myoglobin, is harmful to the kidney and can precipitate acute kidney failure. Signs and symptoms include pain, tenderness, and swelling of the affected muscle, as well as nausea, vomiting, confusion, arrhythmias, coma, anuria, and later disseminated intravascular coagulation (DIC). Treatment is primarily aimed at preventing acute kidney failure with the administration of fluid IV to increase ECF volume, increase the glomerular filtration rate (GFR) and oxygen delivery, and dilute myoglobin and any other toxins.

▮ Nephrotoxic drugs

An example of a nephrotoxic drug is gentamicin, an aminoglycoside antibiotic used to treat severe infections. It is excreted in unchanged form, mostly by glomerular filtration, in the kidney. In renal impairment, gentamicin will accumulate in the kidney, causing destruction of kidney cells (nephrotoxicity). If used, the dosage and treatment period should be minimal and plasma concentration should be closely monitored.

Net ultrafiltration pressure is determined by the Starling equation:

$$GFR = K_f [(P_{GC} - P_{BC}) - (\pi_{GC} - \pi_{BC})]$$

where

K_f = filtration coefficient of the glomerular capillaries. It depends on the membrane permeability of the cells that comprise the glomerular barrier and their surface area.

P_{GC} = glomerular capillary hydrostatic pressure. It is constant throughout the capillary (~45 mm Hg).

P_{BC} = Bowman's capsule hydrostatic pressure. It is analogous to interstitial hydrostatic pressure (P_i) in systemic capillaries and is usually ~10 mm Hg.

π_{GC} = glomerular capillary oncotic pressure. It usually increases along the length of the capillary because as water is filtered out of capillaries, the proteins left behind become increasingly concentrated. It is ~28 mm Hg.

π_{BC} = Bowman's capsule oncotic pressure. It is usually zero because very little protein is filtered under normal conditions.

$$\text{Net ultrafiltration pressure} = (P_{GC} - P_{BC}) - \pi_{GC}$$
$$= (45 \text{ mm Hg} - 10 \text{ mm Hg}) - 28 \text{ mm Hg}$$
$$= +7 \text{ mm Hg}$$

Table 15.2 summarizes the effects of changes in Starling forces on net ultrafiltration pressure and GFR.

Table 15.2 ► Effects of Changes in Starling Forces on Net Ultrafiltration Pressure and Glomerular Filtration Rate (GFR)		
Change in Starling Forces	**Causes**	**Effects**
↑ P_{GC}	Vasodilation of the afferent arteriole Vasoconstriction of the efferent arteriole	↑ net ultrafiltration pressure and GFR
↑ P_{BC}	Constriction of the ureters (e.g., by a ureteral stone)	↓ net ultrafiltration pressure and GFR
↑ π_{GC}	Increase in plasma [protein]	↓ net ultrafiltration pressure and GFR

Regulation of Renal Plasma Flow and Glomerular Filtration Rate

Regulation of GFR is linked to the regulation of RPF, because the flow of plasma to the kidneys influences the rate of filtration.

Autoregulation

Renal autoregulation ensures that RPF and GFR remain almost constant over a wide range of mean arterial blood pressures (BPs) (80–180 mm Hg). As blood pressure (BP) increases within this range, resistance in renal arterioles increases proportionately to minimize large increases in RPF and GFR by limiting changes in glomerular pressure (**Fig. 15.5**).

The two intrarenal mechanisms responsible for renal autoregulation are the myogenic mechanism and the tubuloglomerular feedback mechanism.

Myogenic Mechanism. Increases in renal arterial pressure cause stretching of the afferent arteriolar smooth muscle. This, in turn, causes the smooth muscle to contract, increasing resistance, which decreases RPF (and GFR) to their normal levels.

Tubuloglomerular Feedback Mechanism. Increases in renal arterial pressure increase the GFR and lead to an increased solute load to the macula densa cells in the distal tubule (**Fig. 15.2**). This activates these cells, and they stimulate the adjacent afferent arteriole to constrict. The arteriolar constriction increases resistance, thus decreasing RPF (and GFR) to their normal levels.

Fig. 15.5 ► Autoregulation of renal blood flow (RBF) and glomerular filtration rate (GFR).
Autoregulation of RBF ensures that minimal changes in renal plasma flow (RPF) and GFR occur when the systemic blood pressure (BP) fluctuates between 80 and 180 mm Hg. If the BP falls below ~80 mm Hg, however, renal circulation and filtration will ultimately fail. RBF and GFR can be regulated independently by making isolated changes in the resistances of the afferent and efferent arterioles.

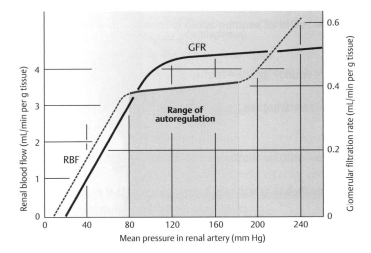

Regulation by Modulation of Arteriolar Resistance

Even in the face of autoregulation, changes in RPF and GFR can occur by local changes in vascular resistance of afferent and efferent arterioles. These resistance changes can be caused by the actions of the autonomic nervous system and various vasoactive humoral agents.

- Sympathetic activation leads to vasoconstriction of both afferent and efferent renal arterioles. This will reduce RPF and GFR. Similarly, decreased sympathetic tone results in decreased renal vascular resistance and increased RPF and GFR.
- Vasoactive humoral agents that act as vasoconstrictors, for example, catecholamines, angiotensin II, ADH (vasopressin), prostaglandins, and endothelin, will reduce RPF and GFR.
- Vasoactive humoral agents that act as vasodilators, for example, atrial natriuretic peptide (ANP), acetylcholine, kinins, and nitric oxide (NO), will increase RPF and GFR.

Note: When resistance is altered in the afferent arteriole only, RPF and GFR change in the same direction. When resistance is altered in the efferent arteriole only, RPF and GFR change in opposite directions. Therefore, GFR tends to decrease less than RPF when sympathetic tone is increased, as both afferent and efferent arterioles are constricted.

15.4 Renal Clearance

Renal clearance measures the efficiency of the kidneys in removing a substance from plasma. It is a useful concept in renal physiology because it can be used to quantitatively measure the intensity of several aspects of renal function (i.e., ultrafiltration, reabsorption, and secretion).

- *Renal clearance* is the volume of plasma from which a given substance is completely cleared by the kidneys and excreted in the urine per unit time. It can be calculated using the following equation (a modification of the Fick equation):

$$C = UV/P$$

where C is the renal clearance (mL/min), U is the urine concentration (mg/mL), P is the plasma concentration (mg/mL), and V is urine output (mL/min).

Example:

Over a 12-hour (720-minute) collection period, a patient produces 504 mL of urine with a creatinine concentration of 0.11 mg/mL. The plasma creatinine is 0.001 mg/mL. The clearance (of creatinine) is

$$C = (0.11\,mg/mL)(504\,mL/720\,min)/(0.001\,mg/mL)$$
$$= 55,440\,mL/720\,minutes$$
$$= 77\,mL/min$$

— A substance that is freely filtered (i.e., that has the same concentration in the filtrate as in plasma) and does not undergo net reabsorption or secretion in the renal tubules (e.g., inulin) will have a clearance value that is equal to GFR.

— A substance that undergoes net reabsorption from the tubules into pericapillary blood (e.g., urea) will have a clearance value that is lower than GFR.

— A substance that undergoes net secretion from pericapillary blood into the tubules (e.g., p-aminohippuric acid [PAH]) will have a clearance value that is higher than GFR.

Measurement of Glomerular Filtration Rate Using Inulin Clearance

Inulin is a nontoxic polysaccharide that is not bound to plasma proteins. It is freely filtered at the glomerulus and is neither reabsorbed nor secreted by renal tubules; therefore, the volume of plasma cleared of inulin per minute is equal to the GFR and may be calculated as follows:

$$GFR = C_{inulin} = [U]_{inulin} \times V/[P]_{inulin}$$

Example:

Urine flow rate = 1 mL/min, plasma inulin = 0.15 mg/mL, urine inulin = 15 mg/mL

$$Inulin\ clearance = GFR = 15\ mg/mL \times 1\ mL/min/0.15\ mg/mL$$
$$= 100\ mL/min$$

Measurement of Glomerular Filtration Rate Using Creatinine Clearance

Creatinine is an end product of skeletal muscle creatine metabolism and has a fairly constant concentration in plasma under normal conditions.

— Creatinine is freely filtered by the glomerulus and is not reabsorbed. However, small amounts are secreted into the renal tubules, so creatinine clearance gives a slightly greater estimate of GFR than inulin clearance. Despite this, creatinine clearance is used to measure GFR rather than inulin clearance because creatinine is endogenously produced.

— There is an inverse relationship between plasma creatinine level and the magnitude of GFR; for example, if GFR decreases to half of normal, and creatinine production remains constant, plasma creatinine will double.

Measurement of Glomerular Filtration Rate Using Urea Clearance

Urea is filtered and reabsorbed. Under conditions when urea reabsorption is approximately a constant fraction of the filtered load, urea clearance can be used to estimate GFR. Plasma levels of urea are used to estimate renal function by the same inverse relationship as serum creatinine. Plasma urea level is expressed as blood urea nitrogen (BUN) concentration. When GFR falls, BUN usually rises in parallel to serum creatinine. However, urea clearance or BUN is usually not a reliable indicator of the magnitude of GFR, as plasma urea concentration varies widely, depending on protein intake, protein catabolism, and variable renal resorption of urea under different states of hydration.

Measurement of Renal Plasma Flow Using *p*-aminohippuric Acid Clearance

PAH is both filtered and secreted into renal tubules; therefore, the renal clearance of PAH is greater than GFR and is not used for its quantitative measurement. However, the renal clearance of PAH can be used to estimate the magnitude of RPF because PAH is completely cleared from the plasma by renal excretion during a single circuit of plasma flow through the kidney. Normally, 85 to 90% of plasma flowing through the kidney reaches the nephron and is cleared of PAH, therefore PAH clearance is a measure of effective renal plasma flow (ERPF) (thus accounting for the 10 to 15% of plasma that does not supply the nephron) and can be calculated using the equation for renal clearance.

$$ERPF = C_{PAH} = [U]_{PAH} \times V/[P]_{PAH}$$

Renal Clearance of Glucose

The renal clearance of glucose is zero at normal plasma glucose concentration (80 mg/100 mL) and up to 300 mg/100 mL because all filtered glucose is reabsorbed by renal tubules. If plasma glucose levels increase above 3 times normal, the renal reabsorptive rate of glucose will reach its maximum tubular transport capacity, and glucose excretion will increase until its clearance approaches GFR. That is, at high plasma glucose concentrations, the majority of excreted glucose comes from its unreabsorbed filtered load.

16 Renal Tubular Transport

Renal excretion of any substance reflects the difference between the rate at which it enters the tubular lumen by filtration and secretion and the rate that it leaves the tubular lumen by reabsorption. For freely filtered solutes, the following expressions relate the various quantities:

Filtered load = glomerular filtration rate (GFR) × plasma concentration (mg/min)
Excretion rate = urine flow rate (U) × urine concentration (mg/min)
Net tubular transport = filtered load – excretion rate (mg/min)

Note: For solutes secreted but not reabsorbed, the net tubular transport is a negative number because more is excreted than filtered.

Example 1: Urea is a solute that is both secreted and reabsorbed. A patient has a GFR of 80 mL/min and plasma [urea nitrogen] of 0.12 mg/mL. The urine flow rate is 1 mL/min, and the urine [urea nitrogen] is 4.8 mg/mL.

Filtered load = 80 mL/min × 0.12 mg/mL = 9.6 mg/min
Excretion rate = 1 mL/min × 4.8 mg/mL = 4.8 mg/min
Net tubular transport = 9.6 mg/min – 4.8 mg/min = 4.8 mg/min (net reabsorption)

Example 2: Sodium is a solute that is reabsorbed but not secreted. For the patient above, the plasma sodium is 1.4 mEq/mL, and urine sodium is 1.12 mEq/mL.

Filtered load = 80 mL/min × 1.4 mEq/mL = 112 mEq/min
Excretion rate = 1 mL/min × 1.12 mEq/mL = 1.12 mEq/min
Net tubular transport = 112 mEq/min – 1.12 mEq/min = 110.9 mEq/min (99% reabsorption)

16.1 Mechanisms of Renal Tubular Transport

The renal epithelial cells use a large array of transport proteins (i.e., primary active transporters, multiporters, and uniporters) to reabsorb or secrete various substances. For the most part, this is *transcellular* transport—solutes travel through the epithelial cells using one transport species in the luminal membrane and a different one in the basolateral membrane. In some cases, secretion or reabsorption is *paracellular* transport (around the cells) where a concentration gradient between the tubular fluid and interstitial fluid drives diffusion through leaky tight junctions.

Na⁺ Transport Mechanisms

The transport of almost all substances is directly dependent on, or strongly influenced by, the reabsorption of Na^+. Na^+ enters tubular cells across the luminal membrane via an array of symporters, antiporters, and, in some places, Na^+ channels. Na^+ is then pumped out of tubular cells across the basolateral membrane primarily by the Na^+–K^+ ATPase and moves by a combination of diffusion and bulk flow into the peritubular capillaries. Some Na^+ also leaks back to the lumen by the paracellular pathway if the limiting gradient is reached (**Fig. 16.1**).

— Water reabsorption is linked to Na^+ because Na^+ is the major osmotic particle in the tubular fluid—in most regions of the tubule, when Na^+ is removed, water follows.
— Cl^-, the major anion in the tubular fluid, follows Na^+ because net transport of cations requires an equal net transport of anions.
— The transport of many other substances is linked to Na^+ via a cascade of symporters and antiporters. The Na^+ gradient set up by the Na^+–K^+ ATPase drives the secondary active transport of other substances (e.g., H^+ ions). The gradient of the other substance then drives the secondary active transport of still other substances.

Fig. 16.1 ▶ Electrochemical Na⁺ gradient.
Na^+–K^+ ATPase pumps Na^+ ions out of the cell while conveying K+ ions into the cell (**1**), thereby producing a chemical Na^+ gradient (**2**). Back diffusion of K^+ (**3**) also leads to the development of a membrane potential (**4**). Both combined result in a high electrochemical gradient that provides the driving force for passive Na^+ influx into tubular cells from the lumen. (ATP, adenosine triphosphate.)

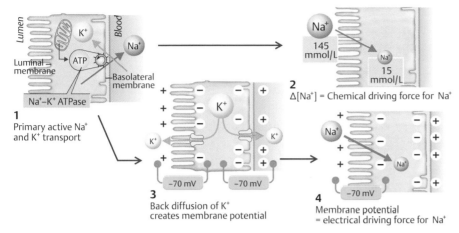

Glucose Transport.   The electrochemical gradient established by the Na^+–K^+ ATPase in the basolateral membrane strongly favors inward movement of the Na^+ from the lumen into tubular cells via Na^+–glucose symporters (SGLTs) in the luminal membrane. As glucose begins to accumulate within the tubular cells, it moves out across the basolateral membrane into peritubular capillary blood via a glucose uniporter (GLUT2).

Transport of Organic Solutes.   A huge number of organic solutes present in the blood at low concentrations are filtered at the glomerulus and then reabsorbed (e.g., amino acids and Krebs cycle intermediates). This process involves either active transport of substances from the lumen into tubular cells via the luminal membrane or active transport of substances from tubular cells into peritubular capillary blood via the basolateral membrane. In most cases, these processes consist of symport or antiport with Na^+ or another organic solute.

Transport of *p*-Aminohippuric Acid and Toxic Substances.   *P*-aminohippuric acid (PAH) moves from peritubular capillary blood into tubular cells across the basolateral membrane via an anion antiporter and secreted from tubular cells into the tubular lumen across the luminal membrane via a multidrug-resistant protein. This transport system is also used by the body to rid itself of organic toxic substances.

16.2 Progress of Renal Tubular Transport

The progress of transport (reabsorption or secretion) for any substance that is freely filtered can be followed by looking at the ratio of its concentration in the tubular fluid, TF_x, to its concentration in the plasma, P_x. This ratio, $[TF]_x/[P]_x$, may rise, fall, or stay constant depending on the transport of the substance and on how much water is reabsorbed. Water is reabsorbed in most regions of the nephron and this has the effect of concentrating any solutes in the tubular fluid even if they are not transported at all. Therefore, we simultaneously have to take into account the reabsorption of water. This is achieved by measuring the [TF]/[P] ratio of inulin. Given that inulin is neither secreted nor reabsorbed, the rise in $[TF]_{inulin}/[P]_{inulin}$ reflects the reabsorption of water. For example, if 75% of the filtered water has been reabsorbed (the original amount of inulin is now dissolved in only 25% of the water), the value of $[TF]_{inulin}/[P]_{inulin}$ will be 4.

— The fractional reabsorption of water can be expressed as

$$1 - 1/([TF]_{inulin}/[P]_{inulin})$$

For example, when 75% of the water is reabsorbed, the expression becomes

$$1 - 0.25 = 0.75$$

— For any solute (other than inulin itself), an expression that corrects the [TF]/[P] ratio for water reabsorption is

$$\text{Fractional reabsorption of solute} \times = 1 - ([TF]_x/[P]_x)/([TF]_{inulin}/[P]_{inulin})$$

— This expression calculates the fraction of a substance that has been reabsorbed. Consider as an example the reabsorption of urea. Suppose at a given point in the nephron the $[TF]_x/[P]_x$ for urea is 1.2 (its concentration is now 20% greater than in plasma), and the $[TF]_{inulin}/[P]_{inulin}$ for inulin is 4 (75% of the water has been reabsorbed). Then the fractional reabsorption of urea is

$$1 - 1.2/4 = 0.7$$

— This tells us that 70% of the urea has been reabsorbed even though its concentration has actually increased. This follows because a slightly smaller fraction of urea has been reabsorbed than water.
— If a solute undergoes net secretion, there will be more of that substance in the tubular fluid than was filtered. Its concentration will rise both because water is reabsorbed and because amounts are added to the fluid. The expression for fractional reabsorption will then result in a negative number. For example, if the fractional reabsorption is –0.5, it means that an amount equal to 50% of the filtered load has been secreted.

16.3 Limitations of Tubular Transport

The capacity of the tubules to secrete and reabsorb is not infinite. Rates reach an upper limit either because the transporters saturate (T_m-limited transport processes) or because the substance leaks back across leaky tight junctions (gradient-limited transport processes).

Saturation of Transporters: T_m-limited Transport Processes

Maximum tubular transport capacity (T_m) is the highest attainable rate of tubular transport of any given solute. T_m is reached due to saturation of the transport carriers and/or saturation of transport sites for a particular substance along the renal tubules.

— Substances with a reabsorptive T_m include glucose, phosphate, and sulfate ions, many amino acids, and Krebs cycle intermediates.
— Substances with a secretory T_m include PAH, uric acid, creatinine, histamine, and drugs (e.g., penicillin and morphine).

Reabsorptive T_m: Glucose.

$$\text{Filtered load of glucose} = GFR \times \text{plasma glucose}$$

— If GFR remains constant, the filtered load of glucose will be proportional to the plasma glucose concentration.
— Glucose is not normally excreted, because filtered glucose is reabsorbed into blood in the proximal tubules by SGLT. As plasma glucose concentration, and consequently the filtered load, increases, renal Na+–glucose carriers become saturated, and the T_m for glucose is reached (400 mg/min). Increases in plasma glucose concentration above T_m will cause glucose to be excreted in urine (**Fig. 16.2**).

Secretory T_m: *p*-Aminohippuric Acid

$$\text{Filtered PAH} = GFR \times \text{plasma [PAH]}$$

■ Osmotic diuresis with diabetes

Glucose is normally completely reabsorbed in the proximal tubule. In diabetes, however, high plasma glucose levels exceed their maximum tubular transport capacity (T_m), causing glucose to pass on to the loop of Henle and distal nephron, where it causes an osmotic diuresis. Net reabsorption of Na+ is also reduced (causing hyponatremia, or low blood [Na+]), because the large amount of tubular water accompanying the glucose also contains large amounts of Na+. These factors explain polyuria (excessive urination), polydipsia (excessive thirst), and dehydration, which are common presenting symptoms in diabetes.

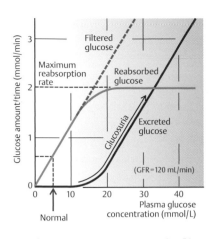

Fig. 16.2 ▶ **Reabsorption of glucose and amino acids.**
Fractional excretion of D-glucose is very low (~0.4%). This
virtually complete reabsorption is achieved by secondary
active transport (Na⁺–glucose symport) at the luminal cell
membrane. About 95% of this activity occurs in the proximal
tubule. If the plasma glucose concentration exceeds 10 to
15 mmol/L, as in diabetes mellitus (normally 5 mmol/L),
glycosuria develops. (GFR, glomerular filtration rate)

- If GFR remains constant, the filtered load of PAH will be proportional to plasma [PAH].
- PAH is normally secreted into urine from peritubular capillary blood by carriers in the proximal tubule. As plasma [PAH] increases, these carriers become saturated, the secretion of
  PAH reaches its T_m, and the renal clearance of PAH decreases toward the value of GFR. The
  amount of PAH that is excreted is therefore the sum of the [PAH] that is filtered across the
  glomerular membrane and the [PAH] that is secreted from peritubular capillary blood into
  urine (**Fig. 16.3**).

Determination of T_m-limited Transport Processes Using Renal Clearance.   Renal
clearance can be used to determine whether or not renal transport of a substance is a T_m-
limited transport process. This is done by constructing a renal titration curve, a combined plot
of the filtered load, the urinary excretion rate, and the transport rate of substance X against the
increasing plasma concentrations of X. If the transport rate becomes constant at high plasma
concentrations of X, then it is a T_m-limited transport process.

Gradient-limited Transport Processes

For some solutes, rates of transport are limited by the concentration gradient between the filtrate and the peritubular capillary blood because the leakiness of the tight junctions does not
allow a large gradient to exist. In this case, the solute diffuses back as fast as it is transported.
- Na⁺ reabsorption along the nephron is an example of a gradient-limited transport process.

Fig. 16.3 ▶ **Secretion and excretion of *p*-aminohippuric acid (PAH).**
The proximal tubule uses active transport mechanisms to secrete numerous waste products. This is
done with carriers for organic anions (e.g., PAH) and organic cations. This makes it possible to raise their
clearance level above that of inulin and thus raise their fractional excretion above 1.0 to eliminate them
more effectively (compare *red* and *blue curves*). Secretion is carrier-mediated and is therefore subject to
saturation kinetics (saturation = maximum tubular transport capacity). The fractional excretion of organic
anions or cations decreases when their plasma concentrations rise (PAH secretion curve reaches a plateau,
and the slope of the PAH excretion curve decreases).

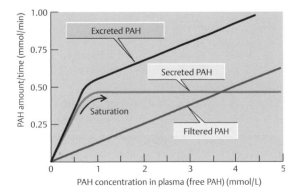

16.4 Renal Handling of Na⁺ and Its Accompanying Solutes

Refer to **Fig. 16.4.**

Proximal Tubule

— About 67% of the filtered load of Na⁺ is reabsorbed from tubular fluid into peritubular capillary blood in the proximal tubule.
— Na⁺ enters the tubular cells from the lumen mostly via the Na⁺–H⁺ antiporter (NHE-3) in the luminal membrane.

Fig. 16.4 ▶ **Na⁺ and Cl⁻ reabsorption.**
In the proximal tubule, Na⁺ ions diffuse passively from the tubular lumen into cells via the electroneural Na⁺–H⁺ antiporter (NHE-3) and various Na⁺ symporters for reabsorption of glucose and other substances (**1**). Because most of these symporters are electrogenic, the luminal membrane is depolarized, and an early lumen-negative transepithelial potential (LNTP) develops (**2**). This LNTP drives Cl⁻ through paracellular spaces out of the lumen and into peritubular capillary blood (**3**). The reabsorption of Cl⁻ lags behind that of Na⁺ and H₂O, so luminal [Cl⁻] rises. As a result, Cl⁻ starts to diffuse down its concentration gradient paracellularly along the middle and late proximal tubule (**4**), thereby producing a lumen-positive transepithelial potential (LPTP) (**5**). The LPTP drives Na⁺ and other cations into peritubular capillary blood. In the thick ascending limb of the loop of Henle (**6**), Na⁺ is reabsorbed via the Na⁺–K⁺–2Cl⁻ symporter (**7**). This symporter is primarily electroneutral, but K⁺ recirculates back into the lumen through K⁺ channels. This hyperpolarizes the membrane, resulting in an LPTP. In the distal convoluted tubule, Na⁺ is reabsorbed via an electroneutral Na⁺–Cl⁻ symporter (**8**). In principal cells of the collecting duct, Na⁺ exits the lumen via Na⁺ channels activated by aldosterone and antidiuretic hormone (ADH) and inhibited by prostaglandins and atrial natriuretic peptide (ANP) (**9**). On the basolateral membrane side, Na⁺ ions exit cells and enter peritubular capillary blood via Na⁺–K⁺ ATPase and an Na⁺–HCO₃⁻ symport carrier. In the latter case, Na⁺ exits the cell as H⁺ is secreted by the cell, resulting in the intracellular accumulation of HCO₃⁻. (ANP, atrial natriuretic peptide; BSC, bumetanide-sensitive cotransporter; TSC, thiazide-sensitive cotransporter)

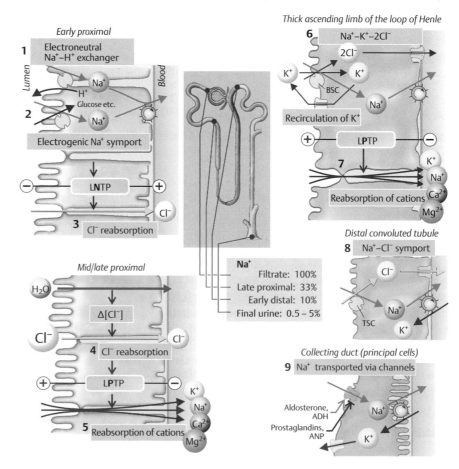

— Na⁺ leaves the tubular cells and enters peritubular capillary blood primarily via the Na⁺–K⁺ ATPase on the basolateral membrane.
— The proximal tubule has a high permeability to water, resulting in water following Na⁺ in equal proportions. The reabsorbed fluid then moves from interstitial space into peritubular capillaries by bulk flow caused by the net balance of hydrostatic and oncotic pressures acting across the capillaries.
— Cl^-, HCO_3^-, glucose, phosphate, sulfate, amino acids, and many other solutes are reabsorbed along with Na^+ and H_2O. Cl^-, the major anion in the filtrate, is mostly passively reabsorbed. The concentration gradient favors the movement of Cl^- from lumen to peritubular capillary blood. This gradient is established by the reabsorption of water.
— In the early part of the proximal tubular lumen, the potential difference across the luminal membrane is slightly negative (−4 mV). In the late part of the proximal tubule, the potential difference becomes slightly positive (+3 mV).
— Organic acids and NH_3 are secreted.
— H^+ ions are actively secreted via the Na^+–H^+ antiporter in the luminal membrane.

Glomerulotubular Balance.   Under steady-state conditions, a relatively constant fraction (two-thirds, or 67%) of the filtered Na^+ and H_2O is reabsorbed from tubular fluid into peritubular capillary blood in the proximal tubule despite variations in GFR. The absolute rate of Na^+ reabsorption in the proximal tubule changes proportionately to the change in GFR (Na^+ and H_2O filtered load). This helps to minimize changes in Na^+ and H_2O excretion that follow changes in GFR.

Loop of Henle

— Twenty-five percent of Na^+ is reabsorbed, along with Cl^- and K^+, in the thick ascending limb via an Na^+–K^+–$2Cl^-$ symporter in the luminal membrane. Net K^+ reabsorption in this segment is very small compared with net reabsorption of Na^+ and Cl^-. K^+ diffuses back to the tubular lumen.
— About 10% of the water is reabsorbed in the thin descending limb, and virtually none in the thick ascending limb. The relatively greater reabsorption of NaCl than water in the loop as a whole dilutes the tubular fluid. The loop of Henle is therefore the primary site for dilution of urine.
— The transepithelial potential difference is 10 mV lumen positive in the thick ascending limb, which helps promote cation reabsorption.

Distal Tubule and Collecting Duct

— Eight percent of Na^+ is reabsorbed into tubular cells, along with Cl^- in the early distal tubule, via an Na^+–Cl^- cotransporter in the luminal membrane. The early distal tubule is also impermeable to water, allowing NaCl to be reabsorbed without water. It is therefore a secondary site for dilution of urine.
— NaCl is reabsorbed along with water in the late distal tubule.
— H^+ is actively secreted into the tubular lumen from peritubular capillary blood against a concentration gradient of 1000:1 via an H^+–ATPase system in luminal membrane of the late distal tubule. Tubular fluid can be significantly acidified in the distal nephron.

Regulation of Na⁺ Reabsorption

Na^+ is the most abundant solute in extracellular fluid (ECF). Consequently, the status of Na^+ balance critically determines the volume of the ECF compartment and the long-term regulation of blood pressure. The renin–angiotensin–aldosterone system is the most important regulator of Na^+ balance. The kidney regulates Na^+ balance by adjusting the amount of Na^+ excretion according to Na^+ intake. Na^+ excretion is the result of two processes: glomerular filtration and tubular reabsorption. The kidneys conserve Na$^+$ by normally reabsorbing 99.4% of filtered Na$^+$. Autoregulation of GFR automatically prevents excessive changes in the rate of Na$^+$ excretion in response to spontaneous changes in blood pressure. In addition, glomerulotubular balance compensates for changes in the filtered load of Na$^+$ due to acute changes in GFR under normal Na$^+$ and volume status. When there is a chronic change in GFR due to a change in the size of the ECF compartment, glomerulotubular balance is abolished, and Na$^+$ excretion varies more directly with GFR, such that Na$^+$ balance and ECF volume are restored.

Renin–Angiotensin–Aldosterone System Regulation

Renin, synthesized in the juxtaglomerular cells of renal afferent arterioles, acts on angiotensinogen (synthesized in the liver) to form angiotensin I in the bloodstream (**Fig. 16.5**). Angiotensin I is converted to angiotensin II by ACE. ACE is located primarily in pulmonary capillary endothelium but is also present in systemic capillary endothelial cells. Angiotensin II stimulates the release of aldosterone from the adrenal cortex. Aldosterone stimulates Na$^+$ reabsorption in the distal tubules and collecting ducts. Angiotensin II is a potent vasoconstrictor and simulates the renal tubules to reabsorb Na$^+$.

Renin release is stimulated by the following:

— ↓volume of the ECF compartment. This is detected by baroreceptors in renal afferent arterioles.
— ↑sympathetic nervous system activation

Fig. 16.5 ▶ **Renin–angiotensin–aldosterone system (RAS).**
If the mean renal blood pressure acutely drops below 99 mm Hg or so, renal baroreceptors will trigger the release of renin, thereby increasing the systemic plasma renin concentration. Renin is a peptidase that catalyzes the cleavage of angiotensin I from angiotensinogen. Angiotensin-converting enzyme (ACE) cleaves two amino acids from angiotensin I to produce angiotensin II. Angiotensin II and aldosterone are the most important effectors of the RAS. Angiotensin II stimulates the release of aldosterone by the adrenal cortex. Both hormones directly and indirectly lead to a renewed increase in arterial blood pressure, and in response, renin release decreases to normal levels. Moreover, both hormones directly inhibit renin release (negative feedback). (GFR, glomerular filtration rate; RBF, renal blood flow)

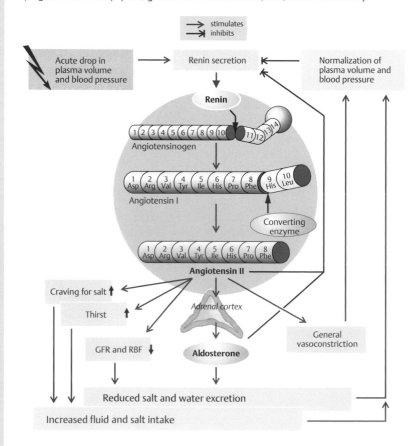

Sympathetic Nervous System and Humoral Regulation

— Increased sympathetic nervous system activation directly stimulates Na^+ reabsorption in the proximal tubule.
— Atrial natriuretic peptide (ANP) is a hormone that is released from atrial myocardial cells in response to stretch of the atria. This stretch is due to an expansion of the ECF compartment. ANP increases Na^+ excretion by
   — ↑ GFR
   — ↓ tubular reabsorption of Na^+ in collecting ducts
   — ↓ renin and aldosterone secretion

Figure 16.6 shows overall regulation of NaCl balance.

Effects of Diuretics on Na^+ Reabsorption

Diuretics decrease the absorption of Na^+ and water by various mechanisms and thereby increase their excretion

Fig. 16.6 ► Regulation of salt (NaCl) balance.
(1) Salt deficit. When there is low blood Na^+ (hyponatremia) (e.g., in aldosterone deficiency) in the presence of a primarily low water content of the body, blood osmolality and therefore antidiuretic hormone (ADH) secretion decrease, thereby transiently increasing the excretion of water. Although the hyposmolality is elevated, the extracellular fluid (ECF) volume, plasma volume, and blood pressure consequently decrease. This, in turn, activates the renin–angiotensin–aldosterone system (RAS), which stimulates thirst by secreting angiotensin II and induces Na^+ retention by secreting aldosterone. The retention of Na^+ increases plasma osmolality, leading to the secretion of ADH and, ultimately, to the retention of water. The additional intake of fluids in response to thirst also helps normalize ECF volume. (2) Salt excess. An abnormally high NaCl content of the body leads to increased plasma osmolality (thirst → drinking), as well as ADH secretion (retention of water). Thus, the ECF volume increases, and RAS activity is decreased. The additional secretion of atrial natriuretic peptide (ANP) leads to increased excretion of NaCl and water. (STN, solitary tract nucleus)

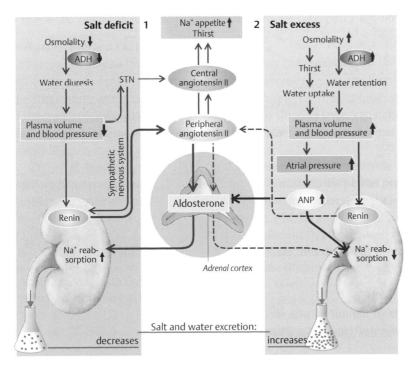

Carbonic Anhydrase Inhibitors. Carbonic anhydrase inhibitors (e.g., acetazolamide) decrease the production and reabsorption of HCO_3^- in tubular cells.

— *Effects:*
  — ↑ HCO_3^- excretion
  — ↓ Na^+ reabsorption because fewer H^+ ions are available for the Na^+–H^+ antiporter.
— *Uses:*
  — Treatment of glaucoma (↓ intraocular pressure via inhibition of aqueous humor formation)

Loop Diuretics. Loop diuretics (e.g., furosemide, bumetanide, and ethacrynic acid) inhibit the Na^+–K^+–$2Cl^-$ symporter in the thick ascending limb of the loop of Henle.

— *Effects:*
  — ↑ excretion of NaCl, K^+, and Ca^{2+}
  — ↓ ability to concentrate urine (by ↓ corticomedullary gradient)
  — ↓ ability to dilute urine (by inhibition of diluting segment of the loop of Henle)
— *Uses:*
  — Pulmonary edema due to left ventricular failure
  — Chronic congestive heart failure
  — Acute oliguria (by maintaining urine formation)
  — Hypertension
  — Acute hypercalcemia (with fluid replacement therapy)

Thiazide Diuretics. Thiazide diuretics (e.g., hydrochlorothiazide) inhibit the Na^+–Cl^- symporter in the early distal tubule.

— *Effects:*
  — ↑ NaCl and K^+ excretion
  — ↓ Ca^{2+} excretion
  — ↓ ability to dilute urine (by inhibition of the cortical diluting segment of the loop of Henle)
— *Uses:*
  — Mild to moderate hypertension
  — Idiopathic hypercalciuria

Potassium-sparing Diuretics. Potassium-sparing diuretics (e.g., spironolactone, triamterene, and amiloride) act on the late distal tubule and collecting duct.

— Spironolactone competitively antagonizes aldosterone, thus inhibiting the synthesis of Na^+ channel proteins and Na^+–K^+ ATPases.
— Triamterene and amiloride directly block Na^+ channels, thus inhibiting Na^+ reabsorption and K^+ secretion.
— *Effects:*
  — Na^+ excretion
  — ↓ K^+ secretion
— *Uses:*
  — Hypertension (spironolactone, triamterene, and amiloride). Triamterene and amiloride are weak diuretics and have little hypotensive action when given alone. However, they are useful when given along with the thiazides to prevent K^+ depletion.
  — Primary aldosteronism (spironolactone)
  — Hyperuricemia, hypokalemia, or glucose intolerance (spironolactone)

Figure 16.7 summarizes the site and mechanism of action of diuretics.

Osmotic diuretics and their role in head injury management

Osmotic diuretics (e.g., mannitol) are solutes that cannot be reabsorbed from the tubular lumen into pericapillary blood and thus increase tubular fluid osmolarity. This increased osmolality causes water retention in the lumen, which is subsequently excreted. The excretion of other electrolytes (e.g., Na^+, K^+, Cl^-, HCO_3^-, Ca^{2+}, Mg^{2+}, and PO_4^{3-}) is also increased due to dilution of their tubular concentration by retained water. Mannitol is widely used to manage head injuries where there is a need for the acute reduction of intracranial pressure (ICP), for example, if the patient shows signs of brain herniation, a situation when brain tissue, cerebrospinal fluid, and blood vessels are moved or pressed away from their usual position. When given as a bolus, an osmotic gradient is set up so that fluid is drawn out of cells, thus decreasing edema (and ICP). Then circulating blood volume increases, and blood viscosity decreases. This has the beneficial effect of increasing cerebral blood flow and oxygen delivery.

Fig. 16.7 ▶ **Site of action of diuretics.**
(ECF, extracellular fluid)

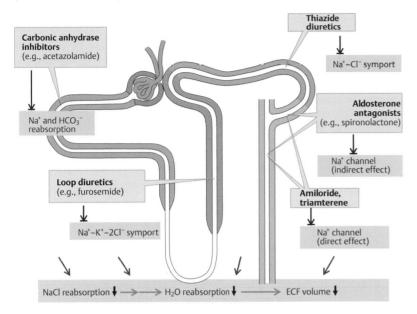

16.5 Renal Handling of K⁺

Refer to **Fig. 16.8.**

Proximal Tubule

- Sixty-seven percent of K^+ is reabsorbed, along with Na^+ and H_2O.
- K^+ is reabsorbed passively by its concentration gradient established by the reabsorption of water.

Thick Ascending Loop of Henle

- Twenty percent of K^+ is reabsorbed along with Na^+ and Cl^- via the Na^+–K^+–$2Cl^-$ symporter in the luminal membrane.

Distal Tubule and Collecting Duct

- K^+ is reabsorbed in the distal tubule via H^+–K^+ ATPase in the luminal membrane of α-intercalated cells. This occurs only on a low-K^+ diet when the body tries to retain as much of the filtered load of K^+ as possible. A diet that is low in K^+ also decreases the number of channels; thus, K^+ secretion is decreased.
- The relative amount of K^+ secretion in the distal tubule depends on dietary intake of K^+, aldosterone levels, acid–base status, tubular flow, and diuretics (**Fig. 16.9**).

Dietary Intake

- At normal and high K^+ diets, K^+ is secreted by principal cells into the cortical collecting ducts by active uptake from the interstitium via Na^+–K^+ ATPase on the basolateral membrane. K^+ is then secreted from tubular cells into the lumen by passive transport, which is driven by the prevailing electrochemical gradient. Because most filtered K^+ is reabsorbed in prior nephron segments, the rate of K^+ excretion is proportional to its secretory rate in the distal nephron.
- A diet that is high in K^+ increases the number of K^+ channels in the luminal membrane, so K^+ secretion is increased.

Fig. 16.8 ► Reabsorption and secretion of K⁺ in the kidney.
Approximately 67% of K⁺ is reabsorbed in the proximal tubule (comparable to the percentage of Na⁺ and H_2O reabsorbed). This type of K⁺ transport is mainly paracellular and therefore passive. Solvent drag (when solute particles are carried along with the water flow or filtration) and the lumen-positive transepithelial potential (LPTP) (**1**) in the middle and late proximal tubule are responsible for it. In the loop of Henle, another 15% of the filtered K⁺ is reabsorbed by trans- and paracellular routes (**2**). The amount of K⁺ secreted is determined in the collecting ducts. Larger or smaller quantities of K⁺ are then either reabsorbed or secreted according to need. The collecting duct contains principal cells (**3**) that reabsorb Na⁺ and secrete K⁺. Accumulated intracellular K⁺ can exit the cell through K⁺ channels on either side of the cell. The electrochemical K⁺ gradient across the membrane in question is decisive for the efflux of K⁺. The luminal membrane of principal cells also contains Na⁺ channels through which Na⁺ enters the cell. This depolarizes the luminal membrane, which reaches a potential of ~−20 mV. The driving force for K⁺ efflux is therefore higher on the luminal side than on the opposite side. Hence, K⁺ preferentially exits the cell toward the lumen (secretion). This is mainly why K⁺ secretion is coupled with Na⁺ reabsorption; that is, the more Na⁺ reabsorbed by the principal cell, the more K⁺ secreted. Type A (α) intercalated cells can actively reabsorb K⁺ in addition to secreting H⁺ due to a H⁺–K⁺ ATPase in the luminal membrane (like the parietal cells in the stomach) (**4**). (CA, carbonic anhydrase)

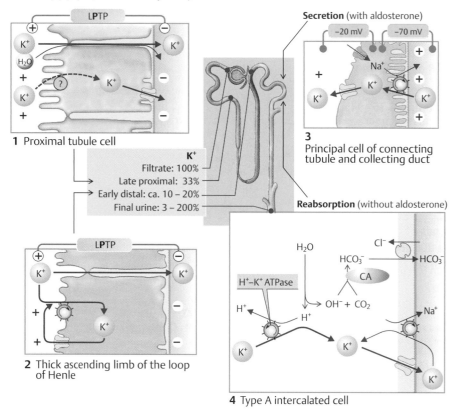

1 Proximal tubule cell

K⁺
Filtrate: 100%
Late proximal: 33%
Early distal: ca. 10 – 20%
Final urine: 3 – 200%

2 Thick ascending limb of the loop of Henle

3 Principal cell of connecting tubule and collecting duct

4 Type A intercalated cell

Aldosterone Levels

— Aldosterone increases K⁺ secretion and stimulates Na⁺ reabsorption into cells in the distal tubule and collecting duct. The intracellular Na⁺ is then pumped out of the tubular cell in exchange for K⁺ by the Na⁺–K⁺ ATPase pump. This increases the intracellular [K⁺], which provides a favorable electrochemical gradient for the passive secretion of K⁺ into the lumen via K⁺ channels. Aldosterone also increases the number of K⁺ channels in the luminal membrane.

Acid–base Status.   Acidosis decreases K⁺ secretion; alkalosis increases K⁺ secretion.

— In *acidosis,* there is an excess of H⁺ ions in the bloodstream. H⁺ ions enter cells in exchange for K⁺ via the H⁺–K⁺ ATPase pump. This decreases the intracellular [K⁺], so K⁺ secretion is decreased.
— In *alkalosis,* there is a deficiency of H⁺ ions in the bloodstream, causing H⁺ to enter the blood from the cell in exchange for K⁺. This increases the intracellular [K⁺], so K⁺ secretion is increased.

Fig. 16.9 ► **Factors that affect K⁺ secretion and excretion.**
An increased K⁺ intake raises the intracellular and plasma [K⁺], which thereby increases the chemical driving force for K⁺ secretion. The intracellular [K⁺] in renal cells rises in alkalosis and falls in acute acidosis. This leads to a simultaneous fall in K⁺ excretion, which again rises in chronic acidosis. This is because acidosis-related inhibition of Na⁺–K⁺ ATPase reduces proximal Na⁺ reabsorption, resulting in increased distal urinary flow, and the resulting hyperkalemia (high blood [K⁺]) stimulates aldosterone secretion. If there is increased urinary flow in the collecting duct (e.g., due to a high Na⁺ intake, osmotic diuresis, or other factors that inhibit Na⁺ reabsorption upstream), larger quantities of K⁺ will be excreted. Aldosterone leads to retention of Na⁺, an increase in extracellular fluid (ECF) volume, a moderate increase in H⁺ secretion, and increased K⁺ secretion. It also increases the number of Na⁺–K⁺ ATPase molecules in the target cells.

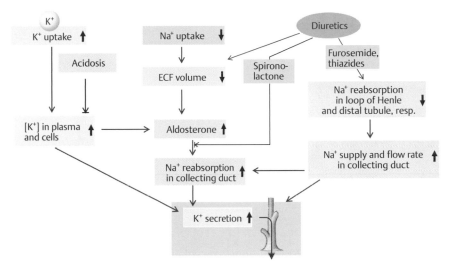

Tubular Flow

- Increased fluid flow in the distal nephron, detected by mechanosensitive elements in principal cells, increases K⁺ permeability and therefore increases K⁺ secretion.

Diuretics

- Thiazide and loop diuretics increase K⁺ secretion because of increased fluid flow past the principal cells. This can cause decreased blood K⁺ levels (hypokalemia) unless they are combined with a K⁺-sparing diuretic, which decreases K⁺ secretion.

16.6 Renal Handling of Urea, Phosphate, Calcium, and Magnesium

Urea

- Fifty percent of the filtered load of urea undergoes passive net reabsorption into peritubular capillary blood in the proximal tubule. The remaining 50% is excreted in urine.
- Urea is secreted from interstitial fluid into tubular fluid in the deep medullary regions of the loop of Henle, thereby restoring the urea content of the tubular fluid.
- The distal tubule, cortical, and outer medullary collecting ducts are relatively impermeable to urea, so no reabsorption occurs in these segments.
- The inner medullary collecting duct is permeable to urea in the presence of antidiuretic hormone (ADH), and urea is reabsorbed a second time. Some of the urea that is reabsorbed is recycled back to the loop of Henle and in doing so sets up the corticopapillary osmotic gradient (see page 179 and **Fig. 17.3**).

Fig. 16.10 ► Reabsorption of phosphate, Ca²⁺, and Mg²⁺.

Inorganic phosphate (P_i) is filtered, and a large part of this is reabsorbed (**1**). P_i is reabsorbed at the proximal tubule (**2,3**). Its luminal membrane contains an Na^+–P_i symporter. P_i excretion rises in the presence of a P_i excess and falls during a P_i deficit. Acidosis also results in the excretion of phosphate and H^+ ions. Hypocalcemia (low blood [Ca^{2+}]) also induces a rise in P_i excretion. Parathyroid hormone (PTH) inhibits P ↑ reabsorption (**3**). Ca^{2+} reabsorption occurs practically throughout the entire nephron (**1,2**) and is paracellular, that is, passive (**4a**) The lumen-positive transepithelial potential (LPTP) provides most of the driving force for this activity. Because Ca^{2+} reabsorption in the thick ascending limb of the loop of Henle (TAL) depends on NaCl reabsorption, loop diuretics inhibit Ca^{2+} reabsorption there. Parathyroid hormone (PTH) promotes Ca^{2+} reabsorption in TAL, as well as in the distal convoluted tubule, where Ca^{2+} is reabsorbed by transcellular active transport (**4b**). Thus, Ca^{2+} influx into the cell is passive and occurs via luminal Ca^{2+} channels, and Ca^{2+} efflux is active and occurs via Ca^{2+} ATPase and via the Na^+–Ca^{2+} antiporter. The majority of Mg^{2+} is subject to paracellular reabsorption in the TAL (**4a**). Another 10% of Mg^{2+} is subject to transcellular reabsorption in the distal tubule (**4b**), probably like Ca^{2+}.

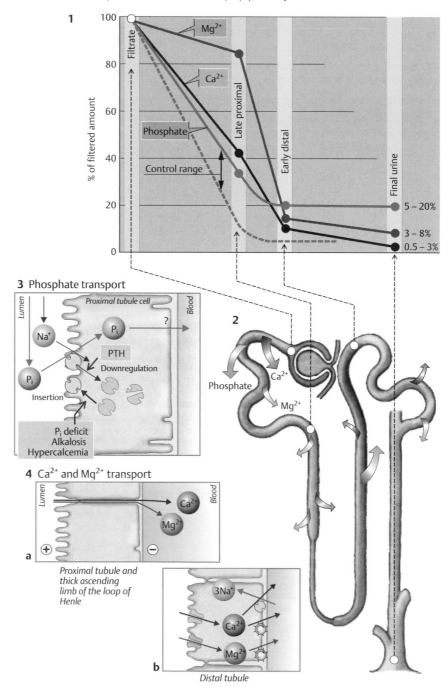

Phosphate

- Eighty-five percent of filtered phosphate is reabsorbed in the proximal tubule by Na^+–phosphate cotransport. The remaining 15% is excreted in urine.
- Parathyroid hormone (PTH) inhibits phosphate reabsorption in the proximal tubule by activating adenylate cyclase and increasing cyclic adenosine monophosphate (cAMP) production (↑ urinary cAMP). This inhibits Na^+–phosphate cotransport.
- Phosphate is an important urinary buffer. It combines with excess H^+ ions, which are excreted as phosphoric acid (H_2PO_4). This is also called *titratable acid* and is discussed further on page 188.

Calcium

- Sixty percent of plasma Ca^{2+} is filtered across the glomerulus.
- Ninety percent of the filtered load of Ca^{2+} is reabsorbed into peritubular capillaries by passive transport in the proximal tubule and in the thick ascending loop of Henle. Although not coupled mechanistically, the amount reabsorbed depends on Na^+ reabsorption because of its effects on driving forces.
- PTH facilitates further active Ca^{2+} reabsorption in the distal tubule by activating adenylate cyclase and ↑ cAMP.
- Loop diuretics (e.g., furosemide) inhibit Na^+ reabsorption (via inhibition of the Na^+–K^+–2Cl^- symporter) in the thick ascending loop of Henle. Because Ca^{2+} reabsorption depends on Na^+ reabsorption, Ca^{2+} reabsorption decreases, and more Ca^{2+} is excreted. Loop diuretics can therefore be used in the treatment of hypercalcemia (with appropriate fluid replacement).
- Thiazide diuretics (e.g., hydrochlorothiazide) decrease Na^+ reabsorption but actually stimulate Ca^{2+} reabsorption. They act in the early distal tubule to inhibit the Na^+–Cl^- symporter. Blocking Na^+ uptake lowers the cytostolic [Na^+] and stimulates Na^+–Ca^{2+} antiport at the basolateral membrane. This increases Ca^{2+} reabsorption. Thiazides are used for the treatment of idiopathic hypercalciuria.

Magnesium

- The thick ascending loop of Henle is the major site for Mg^{2+} reabsorption through tight junctions, although reabsorption also occurs in the proximal and distal tubules.
- Mg^{2+} and Ca^{2+} compete for reabsorption in the thick ascending loop of Henle.
- Hypercalcemia inhibits Mg^{2+} reabsorption; therefore, more Mg^{2+} is excreted.
- Hypermagnesemia inhibits Ca^{2+} reabsorption; therefore, more Ca^{2+} is excreted.

Figure 16.10 shows the reabsorption of phosphate, Ca^{2+}, and Mg^{2+}.

Table 16.1 and **Fig. 16.11** summarize the renal handling of solutes.

Table 16.1 ▶ Summary of the Renal Handling of Solutes				
Substance	**Reabsorption**	**Secretion**	**Effects of Diuretics**	**Effects of Hormones**
Na^+ and Cl^-	67% in proximal tubule 25% in thick ascending loop of Henle 8% in the distal nephron, mostly in the cortical collecting ducts	—	All classes of diuretics ↑ Na^+ excretion	Aldosterone ↑ Na^+ reabsorption in the distal tubule ANP ↓ Na^+ reabsorption in the collecting ducts Angiotensin II ↑ aldosterone secretion and ↑ Na^+ reabsorption in the proximal tubule
K^+	67% in proximal tubule 20% in thick ascending loop of Henle Reabsorption in distal tubule and collecting duct if low K^+	Secretion of K^+ in high K^+ states and acidosis	Osmotic diuretics, loop diuretics, and thiazides ↑ K^+ secretion (unless used with K^+-sparing diuretic)	Aldosterone ↑ K^+ secretion
Urea	50% reabsorbed in proximal tubule Reabsorption in inner medullary collecting duct (if ADH present)	Secreted in the loops of Henle that reach deep medulla	—	ADH ↑ permeability of inner medullary collecting duct to urea
PO_4^{3-}	85% reabsorbed in proximal tubule	—	Osmotic diuretics ↑ PO_4^{2-} secretion	PTH ↓ PO_4^{3-} reabsorption in the proximal tubule
Ca^{2+}	60% in proximal tubule 30% in thick ascending loop of Henle	—	Osmotic diuretics and loop diuretics ↑ Ca^{2+} excretion Thiazide diuretics ↓ Ca^{2+} excretion	PTH ↑ Ca^{2+} reabsorption in the distal tubule
Mg^{2+}	Reabsorption occurs in thick ascending loop of Henle (mainly)	—	Osmotic diuretics ↑ Mg^{2+} excretion	—

Note: The net percentage of a substance that is not reabsorbed is excreted as final urine.
Abbreviations: ADH, antidiuretic hormone; ANP, atrial natriuretic peptide; PTH, parathyroid hormone.

Fig. 16.11 ▶ Summary of the transport of substances along the nephron. (PAH, p-aminohippuric acid)

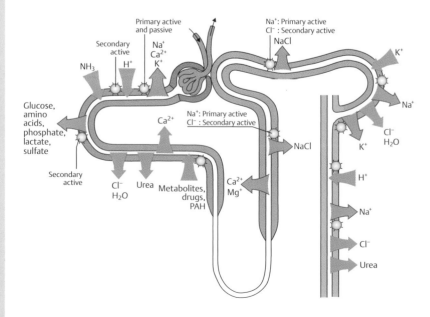

17 Regulation of Water Balance and the Concentration and Dilution of Urine

17.1 Regulation of Water Balance

The regulation of body water depends on the dynamic balance between the rates of water movement into and out of the body. The two major mechanisms responsible for water balance are thirst and antidiuretic hormone (ADH) regulation of urinary water excretion.

— The circulating level of ADH regulates the amount of water reabsorption from distal tubules and collecting ducts into peritubular capillary blood. Therefore, the regulation of renal excretion of water is ultimately determined by factors that influence the rate of synthesis and release of ADH into the blood and its renal action. ADH is a peptide hormone synthesized in specialized hypothalamic neurons and transported within axons to the posterior pituitary, where it is stored until release. Because water gain or deficit significantly affects total solute concentration within body fluids, plasma osmolality is the most important regulator of ADH release (**Figs. 16.6**, p.169, and **17.1**). The amount of ADH released increases with a rise in plasma osmolality via stimulation of hypothalamic osmoreceptors. A decrease in blood volume also stimulates ADH release, because the resulting decrease in atrial pressure relieves the inhibitory effect of atrial baroreceptors on ADH release. ADH binds to V_2 receptors on peritubular membranes of epithelial cells of the distal nephron and, via activation of the

Fig. 17.1 ▶ **Regulation of water balance.**
(**1**) Net water losses (hypovolemia) due, for example, to sweating make extracellular fluid (ECF) hypertonic. Osmolality rises of only 1 to 2% are sufficient to increase the secretion of antidiuretic hormone (ADH) from the posterior pituitary. ADH decreases urinary H_2O excretion. Fluid intake from outside the body is also required. The hypertonic cerebrospinal fluid (CSF) stimulates the secretion of (central) angiotensin II (AT II), which triggers hyperosmotic thirst. (**2**) In water excess, the osmolality of ECF is reduced. This signal inhibits the secretion of ADH, resulting in water diuresis and normalization of plasma osmolality within 1 hour.

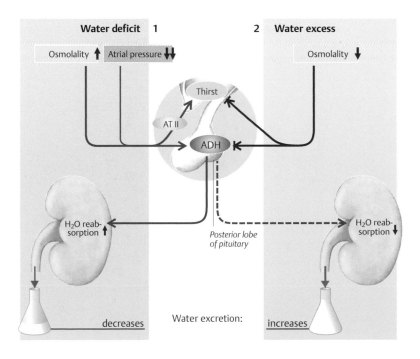

adenylate cyclase enzyme system, increases the permeability of luminal membranes of the epithelial cells to water.

— Water intake is regulated through a thirst center located in the hypothalamus. Thirst is also stimulated by both an increase in plasma osmolality and a decrease in extracellular fluid (ECF) volume, thus working in concert with the ADH mechanism to maintain water balance. Angiotensin also stimulates thirst.

17.2   Concentration and Dilution of Urine

The kidneys are able to produce urine that is either more concentrated or more diluted than plasma by altering the amount of water that is reabsorbed via the regulation of ADH (**Fig. 17.2**).

Production of Concentrated (Hyperosmotic) Urine

Hyperosmotic urine is produced when plasma ADH concentration is high and is facilitated by the corticomedullary osmotic gradient.

Fig. 17.2 ▸ **Water reabsorption and excretion.**
Approximately 65% of the glomerular filtrate is reabsorbed at the proximal tubule. The "driving force" is the reabsorption of NaCl. This slightly dilutes the urine in the tubule, but H_2O immediately follows this small osmotic gradient because the proximal tubule is "leaky." The reabsorption of H_2O can occur through leaky tight junctions or through water channels (aquaporins) in cell membranes. The urine in the proximal tubule therefore remains isotonic. Oncotic pressure in the peritubular capillaries provides an additional driving force for water reabsorption. The more water filtered at the glomerulus, the higher this oncotic pressure. Thus, the reabsorption of water at the proximal tubule is, to a certain extent, adjusted in accordance with the glomerular filtration rate (GFR). Because the descending limb of the loop of Henle has aquaporins that make it permeable to water, the urine in it is largely in osmotic balance with the hypertonic interstitium, the content of which becomes increasingly hypertonic as it approaches the papillae. The urine therefore becomes increasingly concentrated as it flows in this direction. In the thin descending limb, which is only sparingly permeable to NaCl, this increases the concentration of Na^+ and Cl^-. Most water drawn into the interstitium is carried off by the vasa recta. Because the thin and thick ascending limbs of the loop of Henle are largely impermeable to water, Na^+ and Cl^- passively diffuse (thin limb) and are actively transported (thick limb) out into the interstitium. Because water cannot escape, the urine leaving the loop of Henle is hypotonic. Final adjustment of the excreted urine volume occurs in the collecting duct. In the presence of antidiuretic hormone (ADH) via V_2 receptors, aquaporins in the luminal membranes are used to extract water from the urine passing through the increasingly hypertonic renal medulla. This results in maximum antidiuresis. The absence of ADH results in water diuresis. (FE_{H_2O}, fraction excreted)

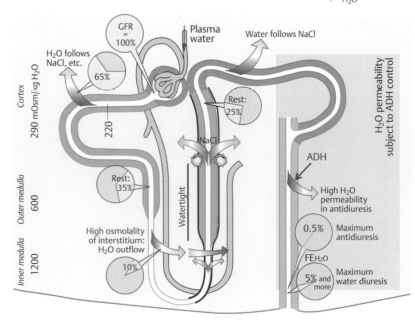

Corticomedullary Osmotic Gradient

The corticomedullary osmotic gradient is a gradient of osmolarity from the corticomedullary border (~300 mOsmol/kg H_2O) to the inner medulla. The maximum value reached in the papilla varies depending on hydration status, but it may reach 1200 mOsmol/kg H_2O in conditions of severe dehydration.

- The gradient is established by NaCl and urea and leads to the reabsorption of water (in the presence of ADH).

Role of NaCl in establishing the corticomedullary gradient.   The corticomedullary osmotic gradient is established by the *countercurrent multiplication system* in the loop of Henle. This is dependent upon the close apposition of the descending and ascending limbs of the loop of Henle in the medulla and the transport characteristics of each limb.

- The descending limb of the loop of Henle is permeable to water but poorly permeable to solutes.
- The thick ascending limb of the loop of Henle actively reabsorbs NaCl but is impermeable to water.

These characteristics allow the ascending limb to dilute the tubular fluid and concentrate the medullary interstitium ("single effect"), creating a horizontal osmotic gradient between tubular fluid in the ascending limb and that in the descending limb. This horizontal osmotic gradient is then multiplied vertically along the length of the descending loop of Henle, generating an osmotic gradient within the tubular fluid. The medullary interstitium is equilibrated with the fluid in the descending limb because this nephron segment is highly permeable to water.

Role of urea in establishing the corticomedullary gradient.   Urea is the other major solute within the medullary interstitium besides NaCl. Filtered urea undergoes passive net reabsorption into proximal tubule cells (**Fig. 17.3**). However, urea concentration at the end of the proximal tubule is approximately twice that of plasma due to water reabsorption. Urea is secreted into the tubular lumen in the deep regions of the loop of Henle. Due to the low permeability of the distal tubule and cortical collecting duct to urea, urea concentration in the tubular fluid remains high as the fluid flows through the medullary collecting duct, which is more permeable to urea. Urea diffuses out of the collecting duct, entering the interstitium and vasa recta, as well as reentering the loop of Henle. Therefore, there is a medullary recycling of urea.

- In the presence of ADH, urea constitutes ~40% of papillary osmolality because ADH increases the permeability of medullary collecting ducts to urea as well as to water.
- In the absence of ADH, < 10% of medullary interstitial osmolality is due to urea.

The medullary recycling of urea thus helps establish an osmotic gradient within the medulla with less energy expenditure (urea transport is passive) and enhances water conservation.

Role of the vasa recta in maintaining the corticomedullary osmotic gradient.   The vasa recta are hairpin blood vessels in the renal medulla formed from efferent arterioles of juxtamedullary glomeruli in apposition to the loop of Henle and collecting ducts. Like other systemic capillaries, they are permeable to solutes and water. Because of their unique countercurrent arrangement, the vasa recta act as a passive countercurrent exchanger system (**Fig. 17.4**). As solutes are transported out of the ascending loop of Henle, they diffuse down their concentration gradients into the descending vasa recta. Thus, blood in the descending vasa recta becomes progressively more concentrated as it equilibrates with the corticomedullary osmotic gradient. In the ascending vasa recta, solutes diffuse back into the medullary interstitium and into the descending vasa recta. In this manner, solutes recirculate within the renal medulla, keeping the solute concentration high within the medullary interstitium. The passive equilibrium of blood within each limb of the vasa recta with the preexisting medullary osmotic

Fig. 17.3 ► Urea in the kidney.
About 50% of the filtered urea leaves the proximal tubule by diffusion. Much of this reenters the tubule by secretion in the loop of Henle. Because the distal tubule and the cortical and outer medullary sections of the collecting duct are only sparingly permeable to urea, its concentration increases in these parts of the nephron. Antidiuretic hormone (ADH) can (via V_2 receptors) introduce urea carriers in the luminal membrane, thereby making the inner medullary collecting duct permeable to urea. Urea now diffuses back into the interstitium (where urea is responsible for half of the high osmolality there) and is then transported by carriers back into the descending limb of the loop of Henle, comprising the recirculation of urea. The nonreabsorbed fraction of urea is excreted (FE_{urea}).

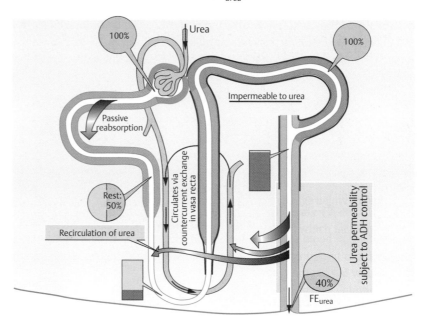

Fig. 17.4 ► Countercurrent systems.
Countercurrent exchange of water in the vasa recta of the renal medulla occurs if the medulla becomes increasingly hypertonic toward the papillae and if the vasa recta become permeable to water. Part of the water diffuses by osmosis from the descending vasa recta to the ascending ones, thereby "bypassing" the inner medulla. Due to the extraction of water, the concentration of all other blood components increases as the blood approaches the papillae. The plasma osmolality in the vasa recta is therefore continuously adjusted to the osmolality of the surrounding interstitium, which rises toward the papillae. Conversely, substances entering the blood in the renal medulla diffuse from the ascending to the descending vasa recta, provided the walls of the vessels are permeable to them. The countercurrent exchange in the vasa recta permits the necessary supply of blood to the renal medulla without significantly altering the high osmolality of the renal medulla and hence impairing the urine concentration capacity of the kidney.

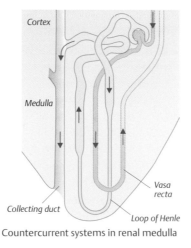

Countercurrent systems in renal medulla

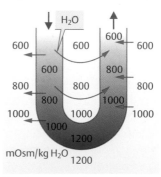

Countercurrent exchange (water) in loop (e.g., vasa recta)

gradient at each horizontal level helps maintain the medullary osmotic gradient necessary for the production of hyperosmotic urine.

Note: If blood flow is rapid in the vasa recta, the osmotic gradient will decrease. This happens when a person takes on a large water load.

Production of Dilute (Hyposmotic) Urine

Hyposmotic urine is produced when plasma [ADH] is low.

— Isosmotic tubular fluid from the proximal tubule enters the descending limb of the loop of Henle. It becomes progressively more concentrated as it moves toward the bend because fluid in the descending limb equilibrates osmotically with fluid within the medullary interstitium. The concentrated fluid at the bend of the loop then becomes progressively more diluted as it flows through the ascending loop of Henle because NaCl is reabsorbed without water. When there is a very low level of ADH, the permeability of the collecting duct to water is very low. Consequently, no water will be reabsorbed in the distal nephron even though salts continue to be reabsorbed. Therefore, the diluted tubular fluid that emerges from the ascending loop of Henle will remain hyposmotic as it flows through the distal nephron and medullary collecting ducts, producing hyposmotic urine (~100 mOsmol/kg H_2O).

17.3   Measurement of Concentrating and Diluting Ability

Free Water Clearance

Free water (solute free water) is produced in the ascending loop of Henle and early distal tubule, where NaCl is reabsorbed without water.

Free water clearance (C_{H_2O}) is defined as the amount of pure water that must be subtracted from or added to the urine (per unit of time) to make that urine isosmotic with plasma. It is used as a measure of the ability of the kidneys to concentrate or dilute urine.

— Urine is C_{H_2O} positive when there is no ADH and free water is excreted (i.e., the urine has excess water).

— Urine is C_{H_2O} negative when ADH is secreted and water is reabsorbed by the late distal nephron and the collecting ducts. The negative C_{H_2O} represents the amount of water that would have to be added back to make the urine isosmotic.

Free water clearance is calculated from the following equation:

$$C_{H_2O} = V - C_{osm}$$

where C_{H_2O} is free water clearance (in mL/min), V is urine flow rate (in mL/min), and C_{osm} is osmolar clearance, which equals V × (urine$_{osm}$/plasma$_{osm}$) (in mL/min).

Example: A patient with a plasma osmolality of 300 mOsmol/kg H_2O produces urine at a rate of 1.2 mL/min. The urine osmolality is 450 mOsmol/kg H_2O.

$$C_{osm} = 1.2 \text{ mL/min} \times [(450 \text{ mOsmol/kg } H_2O)/(300 \text{ mOsmol/kg } H_2O)] = 1.8 \text{ mL/min}$$

$$C_{H_2O} = 1.2 \text{ mL/min} - 1.8 \text{ mL/min} = -0.6 \text{ mL/min}$$

Note: The negative value indicates that water would have to be added to make the urine isosmotic.

Table 17.1 summarizes the hormones that act on the kidney.

Nephrogenic diabetes insipidus

In nephrogenic diabetes insipidus, the kidney is unresponsive to ADH, and thiazide diuretics cause a paradoxical reduction in the excessive urination (i.e., they decrease polyuria). The mechanism for this effect is uncertain, but it is usually attributed to changes in Na$^+$ excretion. Thiazides inhibit NaCl reabsorption in the early segments of the distal tubule but have little effect in the thick ascending limb, which is involved in concentrating the urine. Although all thiazides share this effect, chlorothiazide is most commonly used to treat this condition.

Table 17.1 ▶ Summary of Hormones that Act on the Kidney			
Hormone	**Stimulus for Secretion**	**Mechanism of Action**	**Effects**
ADH	↑ plasma osmolarity ↓ blood volume (major hypovolemia) ↓ blood pressure (severe hypotension)	Acts on V_2 receptors activating adenylate cyclase (cAMP). This causes the insertion of water pores (aquaporins) into membranes	↑ permeability of the collecting duct to water
PTH	↓ plasma [Ca^{2+}] ↑ plasma [PO$_4^{2-}$] ↓ plasma [calcitriol]	Activates adenylate cyclase (↑ urinary cAMP)	↓ PO$_4^{2-}$ reabsorption in the proximal tubule ↑ Ca^{2+} reabsorption in the distal tubule ↑ calcitriol synthesis (via ↑ in 1-α-hydroxylase synthesis and activity)
Aldosterone	↓ blood pressure ↓ blood volume (hypovolemia) ↑ plasma [K$^+$]	Activation of the RAS promotes protein synthesis	↑ Na$^+$ reabsorption in the distal tubule and collecting duct ↑ K$^+$ and H$^+$ secretion in the distal nephron
ANP	Stretch of the atria in response to an increase in atrial pressure	Activates guanylate cyclase (cGMP)	↑ GFR ↓ tubular reabsorption of Na$^+$ in collecting ducts ↓ renin and aldosterone secretion
Angiotensin II	↓ blood pressure ↓ blood volume (hypovolemia)	Angiotensin II binds to angiotensin receptors (mainly AT$_1$ and AT$_2$)	↑ release of aldosterone from adrenal cortex ↑ Na$^+$–H$^+$ exchange and HCO$_3^-$ reabsorption in the proximal tubule ↓ GFR through vascular effects

Abbreviations: ADH, antidiuretic hormone; ANP, atrial natriuretic peptide; cAMP, cyclic adenosine monophosphate; cGMP, cyclic guanosine monophosphate; GFR, glomerular filtration rate; PTH, parathyroid hormone; RAS, renin–angiotensin–aldosterone system.

18 Acid–base Balance

Acid–base balance, or the concentration of H^+ in the extracellular fluid (ECF), is tightly regulated, such that the pH of the arterial blood is maintained within a small range, pH 7.37 to 7.42, or $[H^+]$ = 40 nmol/L. This is important as pH affects many cellular processes, for example, electrochemical gradients necessary for muscle contraction, transport of substances across membranes, bone mineralization, body fluid balance, optimal enzyme functioning, metabolic processes (e.g., gluconeogenesis and glycolysis), and cell division.

The body continuously produces respiratory (or volatile) acids and metabolic (or fixed) acids. It is able to respond very effectively to perturbations in acid (or base) production by utilizing extracellular and intracellular buffers, by altering ventilation rates, and by altering their renal excretion (**Fig. 18.1**).

Volatile Acid (Carbon Dioxide)

Volatile acid (carbon dioxide, CO_2) is produced from cellular aerobic respiration. CO_2 combines with H_2O to form carbonic acid (H_2CO_3), which is catalyzed by the enzyme carbonic anhydrase. H_2CO_3 then rapidly dissociates to form H^+ and HCO_3^-. This reaction sequence can be expressed as follows:

$$CO_2 + H_2O \leftrightarrow H_2CO_3 \leftrightarrow H^+ \text{ and } HCO_3^-$$

Note: CO_2 itself is not an acid, as it does not contain an H^+ ion and therefore cannot act as a proton donor. However, it is designated as a volatile acid, as it has the potential to create H_2CO_3.

Nonvolatile (Fixed) Acids

Fixed acids are produced from protein catabolism (sulfuric acid, H_2SO_4), from phospholipid catabolism (phosphoric acid, H_3PO_4), or from anaerobic carbohydrate catabolism (lactic acid). Fixed acid concentrations may rise during exercise (e.g., lactic acid) or many pathological conditions (e.g., diabetes mellitus may cause ketoacid production).

18.1 Chemical Buffering Systems

Chemical buffers are solutions consisting of a mixture of a weak acid and its conjugate base or a weak base and its conjugate acid. Chemical buffers allow only small pH changes when a strong acid or base is added to them. Chemical buffers are able to resist changes in pH due to the equilibrium between the acid (HA) and its conjugate base (A^-) as expressed by

$$HA \leftrightarrow H^+ + A^-$$

When strong acid is added to a chemical buffer solution consisting of a weak acid and its conjugate base, the equilibrium is shifted to the left, consuming some of the H^+ ions of the strong acid and therefore resisting an increase in $[H^+]$ of the magnitude that would be expected. Similarly, when a strong alkali is added, the equation shifts to the right, and the $[H^+]$ decreases less than would be expected.

Chemical buffering systems act within seconds and are the first line of defense against changes in $[H^+]$ and therefore pH.

Extracellular Fluid Buffers

Bicarbonate

- The most important physiological buffer pair of ECF is HCO_3^-/CO_2. This is due to their high concentration in plasma (24 mM) and its tight regulation: CO_2 by the lungs and HCO_3^- by the kidneys.

Fig. 18.1 ▶ **Factors affecting blood pH.**
Various pH buffers are responsible for maintaining the body at a constant pH. The most important buffer for blood and other body fluids is the bicarbonate/carbon dioxide (HCO_3^-/CO_2) buffer system. The pK_a value determines the prevailing concentration ratio of the buffer base and buffer acid ([HCO_3^-] and [CO_2], respectively) at a given pH (Henderson–Hasselbalch equation). The primary function of the HCO_3^-/CO_2 buffer system in blood is to buffer H^+ ions. However, this system is especially important because the concentrations of the two buffer components can be modified largely independently of each other: by respiration and by the kidney. Hemoglobin in red blood cells, the second most important buffer in blood, is a nonbicarbonate buffer system.

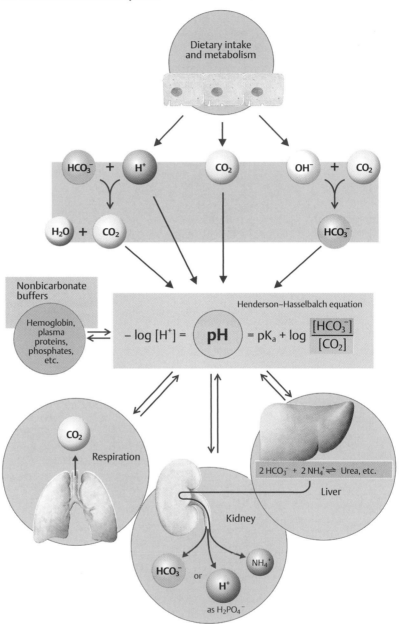

— The pK_a (the negative logarithm of the ionization constant of an acid) of the HCO_3^-/CO_2 buffer pair is 6.1. The buffering power of a buffer system is greatest when its pK_a equals the pH.

Phosphate

— Phosphate contributes little to the buffering capacity of the ECF because of its low concentration. However, phosphate is an important urinary buffer: excess H^+ is excreted as $H_2PO_4^-$ (titratable acid; see page 188).

— The pK_a of the HPO_4^{2-}/$H_2PO_4^-$ buffer pair is 6.8.

Intracellular Fluid Buffers

Organic Phosphate

– Organic phosphates, for example, adenosine monophosphate (AMP), adenosine diphosphate (ADP), adenosine triphosphate (ATP), and phosphorylated proteins, are important buffers within intracellular fluid (ICF).

Intracellular Proteins and Hemoglobin

– Proteins have broad-ranged pK_a values, so they vary in their buffering capacity.
– Hemoglobin, especially deoxygenated hemoglobin, is a strong ICF buffer.

Because all buffer pairs in plasma are in equilibrium with the same $[H^+]$, a change in the buffering capacity of the entire blood buffer system will be reflected by a change in the buffering capacity of only one buffer pair (*isohydric principle*). The acid–base status or pH of ECF can be evaluated by examining only the bicarbonate buffer system.

Calculating pH Using the Henderson–Hasselbalch Equation

The Henderson–Hasselbalch equation is expressed as follows:

$$pH = pK_a + \log [A^-]/[HA]$$

where pH is $-\log_{10} [H^+]$, pK_a is $-\log_{10}$ equilibrium constant, $[A^-]$ is the base form of the buffer system (mM), and [HA] is the acid form of the buffer system (mM).
Note: The pK_a value determines the prevailing ratio of the buffer base and buffer acid at a given pH. Buffer pairs are most effective when the pK_a value is close to the pH (within 1.0 pH).

The pH of ECF containing the HCO_3^-/CO_2 buffer system can be expressed as functions of the buffer pair concentration using the Henderson–Hasselbalch equation:

$$pH = 6.1 + \log [HCO_3^-]/[H_2CO_3]$$

$$pH = 6.1 + \log [HCO_3^-]/[0.03 \times P_{CO_2}]$$

Note: The proportionality constant between dissolved CO_2 and P_{CO_2} (partial pressure of CO_2) is 0.03, and P_{CO_2} is 40 mm Hg.

$$pH = 6.1 + (\log 24 \text{ mmol/L}/0.03 \times 40 \text{ mm Hg})$$

$$pH = 6.1 + \log (20/1)$$

Thus, the maintenance of a normal plasma pH depends on the preservation of the ratio of $[HCO_3^-]$ to $[CO_2]$ in plasma at ~20:1.

18.2 Respiratory Regulation of Acid–base Balance

When arterial pH is made more acidic or alkaline, the rate of alveolar ventilation is altered via changes in signals from respiratory chemoreceptors. The resulting hyper- or hypoventilation will change arterial P_{CO_2} in the direction that will return arterial pH toward normal.

$\downarrow$ pH → $\uparrow$ alveolar ventilation → $\downarrow$ P_{CO_2} → $\uparrow$ $[HCO_3^-]/[CO_2]$ → $\uparrow$ pH

$\uparrow$ pH → $\downarrow$ alveolar ventilation → $\uparrow$ P_{CO_2} → $\downarrow$ $[HCO_3^-]/[CO_2]$ → $\downarrow$ pH

Influence of pH on the excretion of weak acids and bases
The ionization state of weak acids and bases, which include several useful drugs, depends on the pH of their environment. A membrane is more permeable to these substances in their nonionized form. As they become concentrated by the removal of water, they are substantially reabsorbed if they are nonionized. At tubular pH values where they are nonionized, relatively more is reabsorbed, and less is excreted. In contrast, at tubular pH values at which they are charged, they remain in the lumen and are excreted. Most weak acids are nonionized when in their protonated form. When the urine is acidic (low pH), which is the normal situation, this favors passive reabsorption and reduced excretion. Thus, for drugs that are weak acids, a low urinary pH reduces the excretion rate. For many weak bases, the influence of urinary pH is just the opposite. When a filtered weak base becomes protonated at low pH, it usually becomes a cation and is trapped in the tubular lumen, and subsequently excreted. Raising the pH keeps weak bases in their neutral form and allows them to be reabsorbed. Therefore, alkalinizing the urine via dietary manipulation helps the body retain drugs that are weak bases.

18.3 Renal Regulation of Acid–base Balance

The respiratory system alone cannot restore acid–base status to normal. The renal system is needed to restore body HCO_3^- supplies and to eliminate the buffered acids or bases. This is accomplished by the following:

- Reabsorption of virtually all filtered HCO_3^- from the tubular lumen into peritubular capillary blood
- Formation of "new" HCO_3^-
- Excretion of excess H^+ from peritubular capillary blood into the tubular lumen via urinary buffers

Reabsorption of Filtered Bicarbonate

The vast majority of HCO_3^- reabsorption occurs in the proximal tubule.

- H^+ is secreted by the proximal tubule cells into the tubular lumen via an Na^+–H^+ antiporter (NHE-3) (**Fig. 18.2**). The secreted H^+ combines with filtered HCO_3^- in luminal fluid to form CO_2 and H_2O (catalyzed by luminal carbonic anhydrase), which diffuse into proximal tubule cells (**Fig. 18.3**). Within these cells, the reverse reaction occurs. The HCO_3^- produced

Fig. 18.2 ▶ **H⁺ secretion.**
Very large quantities of H^+ ions are secreted into the lumen of the proximal tubule (**1**) by primary active transport via H^+ ATPase and by secondary active transport via an electroneutral Na^+–H^+ antiporter (NHE-3). The luminal pH then decreases from 7.4 (filtrate) to ~6.6. The secreted H^+ ion combines with HCO_3^- to form CO_2 and H_2O. One OH^- ion remains in the cell for each H^+ secreted; OH^- reacts with CO_2 to form HCO_3^- (accelerated by carbonic anhydrase [CA]). HCO_3^- leaves the cell for the blood. (**2**), α-intercalated cells secrete H^+ ions via H^+–K^+ ATPase, allowing luminal pH to drop as low as 4.5. The remaining OH^- in the cell reacts with CO_2 to form HCO_3^-, released through the basolateral membrane via the anion exchanger AE1. (NBC3, Na^+–bicarbonate cotransporter 3)

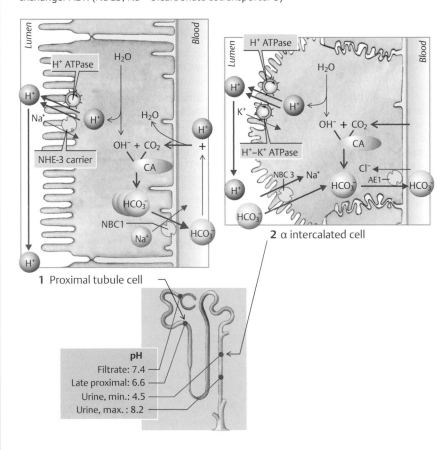

1 Proximal tubule cell

2 α intercalated cell

pH	
Filtrate:	7.4
Late proximal:	6.6
Urine, min.:	4.5
Urine, max.:	8.2

Fig. 18.3 ▶ **HCO₃⁻ reabsorption.**
The H^+ secreted into the lumen of the proximal tubule reacts with ~90% of the filtered HCO_3^- to form CO_2 and H_2O. Carbonic anhydrase (CA) anchored in the luminal membrane catalyzes this reaction. CO_2 readily diffuses into the cell via water channels (aquaporins). CA then catalyzes the transformation of $CO_2 + H_2O$ to $H^+ + HCO_3^-$ within the cell. The H^+ ions are again secreted into the lumen, while HCO_3^- exits through the basolateral membrane via an electrogenic $Na^+–HCO_3^-$ cotransporter. Thus, HCO_3^- is transported through the luminal membrane in the form of CO_2 (main driving force: concentration gradient for CO_2) and exits the basolateral membrane as HCO_3^- (main driving force: membrane potential). (NBC, Na^+–bicarbonate cotransporter)

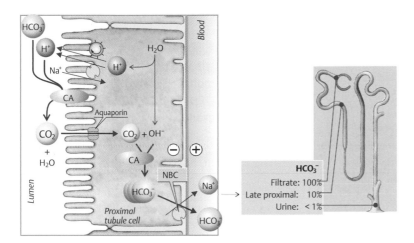

is transported across the basolateral membrane of proximal tubule cells and reabsorbed into peritubular capillary blood. The H^+ ions are secreted again into luminal fluid.

Factors Influencing Bicarbonate Reabsorption

Filtered Load of HCO₃⁻
The rate of HCO_3^- reabsorption from the lumen into peritubular capillary blood increases in proportion to the filtered load of HCO_3^-. It is possible that high concentrations of HCO_3^- in the filtrate raise tubular fluid pH, favoring more H^+ secretion (via H^+–ATPase and Na^+–H^+ antiporter) and thus more HCO_3^- reabsorption (as CO_2 combines with the additional H^+ ions in the lumen and is reabsorbed as HCO_3^-).

Extracellular Fluid Volume
— Expansion of ECF volume results in decreased Na^+ reabsorption from the tubular lumen tubular cells. This decreases Na^+-coupled H^+ secretion (via Na^+–H^+ antiporter) from tubular cells into the lumen and thus decreases HCO_3^- reabsorption.
— Similarly, contraction of ECF volume results in increased Na^+ reabsorption, increased H^+ secretion, and therefore increased HCO_3^- reabsorption. This is known as *contraction alkalosis* and is treated by giving an infusion of NaCl to correct ECF volume contraction and the metabolic alkalosis.

Pco₂
— High arterial P_{CO_2} will increase the rate of HCO_3^- reabsorption from the lumen into tubular cells because the elevated P_{CO_2} causes an increase in cellular H^+ production and secretion.
— Low arterial P_{CO_2} will similarly decrease the rate of HCO_3^- reabsorption from the tubular lumen into peritubular capillary blood.

Note: The dependence of HCO_3^- reabsorption on P_{CO_2} allows the kidney to respond to respiratory acidosis and alkalosis.

Ion Concentrations

— High plasma [Cl^-] decreases HCO_3^- reabsorption.
— High plasma [K^+] decreases H^+ secretion and therefore HCO_3^- reabsorption.

Hormones

— Angiotensin II increases H^+ secretion (via Na^+–H^+ antiporter) and HCO_3^- reabsorption in the proximal tubule; aldosterone has similar effects on the distal nephron. Corticosteroids (e.g., cortisol) will exert mineralocorticoid effects in high concentrations, causing HCO_3^- reabsorption.
— Parathyroid hormone (PTH) decreases HCO_3^- reabsorption.

Formation of "New" Bicarbonate

H^+ produced within tubular cells is secreted into tubular fluid. The secreted H^+ can combine with other non-HCO_3^- buffers in luminal fluid, namely, phosphates and ammonia (NH_3). The H^+ ions remain in luminal fluid as part of the buffer pairs and are later excreted. The HCO_3^- that is formed within renal cells at the same time is then transported across the basolateral membrane into peritubular capillary blood by a Cl^-–HCO_3^- exchanger. Therefore, for every H^+ that is secreted and combines with non-HCO_3^- buffers, a new moiety of HCO_3^- is formed within renal cells and added to body fluids.

Excretion of Excess H+

Excretion of Excess H+ as Titratable Acid

Titratable acid is acid in the urine formed by the protonation of filtered buffer bases. Its amount can be quantified by titrating urine back to pH 7.4 with a strong base (sodium hydroxide [NaOH]). Secreted H^+ ions combine with filtered buffer bases, principally divalent phosphate (HPO_4^{2-}), to form acids, for example, monovalent phosphate ($H_2PO_4^-$), which are then excreted, as shown below.

$$H^+ \text{ (secreted)} + \text{buffer base (filtered)} \rightarrow \text{buffer acid (excreted)}$$

The supply of filtered buffer base available to be protonated is limited and cannot be upregulated. The only means of increasing H^+ excretion on filtered buffers is to lower the urinary pH, which converts a higher fraction of the buffer to the protonated form. However, the kidneys cannot produce urine with a pH lower than ~4.4, because H^+ ions within the tubular lumen leak back into tubular cells. Therefore, when large amounts of H^+ have to be excreted, the kidneys use a different mechanism, namely, ammonium (NH_4^+).

Excretion of Excess H+ as Ammonium

The metabolism of amino acids by the liver produces urea (mostly) and glutamine (**Fig. 18.4**). Urea is excreted, but glutamine is reabsorbed by cells of the proximal tubule and metabolized to NH_4^+ and HCO_3^-. The HCO_3^- is returned to the circulation via peritubular capillaries, and the NH_4^+ is secreted into the luminal fluid. Much of the NH_4^+ is reabsorbed in the thick ascending limb of the loop of Henle and accumulates in the medullary interstitium. It is then secreted again in the medullary collecting ducts and excreted. The overall process is shown below.

$$\text{Glutamine (from liver)} \rightarrow NH_4^+ \text{ (excreted)} + HCO_3^- \text{ (returned to circulation)}$$

Because NH_4^+ is a very weak acid ($pK_a = 9.2$), a small fraction exists as neutral ammonia (NH_3), which is relatively permeable. Thus, some NH_4^+ transport actually consists of NH_3 transport in parallel with proton transport, but the end result is the same.

During times of large acid loads, the hepatic production of glutamine is upregulated, thereby increasing the supply of substrate from which the kidneys synthesize NH_4^+. Large acid loads are therefore excreted chiefly in the form of NH_4^+.

Fig. 18.4 ▶ Secretion and excretion of NH$_4^+$ ↔ NH$_3$.
Excretion of ammonium ions (NH$_4^+$ ↔ NH$_3$) is equivalent to H$^+$ disposal and is therefore an indirect form of
H$^+$ excretion (**1**). For every NH$_4^+$ excreted by the kidney, one HCO$_3^-$ is spared by the liver. This is equivalent
to one H$^+$ disposed because the spared HCO$_3^-$ can buffer an H$^+$ ion. The liver utilizes NH$_4^+$ and HCO$_3^-$ to
form urea. Thus, one HCO$_3^-$ less is consumed for each NH$_4^+$ that passes from the liver to the kidney and
is eliminated in the urine. Before exporting NH$_4^+$ to the kidney, the liver incorporates it into glutamate,
yielding glutamine; only a small portion reaches the kidneys as free NH$_4^+$. In the kidneys, glutamine
enters proximal tubule cells by Na$^+$ symport and is cleaved by mitochondrial glutaminase, yielding NH$_4^+$
and glutamate (Glu$^-$). Glu$^-$ is further metabolized by glutamate dehydrogenase to yield α–ketoglutarate,
producing a second NH$_4^+$ ion (**2**). The NH$_4^+$ can reach the lumen in two ways: it dissociates within the cell
to yield NH$_3$ and H$^+$, allowing NH$_3$ to diffuse into the lumen, where it rejoins the separately secreted H$^+$
ions; or the NHE-3 carrier secretes NH$_4^+$ (instead of H$^+$). Once NH$_4^+$ has reached the thick ascending loop
of Henle, the bumetanide-sensitive Na$^+$–K$^+$–2Cl$^-$ cotransporter (BSC) carrier reabsorbs NH$_4^+$ (instead of
K$^+$) so that it remains in the medulla. Recirculation of NH$_4^+$ through the loop of Henle yields a very high
concentration of NH$_4^+$ ↔ NH$_3$ + H$^+$ toward the papilla (**3**). While the H$^+$ ions are then actively pumped out
into the lumen of the collecting duct (**4**), the NH$_3$ molecules arrive there by nonionic diffusion and possibly
by an NH$_3$ transporter. The NH$_3$ gradient required to drive this diffusion can develop because the low luminal pH (~4.5) leads to a much smaller NH$_3$ concentration in the lumen than in the medullary interstitium,
where the pH is about two pH units higher and the NH$_3$ concentration is consequently about 100 times
higher than in the lumen.

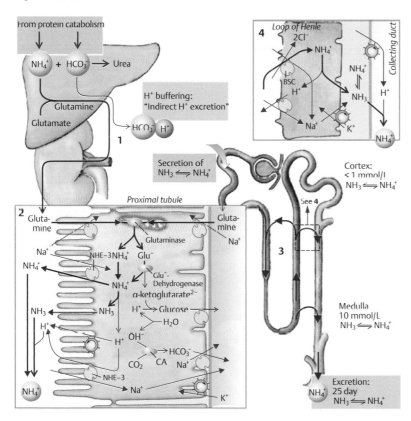

Under normal conditions, all filtered HCO$_3^-$ is reabsorbed, and an additional 40 to 60 mmol of
acid is excreted in the form of titratable acid and NH$_4^+$ per day. This contributes 40 to 60 mmol of
new HCO$_3^-$ to blood and replenishes the HCO$_3^-$ used to buffer the acid produced from metabolism.

— During acidosis the kidneys compensate by excreting more acidic urine (urine is usually
   slightly acidic, about pH 6) still completely reabsorbing HCO$_3^-$ and increasing the excretion
   of NH$_4^+$.

— During alkalosis cell pH rises, providing less driving force for H$^+$ secretion, so less HCO$_3^-$ is
   reabsorbed. The unabsorbed HCO$_3^-$ will alkalinize the urine. Less NH$_4^+$ is formed and less
   acid excreted. Overall, more HCO$_3^-$ will be eliminated from the body, and the body fluid will
   become more acidic.

Quantitation of Renal Tubular Acid Secretion and Excretion

- Total rate of H^+ secretion = rate of HCO_3^- reabsorption + rate of titratable acid excretion + rate of NH_4^+ excretion
- Total rate of H^+ excretion = rate of titratable acid excretion + rate of NH_4^+ excretion = total rate at which "new" HCO_3^- is added to the blood

18.4 Acid–base Disturbances

Arterial pH < 7.36 is acidosis; arterial pH > 7.44 is alkalosis. The causes of such acid–base disturbances are illustrated in **Figs. 18.5** and **18.6.**

- Respiratory disturbances change $[H^+]$ by a primary change in PCO_2 because PCO_2 is regulated by the rate of alveolar ventilation.
- Metabolic disturbances primarily change the $[HCO_3^-]$ by the addition or loss of fixed acids or bases derived from metabolic processes.

The efficiency of compensatory responses is indicated by how close arterial pH is brought back to 7.4. Metabolic disturbances are compensated almost instantaneously. Primary respiratory disturbances require several days for compensation.

Metabolic Acidosis

Metabolic acidosis results from overproduction or abnormal retention of fixed acids, or from loss of bases.

- Arterial $[H^+]$ is increased, and arterial HCO_3^- is decreased.
- Respiratory compensation for the acidosis involves hyperventilation, which reduces the PCO_2.
- Renal compensation involves virtually complete reabsorption of HCO_3^-, which replenishes that used to buffer the excess acid and an increase in the excretion of titratable acid and NH_4^+. The availability of titratable acid is very limited, but the kidneys can greatly increase production of NH_4^+.

Metabolic Alkalosis

Metabolic alkalosis results from loss of fixed acids or from excessive intake or retention of bases.

- Arterial $[H^+]$ is decreased, and $[HCO_3^-]$ is increased.
- Respiratory compensation involves hypoventilation causing a rise in PCO_2. However, the compensatory rise in PCO_2 will tend to increase H^+ secretion from peritubular capillary blood into the tubular lumen and HCO_3^- reabsoption from the tubular lumen into peritubular capillary blood, thereby limiting its effectiveness.
- Renal response involves increased renal excretion of HCO_3^-.

Respiratory Acidosis

Respiratory acidosis results from failure of the lungs to adequately expire CO_2 due to hypoventilation.

- Arterial PCO_2 is increased, which causes an increase in $[H^+]$ and $[HCO_3^-]$.
- There cannot be respiratory compensation for acidosis of respiratory origin.
- Renal compensation includes increased H^+ excretion as titratable acid and NH_4^+, increased HCO_3^- reabsorption, and production of "new" HCO_3^-.

■ **Serum anion gap**

The serum anion gap represents unmeasured anions in serum. These include sulfates, phosphate, citrate, and proteins.

It is calculated as follows: serum anion gap = (Na^+) – $([Cl^-] + [HCO_3^-])$. The normal value of the serum anion gap is 12 mEq/L (range 8–16 mEq/L). The serum anion gap is normal when the loss of HCO_3^- is almost fully compensated for by an increase in Cl^-. This is known as *hyperchloremic metabolic acidosis*. A high serum anion gap indicates acidosis where the loss of HCO_3^- (used to buffer the fixed acid) is not compensated by an increase in Cl^-. Electroneutrality is maintained by an increase in the levels of unmeasured anions (e.g., ketoacids). A low serum anion gap is usually caused by hypoalbuminemia. Albumin is a negatively charged protein; its loss leads to the retention of HCO_3^- and Cl^-; hence, the anion gap is reduced.

Fig. 18.5 ► Causes of alkalosis.

Metabolic activity may cause the accumulation of organic acids (e.g., lactic acid and ketoacids) (**1**). One H^+ is produced per acid. If these acids are metabolized, H^+ disappears again. Consumption of the acids can cause alkalosis. Mobilization of alkaline salts from bone, for example, during immobilization, can cause alkalosis (**2**). Respiratory alkalosis occurs in hyperventilation (e.g., due to damage to respiratory neurons, high altitude, or salicylate poisoning) (**3**). Numerous disorders can lead to metabolic alkalosis. In hypokalemia (low blood $[K^+]$), the chemical gradient for K^+ across cell membranes is increased. In some cells, this leads to hyperpolarization, which drives more negatively charged HCO_3^- from the cell. Hyperpolarization, for example, raises HCO_3^- efflux from the proximal tubule via $Na^+–HCO_3^-$ cotransport (**4**). The resulting intracellular acidosis stimulates the luminal $Na^+–H^+$ exchange and thus promotes H^+ secretion, as well as HCO_3^- production, in the proximal tubule cell. Both processes lead to alkalosis. Aldosterone is released in hypovolemia, stimulating H^+ secretion in the distal nephron (**5**). Thus, the kidney's ability to eliminate HCO_3^- is compromised, resulting in alkalosis. In vomiting of stomach contents, the body loses H^+ (**6**). What is left behind is the HCO_3^- produced when HCl is secreted in parietal cells. Normally, the HCO_3^- formed in the stomach is reused in the duodenum to neutralize the acidic stomach contents and only transiently leads to a weak alkalosis. In liver failure, hepatic production of urea is decreased, the liver uses up less HCO_3^-, and alkalosis develops (**7**). Reduced protein breakdown, for example, as a result of a protein-deficient diet, reduces the metabolic formation of H^+ and thus favors the development of alkalosis (**8**).

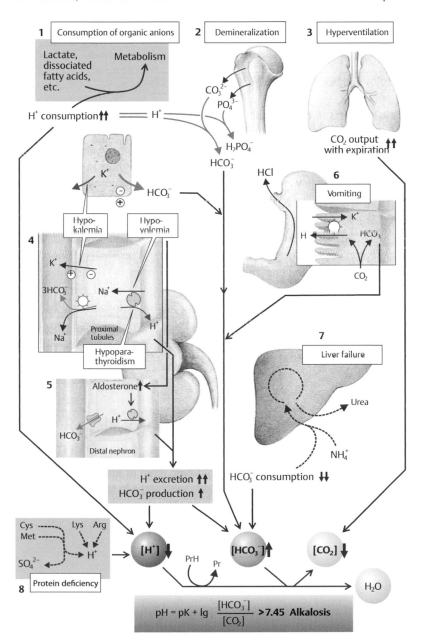

$$pH = pK + lg \frac{[HCO_3^-]}{[CO_2]} > 7.45 \text{ Alkalosis}$$

Renal tubular acidosis type 1

Renal tubular acidosis (RTA) type 1 occurs when the tubules fail to excrete titratable acid and NH_4^+, and there is a failure to create acidic urine. It presents in childhood with increased urination (polyuria), failure to thrive, and bone pain. Adults present with bone pain (osteomalacia), renal stones, constipation, weakness (due to K^+ depletion), or renal failure. Tests will show urine pH > 6 (normally, urine pH ≤ 6), and there is a normal anion gap. Treatment is with HCO_3^- and K^+ replacement.

Renal tubular acidosis type 2

RTA type 2 occurs when there is renal loss of HCO_3^- and serum $[HCO_3^-]$ falls until the filtered load equals the reduced resorptive capacity, allowing the urine to become acidified. It presents in childhood with failure to thrive, metabolic acidosis, and alkaline or slightly acid urine. The anion gap is normal. Treatment is with HCO_3^- replacement.

Renal tubular acidosis type 4

RTA type 4 occurs in diseases associated with aldosterone deficiency (e.g., Addison disease and diabetes mellitus). There is failure to excrete NH_4^+ because hypokalemia (caused by the aldosterone deficiency) inhibits NH_3 synthesis. The anion gap is normal. Treatment (if aldosterone deficient) is with corticosteroid replacement (e.g., fludrocortisone) or a loop diuretic (e.g., furosemide). The underlying disease process should also be treated appropriately.

Acute renal failure

Acute renal failure may occur due to disease of the kidneys themselves, which may be vascular, septic, neoplastic, due to drugs, or due to pregnancy. Extrarenal causes include burns, sepsis, trauma, heart failure, and obstruction. Acute renal failure produces a sharp rise in urea, creatinine, K^+ (hyperkalemia), and Na^+ (hypernatremia) usually with oliguria (low urine output) or anuria (no urine output). There may also be vomiting, confusion, bruising, or gastrointestinal bleeding. A metabolic acidosis (usually with a normal anion gap) will occur due to the failure to excrete H^+ as titratable acid and NH_4^+. Treatment should be aimed at the underlying cause, but hyperkalemia may require urgent correction to avoid cardiac complications.

Fig. 18.6 ▶ **Causes of acidosis.**
Acidosis can develop when there is increased formation or decreased breakdown of organic acids (**1**). One H^+ is produced per acid. Mineralization of bone favors the development of acidosis (**2**). Many primary and secondary diseases of the respiratory system, as well as abnormal regulation of breathing, can lead to respiratory acidosis (**3**). In hyperkalemia (elevated blood K^+ levels), the chemical gradient across the cell membrane is reduced (**4**). The resulting depolarization diminishes the electrical driving force for HCO_3^- transport out of the cell. It slows the efflux of HCO_3^- in the proximal tubules via $Na^+–HCO_3^-$ cotransport. The resulting intracellular acidosis inhibits the luminal Na^+/H^+ exchange and thus impairs H^+ secretion as well as HCO_3^- production in the proximal tubule cells. Both processes lead to acidosis. Hypoaldosteronism reduces the renal excretion of H^+ and HCO_3^- production (**5**). Diarrhea causes a loss of HCO_3^- from the gut (**6**). The liver requires two HCO_3^- ions when incorporating two molecules of NH_4^+ in the formation of urea; thus, increased urea production can lead to acidosis (**7**). A protein-rich diet promotes the development of acidosis as amino acid breakdown generates H^+ ions (**8**).

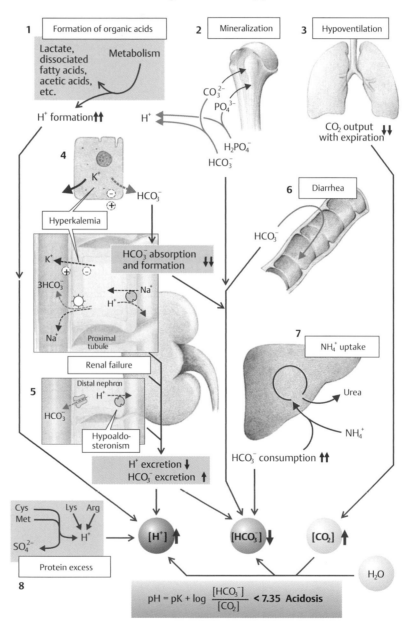

Respiratory Alkalosis

Respiratory alkalosis results from excessive loss of CO_2 due to hyperventilation. Anxious patients presenting with hyperventilation can be treated by having them rebreathe into a paper bag.

— Arterial PCO_2 is decreased, which causes a decrease in $[H^+]$ and $[HCO_3^-]$.
— There cannot be respiratory compensation for alkalosis of respiratory origin.
— Renal compensation includes decreased excretion of H^+ as titratable acid and NH_4^+ and decreased HCO_3^- reabsorption.

Table 18.1 summarizes acid–base disturbances and their compensation/correction.

Table 18.1 ▸ Summary of Acid–base Disturbances			
Acid–base Disorder	Primary Disturbance	Respiratory Compensation	Renal Correction
Metabolic acidosis	↓ HCO_3^-	Hyperventilation (↓ CO_2)	↑ excretion of titratable acid ↑ excretion of $NH4^+$ ↑ production of "new" HCO_3^-
Metabolic alkalosis	↑ HCO_3^-	Hypoventilation (↑ CO_2)	↑ excretion of HCO_3^-
Respiratory acidosis	↑ CO_2	-	↑ excretion of titratable acid ↑ excretion of NH_4^- ↑ reabsorption of HCO_3^- ↑ production of "new" HCO_3^-
Respiratory alkalosis	↓ CO_2	-	↓ excretion of titratable acid ↓ excretion of NH_4^- ↓ reabsorption of HCO_3^-

Review Questions

1. A hyperosmotic solution of urea, a nontoxic substance, is injected intravenously (IV). Urea readily permeates most cell membranes via uniporters. After it has distributed throughout the body fluids,
   A. both intracellular (ICF) and extracellular fluid (ECF) volumes will be decreased.
   B. ECF volume will be decreased.
   C. the osmolality of ICF and ECF will be increased.
   D. the osmolality of the ICF will be decreased.
   E. ICF volume will be increased.

2. Which is the main factor that determines the amount of water, and therefore volume, of the ECF compartment?
   A. Na^+ concentration
   B. Na^+ content
   C. K^+ concentration
   D. K^+ content
   E. Cl^- concentration

3. If a patient hemorrhages 1 L of blood in 5 minutes, you would restore the body fluid balance by giving 1 L of which of the following?
   A. 5% glucose IV
   B. 5% glucose intraperitoneally
   C. 0.9% NaCl IV
   D. 4.5% NaCl IV
   E. Distilled water IV

4. During severe exertion in a hot environment, a person may lose 4 L of hyposmotic sweat per hour. This would result in
   A. decreased plasma volume.
   B. decreased plasma osmolality.
   C. decreased plasma ADH.
   D. decreased plasma aldosterone.
   E. decreased plasma renin.

5. Renal autoregulation of blood flow and glomerular filtration rate (GFR) require
   A. autonomic innervation.
   B. epinephrine.
   C. aldosterone.
   D. ADH.
   E. no action of extrarenal factors.

6. The following data are provided for a substance X. These data are consistent with X being which substance?
   Plasma concentration of X = 0.02 mg/100 mL
   Urine concentration of X = 12 mg/mL
   Urine flow rate = 1 mL/min
   GFR = 125 mL/min
   A. Inulin
   B. Para-aminohippuric acid (PAH)
   C. Sodium ions
   D. Glucose
   E. Alanine

7. A patient with a serum creatinine of 0.8 mg/dL submits a 12-hour urine collection containing 400 mg of creatinine. What is his creatinine clearance in mL/min? (*Note:* Answering this question requires some unit conversions.)
   A. 50 mL/min
   B. 69 mL/min
   C. 100 mL/min
   D. 145 mL/min
   E. 800 mL/min

8. After a patient has a renal transplant, the serum creatinine is 0.6 mg/dL. Six months after the transplant, it rises to 1.2 mg/dL. This last change is most likely due to
   A. a significant decrease in the GFR.
   B. the daily fluctuation of serum creatinine within the normal range.
   C. a rise in creatinine production.
   D. impairment of protein synthesis due to corticosteroids (for immunosuppression).
   E. increase in efferent arteriolar resistance.

9. If ouabain, a drug that inhibits Na^+–K^+ ATPase, is infused into the renal artery,
   A. urine flow will decrease.
   B. there will be an osmotic diuresis.
   C. renal glucose reabsorption will not be affected.
   D. the [Na^+] within proximal tubular cells will decrease.
   E. K^+ reabsorption will increase.

10. Given the data below, what is the rate of excretion of substance X in the urine? Substance X is freely filtered.
    GFR = 125 mL/min
    Plasma concentration of X = 2 mg/mL
    Tubular reabsorption of X = 30 mg/min
    Tubular secretion of X = 60 mg/min
    A. 160 mg/min
    B. 220 mg/min
    C. 250 mg/min
    D. 280 mg/min
    E. 340 mg/min

11. Filtered Cl^- is reabsorbed by a multiporter with K^+ and Na^+ in the
    A. proximal tubule.
    B. proximal tubule and descending limbs of the loop of Henle.
    C. proximal tubule and ascending limb of the loop of Henle.
    D. ascending limb of the loop of Henle.
    E. distal tubule and collecting duct.

12. Which action serves to conserve blood volume?
    A. Suppression of ADH secretion
    B. Increased renin secretion
    C. Increased renal filtration fraction
    D. Increased GFR
    E. Increased renal medullary blood flow

13. A 49-year-old woman complains of weakness, easy fatigability, and loss of appetite. During the past year, she lost 15 lb. Her blood pressure is 80/50 mm Hg, her serum $[Na^+]$ is 130 mEq/L (normal 135–145 mEq/L), and her serum $[K^+]$ is 6.5 mEq/L (normal 3.5 to 5.0 mEq/L). This patient is diagnosed as having Addison disease (decreased levels of adrenal corticosteroids). One possible cause for her hyperkalemia is
    A. decreased K^+ secretion by the cortical collecting duct.
    B. increased K^+ reabsorption by the proximal tubule.
    C. increased Na^+ reabsorption by the collecting duct.
    D. increased volume flow rate of tubular fluid along the nephron.
    E. increased levels of serum aldosterone.

14. During administration of aldosterone, you expect
    A. elevated plasma renin concentration.
    B. reduced 24-hour urinary Na^+.
    C. reduced plasma volume.
    D. reduced blood pressure.
    E. increased urine volume.

15. A 45-year-old man with a blood pressure of 150/90 mm Hq for the past 6 months is stabilized on a diet with 200 mEq Na^+ per day. One day after a dose of a short-acting powerful diuretic drug, you would expect
    A. reduced plasma renin concentration,
    B. elevated urinary Na^+.
    C. reduced plasma volume.
    D. elevated blood pressure.
    E. increased urine volume.

16. A healthy adult with a creatinine clearance of 110 mL/min has a K^+ clearance of 65 mL/min. After administration of a new drug, K^+ clearance increases to 130 mL/min with no change in creatinine clearance. The drug could have produced this change in K^+ clearance by
    A. stimulating K^+ reabsorption in the proximal tubule.
    B. inhibiting aldosterone secretion.
    C. stimulating K^+ reabsorption in the distal nephron.
    D. inhibiting K^+ secretion in the distal nephron.
    E. stimulating K^+ secretion in the distal nephron.

17. The most significant contribution of the loop of Henle to the process of urine concentration and dilution is
    A. the production of hyperosmotic tubular fluid.
    B. serving as the site where ADH controls water reabsorption.
    C. acting as a countercurrent exchanger that establishes the medullary osmotic gradient.
    D. actively reabsorbing Na^+ and Cl^-.
    E. reabsorption of urea.

18. ADH conserves water by
    A. constricting afferent arterioles, thereby reducing the GFR.
    B. increasing water reabsorption by the proximal tubule.
    C. stimulating active reabsorption of solutes in the descending limb of the loop of Henle.
    D. increasing water permeability of the collecting duct.
    E. blocking urea secretion in the loop of Henle.

19. Which of the following actions of a drug is possible if treatment of a patient causes formation of a large volume of urine with an osmolality of 300 mOsm/kg H_2O?
    A. Inhibition of renin secretion
    B. Increase of ADH secretion
    C. Increased permeability to water in distal nephron and collecting ducts
    D. Decreased active Cl^- reabsorption by the ascending limb of the loop of Henle
    E. Inhibition of aldosterone secretion

20. A 60-year-old man was brought unconscious to the emergency room. A neighbor states that the patient has not felt well over the past week and looks as though he has lost weight lately. Which diagnosis fits if his plasma Na^+ and plasma osmolality are reduced, and his kidneys excrete concentrated urine?
   A. Dehydration
   B. Excessive production/release of ADH
   C. Water intoxication
   D. Diabetes insipidus
   E. Diabetes mellitus

21. The kidney excretes acid loads mainly by
   A. excreting filtered HCO_3^-.
   B. secreting titratable acid
   C. a high urinary content of free H^+.
   D. raising urine pH via H^+ secretion.
   E. excreting NH_4^+.

For questions **22** and **23**, a patient suffering from chronic lung disease has an arterial pH of 7.37 and an arterial P_{CO_2} of 54 mm Hg.

22. His condition is a primary
   A. respiratory acidosis.
   B. respiratory alkalosis.
   C. metabolic acidosis.
   D. metabolic alkalosis.
   E. normal acid–base status.

23. One would expect that his arterial
   A. plasma HCO_3^- is lower than normal.
   B. plasma HCO_3^- is higher than normal.
   C. plasma H^+ is lower than normal.
   D. plasma dissolved CO_2 concentration is lower than normal.
   E. total CO_2 is lower than normal.

24. Renal compensation for respiratory acidosis
   A. is a result of decreased plasma P_{CO_2}.
   B. includes hyperventilation.
   C. requires renal generation of HCO_3^-.
   D. is a rapid process that keeps up in time with changes in P_{CO_2}.
   E. does not affect urinary $[NH_4^+]$ excretion.

25. Which of the following values characterizes uncompensated respiratory alkalosis?

	pH	P_{CO_2} (mm Hg)	$[HCO_3^-]$ (mM)
A.	7.40	60	37
B.	7.25	66	28
C.	7.51	29	22
D.	7.60	42	37
E.	7.25	35	15

26. The primary factor that usually determines the long-term (over days) output of urine via the kidney is
   A. arterial blood pressure.
   B. cardiac output (CO).
   C. plasma concentration of antidiuretic hormone (ADH).
   D. central venous pressure.
   E. the volume of total fluid intake.

Answers and Explanations

1. **C.** The urea distributes at equal concentrations in the ICF and ECF and increases osmolality in both compartments (D) (p. 156).
   A,B,E Assuming the injected volume is not significant, the volumes of the compartments do not change because there is no shift of water from one to the other.

2. **B.** Because the osmolality of the body fluids is held nearly constant, the status of Na^+ balance determines the volume of the ECF. Volume = amount of solute/osmolality (p. 155).
   A Na^+ concentration is normally controlled within a narrow range even as Na^+ content varies considerably.
   C,D Potassium is the major intracellular electrolyte.
   E Chloride is distributed close to its equilibrium between intra- and extracellular compartments. Because of the negative membrane potential, intracellular chloride concentration is much lower than extracellular chloride concentration.

3. **C.** Hemorrhage is isosmotic volume depletion. Isotonic NaCl (0.9%) restores both water and ions (p. 156).
   A,B Glucose is metabolized, so it does not restore ion concentrations.
   D A large amount of concentrated NaCl would cause a shrinkage of cells.
   E Water alone is not given IV because it causes lysis of red blood cells.

4. **A.** Sweat contains K^+, Na^+, and Cl^-. Therefore, there is a net loss of water, K^+, Na^+, and Cl^-. Plasma volume would be smaller (p. 156).
   B,C Because the loss of water is relatively greater than the loss of solute, there would be an increase in plasma osmolality stimulating the release of ADH.
   D,E Net Na^+ and volume loss will result in increased plasma levels of renin and aldosterone.

5. **E.** Renal autoregulation is an intrinsic property of the kidney. Afferent arterioles change their resistance proportionally to blood pressure, so renal blood flow and glomerular filtration rate change only slightly (p. 158).
A–D The other answer choices are not requirements for GFR and renal autoregulation.

6. **B.** Clearance of X = (1 mL/min × 12 mg/mL)/0.02 mg/mL = 600 mL/min. Because the volume of plasma being cleared of X is greater than the GFR, the substance must have been secreted into the tubule and so could be PAH (pp. 159, 161).
A The clearance of insulin is equal to its GFR.
C–E These are reabsorbed.

7. **B.** Clearance = excretion rate/plasma concentration. Creatinine excretion/min = 400 mg/(12 h × 60 min/h) = 0.55 mg/min. Plasma creatinine concentration = 0.8 mg/dL = 0.008 mg/mL. Creatinine clearance = 0.55 mg/min/0.008 mg/mL = 69 mL/min (pp. 159–160).

8. **A.** Deterioration in the GFR by about half doubles serum creatinine (p. 160).
B The normal range of daily fluctuation is much less than a change by a factor of 2.
C Creatinine production will rise only if there is an increase in muscle mass. For serum creatinine to double, the person must have doubled his muscle mass, an unlikely situation.
D Serum creatinine is not related to protein synthesis.
E An increase in resistance of efferent arterioles increases the GFR, which would lower the serum creatinine.

9. **B.** Ouabain, a glycoside poison, binds to and inhibits the action of the Na^+–K^+ pump in the cell membrane. Blocking reabsorption of Na^+ will leave excess Na^+ and water in the proximal tubule, which will be passed along to the rest of the nephron and produce an osmotic diuresis (p. 162).
A Urine flow will increase.
C This will stop reabsorption of organic substances, including glucose, that depend on secondary active transport.
D The concentration of $[Na^+]$ within proximal tubular cells will increase because it is not pumped out.
E Reabsorption of K^+ in the proximal tubule will decrease because it depends on the reabsorption of water to increase its luminal concentration.

10. **D.** Amount excreted = filtered load + amount secreted – amount reabsorbed. Filtered load = GFR × plasma concentration. Thus, 125 mL/min × 2 mg/mL + 60 mg/min – 30 mg/min = 280 mg/min.

11. **D.** The Na^+–K^+–$2Cl^-$ (NKCC) multiporter is expressed abundantly in the ascending limb of the loop of Henle (p. 166).
A–C Some Cl^- is reabsorbed actively in the proximal tubule via a different process.
E Cl^- is not reabsorbed actively in the collecting ducts.

12. **B.** Renin stimulates production of angiotensin II and aldosterone, which reduce salt and water excretion (p. 168).
A Less ADH secretion increases excretion of water.
C,D Increased GFR or increased filtration fraction lead to greater excretion, and therefore decreased blood volume.
E Increased medullary blood flow reduces medullary osmolality and leads to greater water excretion.

13. **A.** K^+ secretion is decreased by a decrease in plasma aldosterone, causing retention of K^+ and hyperkalemia (p. 172).
B Urinary K^+ excretion is determined mostly by K^+ secretion into the cortical collecting duct, not by proximal K^+ reabsorption.
C Reduced aldosterone also decreases Na^+ reabsorption.
D Low blood pressure and loss of volume will reduce flow rate.
E Aldosterone is a corticosteroid, and its secretion is decreased in Addison disease.

14. **B.** Aldosterone causes Na^+ retention, so urine Na^+ is reduced (p. 168).
A Renin output is slightly suppressed by hypervolemia, or high blood volume.
C,D Plasma volume expands due to Na^+ retention, and blood pressure rises.
E Aldosterone, by causing increased Na^+ reabsorption, reduces water excretion.

15. **C.** The diuretic causes loss of Na^+ and contraction of plasma volume, resulting in slightly reduced blood pressure (not D) (p. 169).
A,B,E Once the diuretic is no longer acting, the kidneys reabsorb more Na^+. Urinary Na^+ and water excretion now decrease, and the renin level rises.

16. **E.** A rise in K^+ clearance without a change in creatinine clearance indicates increased secretion of K^+, not a decrease (D). K^+ is secreted in the distal nephron, not reabsorbed in the proximal tubule or distal nephron (A,C) (pp.171–173).
B Less aldosterone decreases K^+ secretion in the distal nephron.

17. **D.** The loop of Henle as a whole actively removes salt from the tubular fluid and deposits it in the outer medullary interstitium (pp. 178–179).
A The tubular fluid that leaves the loop of Henle is hyposmotic.
B ADH controls water reabsorption in the collecting ducts.
C Countercurrent exchange occurs in the vasa recta.
E Urea is secreted into the loop of Henle (urea recycling).

18. **D.** ADH increases permeability of the collecting duct to water (p. 178).
A,B ADH at normal levels has minimal effect on arterioles or the proximal tubule.
C Although the main action of ADH involves water reabsorption, it also stimulates salt reabsorption in the thick ascending limb.
E By stimulating urea reabsorption in the medullary collecting ducts, ADH promotes urea secretion in the loop of Henle.

19. **D.** Decreased Cl^- reabsorption by the NKCC multiporter in the loop of Henle decreases dilution of the tubular fluid, reduces net solute reabsorption, and diminishes the corticomedullary osmotic gradient. The result is a diuresis with a nearly isosmotic urine (p. 179).
A,E Inhibition of renin and aldosterone secretion inhibits Na^+ reabsorption, but not in the loop of Henle, so urine could still be hyperosmotic.
B Increase of ADH secretion increases water reabsorption in the distal nephron and medullary collecting ducts, producing a concentrated urine.
C Increased permeability to water produces a concentrated urine.

20. **B.** Excessive ADH promotes excessive water reabsorption, so urine is concentrated, and plasma Na^+ and osmolarity are reduced (p. 178).
A In dehydration, both plasma Na^+ and osmolarity are elevated.
C,D In either diabetes insipidus or water intoxication, the urine is dilute.
E In diabetes mellitus, the plasma would be hyperosmotic due to the excess glucose.

21. **E.** Large acid loads are excreted primarily in the form of $NH4^+$ (p. 188).
A When excreting acid loads, the kidneys reabsorb all filtered bicarbonate.
B Titratable acid is formed by protonating filtered base, not by secretion.
C Even with a low urinary pH, the amount of free, unbuffered H^+ is trivial.
D H^+ secretion lowers urine pH.

22. **A.** A high P_{CO_2} could be either a primary respiratory acidosis or compensation for a primary metabolic alkalosis. Because the pH is toward the low end of normal, this indicates a respiratory acidosis that is well compensated (therefore B–D incorrect) (p. 190).
E Although the pH is toward the low end of normal, the P_{CO_2} is above normal therefore the patient's acid–base status is not normal.

23. **B.** Elevation of P_{CO_2} causes a compensating increase in plasma HCO_3^- concentration, not a decrease (A) (p. 190).
C Plasma H^+ is on the high side of normal.
D Physically dissolved CO_2 is what determines P_{CO_2} and is high.
E Total CO_2 (dissolved CO_2 + bicarbonate) is high.

24. **C.** Raising plasma $[HCO_3^-]$ requires renal generation of HCO_3^- (p. 190).
A,B By definition, a respiratory acidosis is a high P_{CO_2} that exists because of hypoventilation.
D Renal compensation requires hours to days.
E The generation of new bicarbonate requires increased acid excretion, which occurs primarily as NH_4^+.

25. **C.** Uncompensated respiratory alkalosis is characterized by an alkaline pH, reduced P_{CO_2}, and slightly reduced HCO_3^- (pp. 192–193).
A This is compensated respiratory acidosis.
B This is respiratory acidosis.
D This is metabolic alkalosis.
E This is metabolic acidosis.

26. **E.** In a steady state, fluid intake must equal fluid output, and regulatory systems adjust output to match input.
A Urine output is not affected by the normal range of arterial pressures.
B CO has no direct influence on urine output.
C ADH is an important regulator that keeps plasma osmolality constant in the face of variable fluid intake.
D Central venous pressure is normally controlled within a small range, so it has little effect on urine output.

19 Structure and Regulation of the Gastrointestinal Tract

19.1 Structure

The gastrointestinal (GI) system, or alimentary canal, consists of the mouth, pharynx, esophagus, stomach, small intestine, cecum, colon, rectum, and associated secretory organs and glands (salivary glands, exocrine pancreas, gallbladder, and liver). **Figure 19.1** shows the GI organs and

Fig. 19.1 ▶ **Functions of gastrointestinal (GI) organs.**
The GI system extends from the mouth to the rectum and includes secretory organs and glands. The function of each structure and the transit time of the food bolus or chyme from food intake are shown.

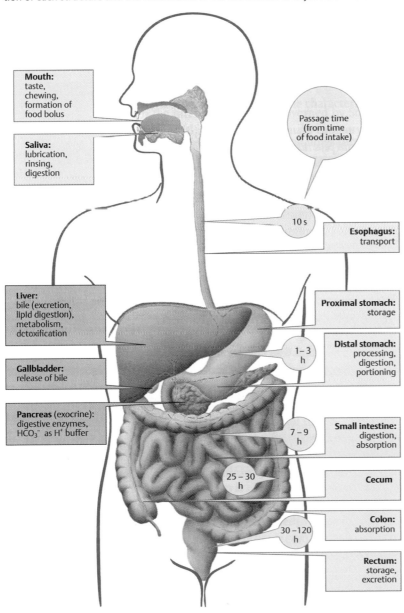

Mouth:
taste,
chewing,
formation of
food bolus

Saliva:
lubrication,
rinsing,
digestion

Passage time
(from time
of food intake)

10 s

Esophagus:
transport

Liver:
bile (excretion,
lipid digestion),
metabolism,
detoxification

Proximal stomach:
storage

Distal stomach:
processing,
digestion,
portioning

1–3
h

Gallbladder:
release of bile

Pancreas (exocrine):
digestive enzymes,
HCO_3^- as H^+ buffer

7–9
h

Small intestine:
digestion,
absorption

25–30
h

Cecum

Colon:
absorption

30–120
h

Rectum:
storage,
excretion

(After Kahle, Leonhardt & Platzer)

Embryology of the gastrointestinal tract

The GI tract and its associated organs (liver, gallbladder, and pancreas) develop from endoderm. The peritoneum, which covers the GI tract, develops from mesoderm. Note that the pancreas, duodenum (descending, horizontal, and some of the ascending part), ascending and descending colon, cecum (variable portions), and upper two-thirds of the rectum are secondarily retroperitoneal; that is, they were intraperitoneal when formed but became retroperitoneal during fetal development.

their individual general functions. However, their overall function is to obtain nutrients and water from the external environment.

Layers of the Gastrointestinal Tract

Working from the lumen outward, the layers of tissue found in most of the GI tract are the mucosa, submucosa, muscularis propria, and serosa (**Fig. 19.2**).

Mucosa

Mucosa is composed of epithelium, lamina propria, and muscularis mucosae.

— Stratified squamous epithelium is found in the esophagus and rectum. Columnar epithelium is found in the rest of the GI tract.

The mucosa in the stomach forms prominent folds (rugae) that serve to increase its surface area. The mucosa in the small intestine has villi and microvilli that serve to increase its surface area.

— The lamina propria is a layer of connective tissue.
— The muscularis mucosae is a layer of smooth muscle. Contraction of the muscularis mucosae alters the surface area for absorption or secretion.

Submucosa

The submucosa is a layer of loose connective tissue with collagen and elastin fibers.

Muscularis Propria (Externis)

The muscularis propria (externis) is a muscular layer composed of circular muscle and longitudinal muscle.

— These muscles are responsible for peristalsis.

Serosa (Adventitia)

The serosa is a fibrous layer that is continuous with the peritoneal lining.

Irritable bowel syndrome

IBS is a chronic idiopathic condition. Symptoms include abdominal pain, bloating, and cramps that are associated with bowel habit alteration in the form of constipation or diarrhea. Treatment is guided by the symptoms and their severity. Mild IBS may respond to dietary changes. Drugs may be called for in patients with moderate to severe symptoms. Antispasmodics (e.g., hyoscyamine and dicyclomine), laxatives (docusate, bisacodyl, senna, or osmotic agents), and loperamide are standard. In severe cases with diarrhea, alosetron, a potent and selective antagonist of the 5-HT$_3$ receptor that decreases intestinal motility and pain, may be used with caution, as it can lead to severe constipation. (*Note:* IBS is not associated with pathophysiological changes in gut structure and is diagnosed only when all else has been excluded.)

Fig. 19.2 ► **Wall structure and enteric nervous system of the GI tract.**
The enteric plexus is the portion of the autonomic nervous system that specifically serves all organs of the GI tract. The submucosal plexus is located between the submucosal layer and circular muscle. The myenteric plexus is located between the circular muscle and the longitudinal muscle. The enteric plexus is subject to modulation by the parasympathetic and sympathetic nervous systems.
From *Thieme Atlas of Anatomy, Head and Neuroanatomy,* © Thieme 2007, Illustration by Karl Wesker.

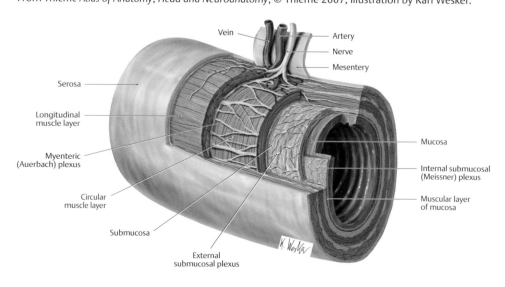

Immune Function of the Gastrointestinal Tract

The mass of immunocompetent lymphoid cells and tissue in the GI tract is equivalent to that found in the rest of the body. Immunocytes can be found in the mucosa and submucosa and within Peyer patches in the terminal ileum. These immunocytes include T and B lymphocytes, plasma cells, mast cells, macrophages, and eosinophils. When they encounter an antigen, they respond by secreting inflammatory mediators (e.g., histamine, prostaglandins, leukotrienes, and cytokines). These stimulate GI motility and the secretion of water and electrolytes, thus promoting expulsion of the offending antigen.

Table 19.1 summarizes the immune defenses of the GI tract.

Table 19.1 ► Immune Defenses of the Gastrointestinal (GI) Tract		
Regions/Organ of the GI Tract	**Immune Defense**	**Mechanism**
Mouth	Saliva contains mucins, immuno-globulin A (IgA), and lysozyme	Mucins prevent the penetration of pathogens. IgA facilitates antibody-dependent cell-mediated cytotoxicity, degranulation of eosinophils and basophils, and phagocytosis of antigens by macrophages, and it triggers the respiratory burst by polymorphonuclear leukocytes. Lysozyme damages bacterial cell walls.
Stomach	Gastric acid	Low pH is bactericidal.
Terminal ileum	Peyer patches absorb ingested antigens (via surface M cells) and pass them to plasma cells. Plasma cells then secrete IgA on subsequent exposure to the antigen.	Mechanism of IgA stated above
Intestines (generally)	Intestinal flora	Normal commensal flora inhibits the proliferation of pathogenic bacteria.
Liver (Kupffer cells)	Macrophages that secrete IgA	Mechanism of IgA stated above

19.2 Regulation of the Gastrointestinal Tract

Neural Control

The GI system is richly innervated by both extrinsic and intrinsic nerves. Extrinsic nerves are those that comprise the autonomic nervous system (ANS). Intrinsic nerves are those that comprise the enteric nervous system, a subdivision of the peripheral nervous system. The enteric nervous system relays information from the ANS and is also able to directly and independently regulate many GI functions.

Extrinsic Innervation (Autonomic Nervous System)

Extrinsic nerves provide for long reflexes, which coordinate activities at widely separated sites along the GI tract.

Parasympathetic division of the autonomic nervous system. In general, the parasympathetic division stimulates GI functions.

— The parasympathetic neural supply for the gut comes from the vagus and pelvic nerves.

  — The vagus nerve innervates the esophagus, stomach, pancreas, and proximal colon. Vagovagal reflexes are long reflexes in which both the afferent and efferent components are mediated by the vagus nerve.
  — The pelvic nerves innervate the distal colon, rectum, and anus.

— Preganglionic parasympathetic fibers synapse with nerves of the intrinsic (enteric) nervous system (the submucosal and myenteric plexus).
— Fibers then project from cell bodies in the ganglia of these plexuses to secretory cells, endocrine cells, and smooth muscle cells in the GI system.

Sympathetic division of the autonomic nervous system.   In general, the sympathetic division inhibits GI functions.

— Sympathetic preganglionic cholinergic nerves synapse in the prevertebral ganglia.
— Sympathetic postganglionic nerves innervate the GI tract from the celiac, superior mesenteric, and superior and inferior hypogastric plexus.
— Postganglionic sympathetic fibers synapse in the submucosal and myenteric plexus of the intrinsic nervous system.
— Some postganglionic fibers synapse directly with blood vessels and smooth muscle.
— Fibers then project from cell bodies in the ganglia of these plexus to secretory cells, endocrine cells, and smooth muscle in the GI tract.

Intrinsic Innervation (Enteric Nervous System)

The enteric nervous system possesses all the elements necessary for short reflex regulation of GI functions, that is, modification of motility and secretory activity by afferent and efferent nerves entirely within the GI tract. It is able to do this without modulation from the extrinsic nervous system, with the exception of the proximal esophagus and external anal sphincter.

Submucosal plexus (Meissner plexus)

— This is located in the submucosal layer between the mucosa and circular muscle in the wall of the GI tract.
— It principally controls GI secretions and blood flow.

Myenteric plexus (Auerbach plexus)

— This is located between the longitudinal and circular layers of smooth muscle in the wall of the GI tract.
— It principally controls the motility of GI smooth muscle.

Hormonal Control

GI hormones are released from endocrine cells in the mucosa of certain regions of the GI system, particularly the antrum of the stomach and the upper small intestine.

The major hormones secreted by the GI system are gastrin, secretin, glucagon-like peptide 1 (GLP-1), cholecystokinin, and glucose-dependent insulinotropic peptide (GIP). The control of their release and their actions are summarized in **Table 19.2.**

Table 19.2 ▶ Hormones in the Gastrointestinal Tract

Hormone and Site of Secretion	Stimulating Factors	Inhibiting Factors	Action(s)
Gastrin G cells in antrum of stomach	Distention of the stomach Peptides and amino acids in the stomach Vagal activation (via GRP)	H^+ in the lumen of the stomach Somatostatin	↑ H^+ and histamine secretion ↑ gastric mucosal growth ↓ gastric emptying
Secretin S cells of duodenum	H^+ and fatty acids in duodenum	—	↑ pancreatic growth, pancreatic enzyme, and HCO_3^- secretion ↑ biliary HCO_3^- secretion ↓ H^+ secretion by gastric parietal cells ↓ gastric emptying by decreasing gastric motility and increasing pyloric sphincter tone ↓ gastric mucosal growth
Glucagon-like peptide 1 (GLP-1) L cells of the distal intestine	Glucose and fructose Amino acids Fatty acids	—	↑ insulin secretion ↑ satiety ↓ gastric emptying ↓ H^+ secretion by gastric parietal cells
Cholecystokinin I cells of duodenum and jejunum	Fatty acids and amino acids in the duodenum	—	↑ pancreatic growth ↑ pancreatic enzyme secretion (major effect) ↑ HCO_3^- secretion (minor effect) ↑ gallbladder emptying (by causing sphincter of Oddi to relax and the gallbladder to contract) ↓ gastric emptying
Glucose-dependent insulinotropic peptide (GIP) K cells of duodenum	Fatty acids, amino acids, and glucose	—	↑ insulin secretion ↓ H^+ secretion by gastric parietal cells ↓ gastric motility ↓ gastric emptying

Abbreviation: GRP, gastrin-releasing peptide.

Paracrine Control

Paracrine substances are signaling molecules released by cells in the GI mucosa that diffuse through interstitial fluid to nearby target cells, where they exert their effects.

The two key paracrine substances of the GI system are histamine and somatostatin. The control of their release and their actions are summarized in **Table 19.3.**

Table 19.3 ▶ Paracrine Substances in the Gastrointestinal Tract

Paracrine Substance and Site of Secretion	Stimulating Factors	Inhibiting Factors	Action(s)
Histamine Enterochromaffin (H) cells in fundus of stomach	Gastrin Vagal activity	—	↑ H^+ secretion by gastric parietal cells
Somatostatin D cells throughout the GI system	H^+ in the lumen of the GI tract Calcitonin gene-related peptide (cGRP)	Vagal activity	↓ gastrin and histamine release, indirectly causing ↓ H^+ secretion

Neurocrine Control

Neurocrine substances are peptides synthesized by neurons of the GI tract. They are released from an axon following an action potential and diffuse across the synaptic cleft to their target tissue.

The three neurocrine substances in the GI tract are vasoactive intestinal peptide, gastrin-releasing peptide (bombesin), and enkephalins. The control of their release and their actions are summarized in **Table 19.4.**

Table 19.4 ▶ Neurocrine Substances in the Gastrointestinal Tract			
Neurocrine Substance and Site of Secretion	**Stimulating Factors**	**Inhibiting Factors**	**Action**
Vasoactive intestinal peptide (VIP) Secreted by neurons in the mucosa and smooth muscle of the GI tract	Distention of the stomach and small intestines Vagal activity	Sympathetic activity	↓ lower esophageal sphincter tone Relaxes proximal muscles of stomach, allowing for entrance of food ("receptive relaxation") ↑ water and electrolyte secretion in intestine
Gastrin-releasing peptide (bombesin) Released from postganglionic fibers of the vagus nerve that innervates the G cells of the stomach	—	—	↑ gastrin release from G cells
Enkephalins Secreted by neurons in the mucosa and smooth muscle of the GI tract	—	—	↓ motility and secretion in the small intestine ↑ contraction of colonic smooth muscle (due to suppression of the net inhibitory action of enteric neurons). This makes the colon more rigid and less able to propel luminal contents aborally. It also increases colonic transit time and decreases the secretion of water.

20 Gastrointestinal Motility

Gastrointestinal (GI) motility is the contractile activity responsible for mixing GI contents and for controlled propulsion of the contents distally.

Types of Muscle in the Gastrointestinal System

Striated muscle is found in the mouth, pharynx, upper esophagus, and external anal sphincter.

The remaining muscle tissue in the GI system is either circular or longitudinal smooth muscle, except in the stomach and gallbladder. The stomach has a layer of oblique muscle in addition to having circular and longitudinal muscle. The gallbladder is composed of smooth muscle, but the fibers are arranged in a reticulum (mesh).

Types of Contractions of Gastrointestinal Smooth Muscles

Phasic Contractions. Phasic contractions are cyclic contractions that permit mixing and propelling of GI contents. These types of contractions are important in the stomach and small intestine.

Tonic Contractions. Tonic contractions are continuous contractions that relax only under neural stimulation. The upper region of the stomach and the sphincters that control the flow of GI contents from one region of the GI tract to another demonstrate tonic contraction.

Slow Waves

Slow waves are oscillating waves of membrane depolarization that are not sufficient to completely depolarize the membrane and stimulate contractions, but that make it possible for contractions to be stimulated more easily by raising the membrane potential closer to threshold (making it less negative) (**Fig. 20.1**).

— Slow waves that act as pacemakers originate in the interstitial cells of Cajal.
— They are a feature of smooth muscle cells in the stomach, small intestine, and large intestine.

Frequency of Slow Waves

Slow waves determine the basic frequency of contractions throughout the GI system by facilitating membrane depolarization at a particular rate.

— Slow-wave frequency is modified by neural or hormonal activity. These influence the amplitude of the oscillation more than the frequency.
— The rate of slow waves is lowest in the stomach (3/min) and highest in the duodenum (12/min).

Fig. 20.1 ▶ **Slow waves and spikes.**
The intestine contains pacemaker cells (interstitial cells of Cajal) that communicate with smooth muscle cells. The membrane potential of smooth muscle cells oscillates between 10 and 20 mV every 3 to 15 seconds, producing slow waves (**1**). These slow waves make the membrane potential less negative and, when the threshold potential is reached (~40 mV), a series of action potentials (spike bursts) are fired, resulting in muscle contraction (**2**). Muscle spasms occur if the trough of the wave also rises above the threshold potential (**3**). The smooth muscle cells are nonexcitable when they are depolarized (~−20 mV) or hyperpolarized (~−70 mV).

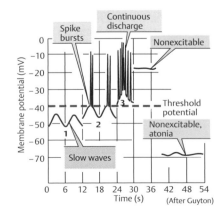

(After Guyton)

Fig. 20.2 ► **Peristaltic reflex.**
Stretching of the intestinal wall during the passage of a bolus triggers a reflex that simultaneously contracts the circular muscles behind the bolus and relaxes the circular muscles in front of it. At the same time, longitudinal muscles behind the bolus are relaxed, and those in front of it are contracted. This propels the bolus in an aboral direction.

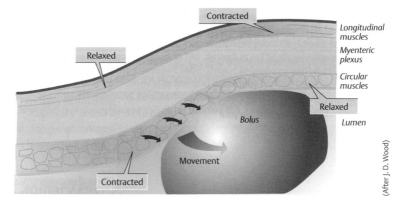

Peristalsis

Peristalsis is a sequential contraction of muscles in the GI system in response to stretching of the gut wall. The lumen is constricted immediately behind the bolus by the contraction of circular muscles, and there is relaxation of longitudinal muscles. Immediately in front of the bolus, there is relaxation of the circular muscles and contraction of the longitudinal muscles (Bayliss and Starling's law of the intestines). This creates a pressure gradient that forces the bolus to move distally.

This coordinated motor activity is called the *peristaltic reflex* and is mediated by the enteric nervous system (**Fig. 20.2**).

20.1 Chewing and Swallowing

Chewing

Food enters the GI system via the mouth, where mastication (chewing) begins the process of mechanical breakdown.

Saliva lubricates the food and contains some digestive enzymes, for example, salivary lipase (very minor action) and salivary amylase.

The chewed food mixes with saliva to form a bolus. The tongue propels the bolus of food to the back of the oral cavity, initiating a swallowing reflex.

Swallowing

Swallowing Reflex

The initial phase of swallowing is voluntary; the later phases are involuntary. Pressure receptors in the pharynx are stimulated by the presence of the bolus to send afferent signals, via the vagus and glossopharyngeal nerves, to the swallowing center in the medulla. Efferent signals are sent back from the medulla to striated muscles in the pharynx, larynx, and upper esophagus

■ Muscles of mastication

The muscles of mastication are the masseter, the temporalis, and the medial and lateral pterygoids. They are innervated by the mandibular division of the trigeminal nerve (cranial nerve [CN] V).

and to smooth muscles in the lower esophagus and stomach, coordinating a sequence of involuntary events that bring food from the mouth to the stomach via the pharynx and esophagus.

Pharyngeal phase of swallowing.   The act of the tongue pushing the bolus into the pharynx elevates the soft palate and uvula to seal off the nasal cavity. Respiration is momentarily inhibited, the larynx is raised, and the epiglottis seals over the glottis so that food cannot enter the trachea. Peristalsis of the striated muscles of the pharynx force the bolus through the pharynx toward the esophagus. At the same time, the upper esophageal sphincter (UES) relaxes, allowing the bolus to pass, then closes to prevent regurgitation. Pressure in the resting esophagus parallels intrapleural pressure and thus is subatmospheric during inspiration. Therefore, closure of the UES also prevents entry of air into the esophagus during inspiration.

Esophageal phase of swallowing.   The esophagus is gated on both ends by sphincters.

— The *UES* is composed of striated muscle and is normally closed due to elasticity of the sphincter and tonic neural excitation.
— The *lower esophageal sphincter (LES)* is composed of smooth muscle. It is closed under resting conditions due in part to tonic myogenic contraction (i.e., the contraction is due to an intrinsic property of the myocyte cell, not due to neural stimulation). The LES relaxes at the start of the swallow, several seconds before the wave of peristalsis in the esophagus approaches the LES due to vagal (noncholinergic) inhibition, allowing the bolus to pass into the stomach. It then closes to prevent reflux of gastric contents into the esophagus.

Table 20.1 lists the factors that modulate the tone of the LES.

Table 20.1 ► Tone of the Lower Esophageal Sphincter (LES)*	
Factors that Increase LES Tone	**Factors that Decrease LES Tone**
Acetylcholine	Nitric oxide
↑ intraabdominal and intragastric pressure	VIP
Gastrin	CCK
Motilin	GIP
Protein-rich food	β adrenergic receptor agonists
	Secretin
	Progesterone
	Prostaglandin E
	Fat-rich food

Abbreviations: CCK, cholecystokinin; GIP, gastrointestinal polypeptide; VIP, vasoactive intestinal peptide.
* Factors that increase LES tone will close the LES, and factors that decrease LES tone will open the LES.

In an upright posture, swallowed liquids can simply flow down the esophagus due to the force of gravity (although gravity is not essential for swallowing). The passage of semisolid food down the esophagus requires peristalsis. Peristalsis in the esophagus is coordinated by the extrinsic and intrinsic nervous systems.

— *Primary peristalsis* automatically occurs when a bolus enters the esophagus. It is coordinated by extrinsic (autonomic nervous system) innervation of striated muscle.
— *Secondary peristalsis* is triggered by local distention of the esophagus. It moves any food left behind, and it returns material refluxed up from the stomach. It is coordinated by intrinsic (enteric nervous system) innervation of smooth muscle (**Fig. 20.3**).

Gastroesophageal reflex disease

Gastroesophageal reflux disease (GERD) occurs when stomach acid continuously refluxes into the esophagus. It may be caused by elevated intraabdominal pressure (e.g., due to obesity, big meals, or tight clothing), reduced LES tone (e.g., due to pregnancy, hiatus hernia, achalasia, fatty meals, and smoking), and by tricyclic and anticholinergic drugs. GERD causes pain, heartburn, and inflammation because the esophagus lacks the protective lining of the stomach. The pain of GERD radiates to the back and is worsened by stooping and ingesting hot drinks. Treatment is with antacids (e.g., calcium carbonate), H_2-receptor antagonists (e.g., cimetidine), or proton pump inhibitors (e.g., omeprazole). Medication to strengthen the LES, known as prokinetic drugs (e.g., metoclopramide), may also be used. If medications alone do not control symptoms, surgery to tighten the LES may be necessary.

Achalasia

Achalasia is a pathological condition caused by a deficiency of inhibitory myenteric neurons in the lower part of the esophagus. The LES fails to relax during swallowing, and peristalsis is absent. Food therefore accumulates above the LES, causing an increased risk of aspiration pneumonia. Symptoms of achalasia include regurgitation of food, chest pain, difficulty swallowing liquids and solids, cough, and weight loss. Drug treatment is aimed at reducing the tone of the LES. This may be achieved with Botox injections (temporary action) or by administration of long-acting nitrates or Ca^{2+} channel blockers. Surgical treatment includes esophagomyomectomy to reduce LES tone and dilation of the esophagus.

Fig. 20.3 ▶ **Esophageal motility.**
Esophageal motility is usually checked by measuring the pressure in the lumen, for example, during a peristaltic wave (**1,2**). The resting pressure within the lower esophageal sphincter (LES) is normally 20 to 25 mm Hg. During receptive relaxation, esophageal pressure drops to match the low pressure in the proximal stomach (**3**), indicating opening of the LES. The LES opens due to a vagovagal reflex mediated by myenteric neurons releasing vasoactive intestinal peptide and nitric oxide.

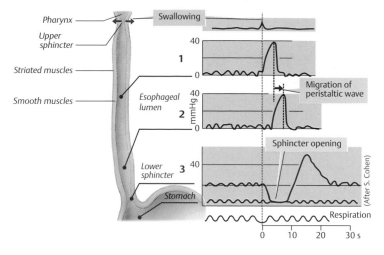

20.2 Gastric Motility

In the stomach, chewed food is converted into a thick liquid called *chyme* by the addition of water and electrolytes (e.g., Na^+, K^+, and Ca^{2+}). It is then mixed with acid and pepsin to begin the digestion process. Chyme is stored for variable amounts of time and then released in a slow, controlled fashion into the duodenum.

Functional Gastric Anatomy

— The *proximal stomach* is made up of the fundus and the proximal part of the corpus (body) (**Fig. 20.4**). The proximal stomach can easily expand to allow for increased volume of stomach contents.

— The *distal stomach* is made up of the distal part of the corpus, the antrum, and the pylorus (gastroduodenal junction [GDJ]), where the stomach and small intestine meet.

Fig. 20.4 ▶ **Anatomy of the stomach.**
The fundus of the stomach is continuous with the body (corpus), followed by the antrum. The pylorus connects the stomach to the duodenum. The mucosa of the fundus and corpus contain chief cells that produce pepsin, as well as parietal cells that produce gastric acid. The mucosa also contains endocrine cells that produce hormones (e.g., gastrin), mucous neck cells that produce protective mucus during feeding, and goblet cells that produce mucus continuously. The stomach is divided functionally into a proximal and distal segment. The proximal stomach mainly serves as a reservoir for food and propels gastric contents to the distal stomach, which is mainly concerned with mixing and pulverizing.

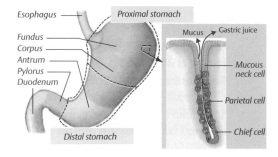

Receptive Relaxation

Receptive relaxation is a vagovagal reflex that causes the muscles of the proximal stomach to relax, which facilitates entry of the bolus into the stomach. It allows the stomach to expand without increasing intragastric pressure. This property is known as *accommodation*.

Receptive relaxation is mediated by vasoactive intestinal peptide (VIP).

Mixing and Digestion

Slow-wave depolarizations and peristaltic contractions usually arise in the corpus and spread distally, forcing chyme toward the pyloric sphincter at the GDJ (**Fig. 20.5**). Contractions are weak in the body of the stomach, where smooth muscle layers are relatively thin. Contractions are more forceful in the more muscular antrum. As the antrum begins to contract, some chyme may pass through the GDJ into the duodenal bulb. As antral contractions continue (antral systole), the pyloric sphincter closes, and the bulk of the chyme is retropulsed back into the body of the stomach; the process then repeats. Peristaltic contractions and antral systole function to mix gastric contents and to break up digestible solids, allowing them to be suspended in, and emptied along with, the liquid.

— Gastric peristalsis is stimulated by long and short cholinergic reflexes initiated by distention and by elevated serum gastrin. Elevated serum gastrin not only stimulates contractions but also increases the frequency of slow-wave depolarizations.

Migrating Myoelectric Complex

During fasting, the stomach exhibits regular contractile activity every 90 to 120 minutes called *migrating myoelectric complex (MMC)*. The contractions begin slowly and increase in strength, culminating in powerful peristaltic contractions. The pyloric sphincter remains open, allowing the contractions to remove large (> 1 mm) nondigestible solids left behind in the stomach and small intestine. MMC also removes mucus, sloughed cells, and bacteria from the small intestine, helping to prevent bacterial overgrowth.

— MMC is controlled by motilin.

Fig. 20.5 ▶ **Mixing in the distal stomach.**
Peristaltic contractions, originating in the corpus, drive stomach contents toward the pylorus. When the pyloric sphincter at the gastroduodenal junction closes, chyme is propelled back into the body of the stomach. This contributes to the breaking up of food particles, along with mixing and digestion of stomach contents.

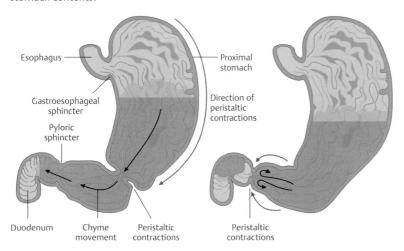

Esophagus

Gastroesophageal sphincter

Pyloric sphincter

Duodenum   Chyme movement   Peristaltic contractions

Proximal stomach

Direction of peristaltic contractions

Peristaltic contractions

Gastric Emptying

Chyme that leaves the stomach enters the duodenum as fluid and small solids (< 0.25 mm in diameter). While peristaltic contractions of the distal stomach aid in gastric emptying, elevation of intragastric pressure due to the sustained contractions of the proximal stomach contributes to emptying of gastric contents.

Rate of Gastric Emptying

— The rate of emptying is faster for liquids, small solids, and carbohydrates. It also increases with increased volume of chyme.

— The rate of emptying is slower for proteins and fats. It is also slower with increased acidity or osmolarity.

Control of Gastric Emptying: Enterogastric Reflex

Gastric emptying is principally regulated by the duodenum through the *enterogastric reflex*. The enterogastric reflex limits the amount of chyme that enters the duodenum at any one time so that digestion and absorption can proceed optimally. This reflex is mediated by hormonal and neural inputs.

Hormonal mediation of the enterogastric reflex

— Acidic chyme in the duodenum stimulates the release of secretin, which reduces gastric motility and increases the tone of the pyloric sphincter.

— The products of lipid digestion stimulate the release of cholecystokinin and glucose-dependent insulinotropic peptide (GIP), which also reduces gastric motility.

— The products of protein digestion stimulate the release of gastrin, cholecystokinin, and GIP, which all slow gastric emptying.

Neural mediation of the enterogastric reflex

— Distention of the duodenum activates duodenal stretch receptors that act via local neuronal circuits to slow gastric emptying.

20.3 Small Intestine Motility

The small intestine is the major site for digestion and absorption of nutrients from food. It receives 7 to 10 L of chyme per day from the stomach, to which various secretions are added.

The basic frequency and patterns of contractions in the small intestine are determined by slow-wave depolarizations. The frequency of slow-wave depolarization is highest in the proximal small intestine, ~12/min, and gradually decreases distally to become ~8/min in the terminal ileum. As in the stomach, the slow waves are myogenic and are not themselves sufficient to elicit contractions. Action potentials, superimposed upon the plateau of the slow wave, cause contraction of smooth muscles of the small intestine.

Entry into the Small Intestine

Contractions of the first section of the duodenum are coordinated with gastric contractions. When the pyloric sphincter is open, the bulb (opening) of the duodenum is relaxed, facilitating emptying of stomach contents and minimizing reflux of duodenal contents into the stomach. Following closure of the pyloric sphincter, the duodenal bulb contracts, moving the contents distally. When the small intestine is distended, stretch receptors are activated. Enteric neurons cause contractions in the circular muscle proximal to the distention and relaxation of longitudinal muscles. They also cause relaxation of the circular muscles and contraction of the longitudinal muscles distal to it. This activity is known as Bayliss and Starling's law of the intestine.

— Overdistention of one portion of the small intestine results in inhibition of motility in the rest of the small intestine. This is known as the *intestino-intestinal reflex*.

Mixing of Chyme and Motility in the Small Intestine

Segmentation Contractions

— Segmentation contractions mix intestinal contents with digestive enzymes and bile salts. Contraction of an intestinal segment propels the chyme in a proximal and distal direction. Relaxation then causes the chyme to occupy the same segment as before; thus, there is no forward propulsion of chyme.

Peristaltic Contractions

— Peristaltic contractions are directed aborally (away from the mouth) and progressively move chyme through the small intestine.
— They can be initiated anywhere along the length of the small intestine and are propagated for distances of a few centimeters.

Regulation of Motility in the Small Intestine

Coordination of stomach contractions and contractions within the small intestine is mediated by the enteric nervous system and hormones.

— Small intestine motility is stimulated by parasympathetic neural activity, serotonin, gastrin, cholecystokinin, and motilin.
— Small intestine motility is inhibited by sympathetic neural activity, epinephrine, secretin, and glucagon.

Emptying of the Small Intestine

Gastroileal Reflex

Increased peristalsis in the ileum and relaxation of the ileocecal sphincter in response to the presence of food in the stomach is referred to as the *gastroileal reflex*. This reflex results in chyme leaving the small intestine and entering the large intestine.

— The gastroileal reflex is mediated by increased gastrin levels or through extrinsic neural reflexes.

20.4 Vomiting

Vomiting is a mechanism for the rapid evacuation of gastric and duodenal contents. It is often preceded by nausea and increased salivation (water brash).

— The vomiting center in the medulla receives stimuli from visual pathways, motion sensors (CN VIII), vagal afferents from the GI tract, and from the chemoreceptive trigger zone in the fourth ventricle (**Fig. 20.6**).

Vomiting Reflex

The vomiting reflex starts with a single retrograde peristaltic contraction beginning in the middle of the small intestine that propels intestinal contents through a relaxed GDJ into the stomach. Inspiration occurs against a closed glottis, lowering intraesophageal pressure. The duodenum

Fig. 20.6 ► Vomiting.
Certain stimuli activate the vomiting center in the medulla oblongata, which receives input from chemosensors in the chemoreceptor trigger zone in the fourth ventricle. Vomiting is preceded by retching and increased salivation. During the act of vomiting, the diaphragm remains in the inspiratory position. The duodenum contracts, pushing intestinal contents into the stomach. The abdominal muscles contract, which then forces stomach contents into the esophagus. Relaxation of the upper esophageal sphincter results in ejection of stomach contents through the mouth.

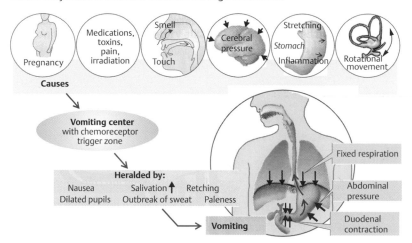

and antrum contract to prevent movement of chyme back into the small intestine. The abdominal muscles then forcibly contract (Valsalva maneuver), increasing intraabdominal pressure, which creates more pressure in the stomach than in the esophagus. This forces gastric contents into the esophagus. The larynx and hyoid bone are drawn forward, decreasing the tone of the UES and leading to the gastric and esophageal contents being expelled via the oral cavity.

20.5 Large Intestine Motility

The colon receives 0.5 to 1.0 L of chyme per day. Chyme moves very slowly from the cecum through the ascending, transverse, descending, and sigmoid parts of the colon. The arrangement of pacemakers in the colon means that retrograde movement is frequently seen, in contrast to the small intestine. This partly explains why colon transit time is longer than small intestine transit time. Chyme is mixed by low-frequency, long-duration contractions called *haustrations,* and fecal material is propelled through the colon by peristalsis.

In the colon, the predominant function of the enteric nerve plexus is to inhibit the innate desire of colonic smooth muscle to contract. The effect of removing this inhibition is highlighted in Hirschsprung disease.

Mass Movement

Periodically, haustrations cease, and colonic contents move distally due to large-scale peristaltic contractions. This mass movement occurs one to three times daily usually during or shortly after ingestion of the first meal of the day. This is most evident in infants. Stimulation of mass movement upon ingestion of a meal is called the *gastrocolic reflex* and may be triggered by increased gastrin or by extrinsic neural reflexes.

Defecation

The rectum normally contains little fecal material and is contracted to prevent entry of material from the colon. Occasionally, after a meal, mass movement shifts some of the colonic contents into the rectum. The resultant distention elicits the *rectosphincter reflex* that relaxes the internal anal sphincter and elicits the urge to defecate. If conditions are not appropriate for defecation, the external anal sphincter is voluntarily contracted, preventing expulsion of fecal material. When defecation is prevented, the rectum slowly pushes the material back into the sigmoid colon, and the internal anal sphincter regains its tone. Thus, the urge to defecate may be suppressed until the next mass movement causes the movement of fecal material from the sigmoid colon back into the rectum.

The act of defecation is partly involuntary and partly voluntary:

— Involuntary responses include contraction of smooth muscles of the distal colon and relaxation of the internal anal sphincter, mediated by the enteric nervous system.
— Voluntary movements include relaxation of the external anal sphincter (which is skeletal muscle that is innervated by the somatic nervous system) and voluntary contraction of abdominal muscles (Valsalva maneuver) to increase intraabdominal pressure to move feces out of the body.

Table 20.2 summarizes the GI reflexes that affect motility.

Table 20.2 ▶ Summary of Gastrointestinal Reflexes		
Reflex*	**Mechanism**	**Function**
Receptive relaxation	This is a vagovagal reflex that causes the muscles of the proximal stomach to relax, which facilitates entry of the bolus into the stomach	It allows the stomach to expand without increasing intragastric pressure (accommodation)
Enterogastric reflex	Entry of chyme into the duodenum inhibits further gastric emptying	Controls amount of chyme entering the duodenum, allowing for optimal digestion and absorption of nutrients
Intestino-intestinal reflex	Overdistention of one portion of the small intestine inhibits motility in the rest of the small intestine	Decompensatory
Gastroileal reflex	Food in the stomach causes increased peristalsis in the ileum and relaxation of the ileocecal sphincter	Promotes the emptying of chyme from the ileum into the colon
Gastrocolic reflex	Increased motility and secretion in the stomach results in increased colonic activity	This reflex begins the process of defecation.
Rectosphincter reflex	Distention of the rectum by colonic contents (following mass movement) causes relaxation of the internal anal sphincter and elicits the urge to defecate	Defecation
* Reflexes mediated by the enteric nervous system.		

21 Gastrointestinal Secretion

Secretion is the process of releasing a substance from a cell or gland. In the gastrointestinal (GI) tract, it mainly occurs in the mouth, stomach, and small intestine and is important for the digestion and absorption of food.

21.1 Salivary Secretion

Formation of Saliva

Saliva is formed by three pairs of salivary glands: the parotid, the submandibular, and the sublingual glands.

— The average adult secretes ~1 L of saliva per day: 250 mL is secreted by the parotid gland, 700 mL by the submandibular gland, and 50 mL by the sublingual gland.

Structure of the Salivary Glands

Acinar (berry-like) cells secrete primary saliva, which has a similar composition to plasma.

    While passing through the ducts before being excreted into the mouth, primary saliva is modified into secondary saliva (**Fig. 21.1**):

— Na^+ and Cl^- content is progressively lowered by reabsorption in the ducts.
— K^+ and HCO_3^- content is increased by secretion in the ducts.

When saliva secretion rates are elevated, primary saliva is not modified as much before being secreted because there is less time available for secretion and absorption of ions. As a consequence, saliva secreted in high volumes has an electrolyte composition closer to primary saliva but is always hypotonic with respect to plasma.

Composition and Functions

Saliva is alkaline during stimulated secretion because of its elevated HCO_3^- content. This neutralizes acid produced by bacteria in the oral cavity and prevents dental caries.

    Saliva has a higher concentration of K^+ and lower concentrations of Na^+ and Cl^- relative to plasma at normal secretion rates. It also contains the following substances:

— Ptyalin, an α-amylase, starts the process of digestion of complex carbohydrates.
— Lingual lipase starts the process of digestion of triglycerides.
— Lysozyme, immunoglobulin A (IgA), and lactoferrin help prevent bacterial overgrowth in the oral cavity.

■ Sjögren syndrome

Sjögren syndrome is an autoimmune disease causing keratoconjunctivitis sicca (diminished tear production) and xerostomia (dry mouth). It is also associated with rheumatoid arthritis (in 50% of cases) and lupus. Lymphocytes and plasma cells infiltrate secretory glands and cause injury. Diminished tear production causes dry, itchy, gritty eyes; diminished saliva production makes swallowing difficult and increases the likelihood of development of dental caries. Rheumatoid arthritis causes joint pain, swelling, and stiffness. Treatment for dry eyes involves the use of artificial tears. Dry mouth may be relieved by artificial saliva, taking frequent sips of water, and chewing gum to stimulate saliva flow. If this is insufficient, pilocarpine, an anticholinergic drug, may be used to stimulate saliva production. Nonsteroidal antiinflammatory drugs (NSAIDs), hydroxychloroquine (an antimalarial drug), and immunosuppressants (e.g., methotrexate and cyclosporine) are used for rheumatoid arthritis.

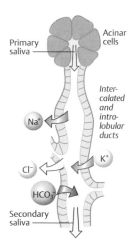

Fig. 21.1 ▶ **Saliva secretion.**
Primary saliva is produced in acinar cells and has a composition similar to plasma. Primary saliva is modified in the excretory ducts, forming secondary saliva. As saliva passes through the excretory ducts, Na^+ and Cl^- are reabsorbed, and K^+ and HCO_3^- are secreted into the lumen. The saliva becomes hypotonic as Na^+ and Cl^- reabsorption is greater than K^+ and HCO_3^- secretion, and the ducts are relatively impermeable to water. At high flow rates, this modification process lags, and the composition of secondary saliva becomes similar to that of primary saliva.

Fig. 21.2 ▸ **Stimulation of salivary secretion.**
Many factors stimulate saliva secretion, including the taste and smell of food, tactile stimulation of the buccal mucosa, mastication, and nausea. Conditioned reflexes also play a role; for example, the clattering of dishes when preparing a meal can induce saliva secretion. Saliva secretion is stimulated via the sympathetic and parasympathetic nervous system.

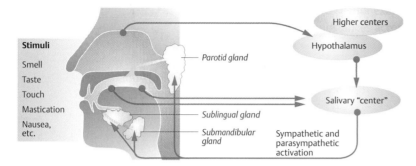

Regulation of Saliva Secretion

Saliva secretion is regulated by both the parasympathetic and sympathetic nervous systems. Both systems increase saliva production, although the parasympathetic system causes a much greater volume response.

Parasympathetic Regulation

Taste, smell, mastication, nausea, and conditioned reflexes stimulate the facial nerve (cranial nerve [CN] VII) and the glossopharyngeal nerve (CN IX), causing the release of acetylcholine (**Fig. 21.2**).

- Acetylcholine interacts with muscarinic (M_3) receptors on acinar cells to stimulate second messengers inositol 1,4,5-triphosphate (IP_3) and Ca^{2+}, resulting in an increased volume of saliva, increased [HCO_3], and increased O_2 consumption (**Fig. 21.3**).
- Acetylcholine also releases kallikrein, which activates bradykinin, a potent vasodilator. Vasodilation improves blood flow to the salivary glands.

Parasympathetic stimulation also causes myoepithelial cells to contract, facilitating movement of saliva from the ducts into the oral cavity.

Sympathetic Regulation

Sympathetic activity causes the release of norepinephrine.

- Norepinephrine binds to β_2-adrenergic receptors on acinar cells to stimulate the second messenger cyclic adenosine monophosphate (cAMP). This briefly produces small increases in the volume of saliva. It also stimulates an increased concentration of mucus and facilitates the movement of saliva from the salivary ducts into the oral cavity.

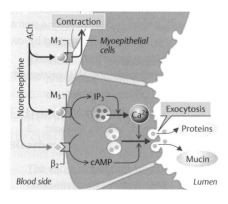

Fig. 21.3 ▸ **Regulation of saliva secretion in acinar cells.**
Parasympathetic stimulation of the facial and glossopharnygeal nerves cause the release of acetylcholine (ACh). ACh binds to M_3 muscarinic receptors and activates inositol 1,4,5-triphosphate (IP_3). This increases cytostolic [Ca^{2+}] and increases the conductivity of the luminal anion channels, resulting in the production of watery saliva and increased exocytosis of salivary enzymes. Binding of ACh to M_3 receptors also causes the contraction of myoepithelial cells and emptying of the acini. Sympathetic stimulation causes the release of norepinephrine, which acts upon β_2 receptors. This causes an increase in cyclic adenosine monophosphate (cAMP) and the secretion of highly viscous saliva, which has a high concentration of mucin.

21.2 Secretion in the Stomach

Functional Anatomy of the Gastric Mucosa

The gastric mucosa contains several gastric glands, many of which open into a common outlet (gastric pits) on the surface of the mucosa (**Fig. 21.4A,B**). Within each gland there are different cell types (**Fig. 21.4B**), each secreting a unique substance or substances. Parietal cells secrete hydrochloric acid (HCl) and intrinsic factor (IF), chief cells secrete pepsinogens, neuroendocrine cells secrete gastrin (G cells) and somatostatin (D cells), and enterochromaffin cells secrete histamine. These secretions mix with the mucus produced by neck cells.

Fig. 21.4 ▶ **(A) Gastric mucosa. (B) Structure of the gastric glands.**
From *Thieme Atlas of Anatomy*, *Neck and Internal Organs*, © Thieme 2006, Illustration by Markus Voll.

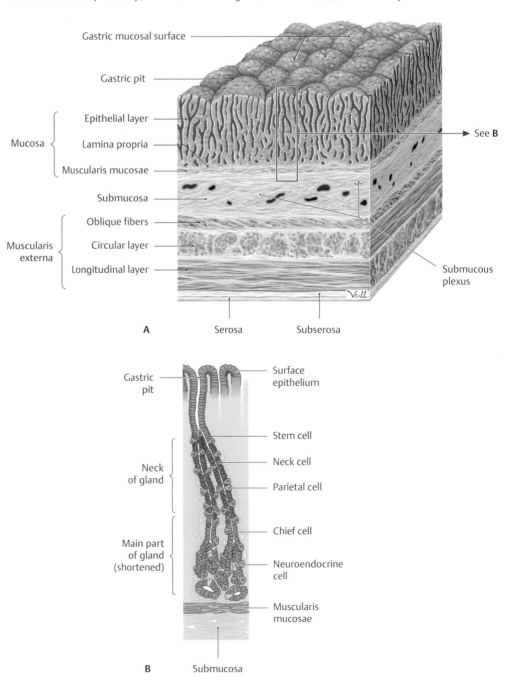

Gastric Secretions

Hydrochloric Acid

HCl is secreted by parietal cells in the stomach.

- *Basal acid output (BAO)* is the baseline amount of HCl produced in a given period in the absence of any stimulatory factors. It is usually < 10 mmol/h.
- *Maximal acid output (MAO)* is the amount of HCl produced in a given period when stimulants (e.g., histamine) are administered. It is usually < 50 mmol/h.

HCl functions to break up cells, to denature proteins for easier digestion, to kill many ingested bacteria, and to convert pepsinogen to pepsin.

Intrinsic Factor

IF is a glycoprotein secreted by parietal cells when stimulated by acetylcholine.

IF binds vitamin B_{12}, following its liberation from R protein by trypsin in the small intestine (see page 233). The IF–B_{12} complex is essential for absorption of vitamin B_{12} in the terminal ileum.

Lack of IF leads to pernicious anemia, which is a frequent complication of achlorhydria (absence of gastric acid secretion due to parietal cell failure).

Pepsin

- Pepsinogen is secreted by gastric chief cells into the lumen and is converted to its active form, pepsin, in the presence of HCl.
- Pepsin is an enzyme that breaks down proteins. It has an optimum pH value of ~ 2.

Gastrin

- Gastrin is secreted by G cells in the antrum of the stomach.
- Gastrin increases H^+ and histamine secretion in response to the presence of food in the stomach.
- Gastrin inhibits gastric emptying via enterogastric reflex (see page 209 and 210).

Mucus

Mucus is secreted by mucous neck cells in the epithelium of the stomach. It is useful for lubrication and protection of gastric epithelial cells from the effects of the strongly acidic environment in the lumen. Mucous neck cells also secrete HCO_3^-, which becomes trapped in the mucous layer and further contributes to epithelial protection. Mucus and HCO_3^- combine to form the *gastric mucosal barrier.*

Acid Secretion in the Stomach

Mechanism of Acid Secretion in Parietal Cells

The following sequence of events results in HCl secretion into the lumen of the stomach and secretion of HCO_3^- into the bloodstream (**Fig. 21.5**):

- CO_2 derived from cellular metabolism combines with water within cells to form H_2CO_3. This is catalyzed by carbonic anhydrase. H_2CO_3 then dissociates to form HCO_3^- and H^+.
- The luminal membranes of parietal cells contain an H^+–K^+ ATPase, which actively secretes H^+ into the lumen of the stomach in exchange for K^+.
- Simultaneously, HCO_3^- is secreted into the bloodstream in exchange for Cl^-. This results in accumulation of Cl^- within the cell and then to passive secretion of Cl^- into the lumen of the stomach.

Extraction of CO_2 from the bloodstream and the secretion of HCO_3^- into it causes venous blood leaving the actively secreting stomach to be more alkaline than arterial blood ("alkaline tide").

▬ Pernicious anemia

Pernicious anemia (a type of megaloblastic anemia) is a disease resulting from malabsorption of vitamin B_{12}. It is caused by an autoimmune reaction to gastric parietal cells that results in a lack of intrinsic factor (IF). Vitamin B_{12} deficiency causes inhibition of DNA synthesis in red blood cell production. Anemia may be asymptomatic, or there may be any of the following signs and symptoms: fatigue; pallor (seen most readily by inspection of the conjunctiva or mucous membranes); dizziness, particularly upon standing (postural hypotension); headache; shortness of breath; coldness of the hands and feet; palpitations; and glossitis (swelling and soreness of the tongue). In severe cases, anemia can cause chest pain (angina due to hypoxia of cardiac muscle) and heart failure (as the heart has to work harder to oxygenate tissues). Treatment is with vitamin B_{12} intramuscular injections given on a monthly basis for life. This allows vitamin B_{12} to enter the body but bypass intestinal absorption, which fails in the absence of IF. Note, pernicious anemia takes time to develop because the liver stores vitamin B_{12}.

▬ Gastric blood supply and mucosal protection

The stomach has a rich mucosal blood supply, supplied by the right and left gastric arteries, the right and left gastro-omental arteries, and the short gastric arteries. It is drained by corresponding veins that empty into the portal vein (some indirectly). This rich blood supply contributes to gastric mucosal protection by ensuring that if gastric H^+ ions manage to penetrate the epithelium, then they are rapidly removed.

Fig. 21.5 ▶ HCl secretion by parietal cells.
The H^+–K^+ ATPase pump in the luminal membrane of parietal cells drives H^+ into the lumen in exchange for K^+ (**1**). The K^+ circulates back into the lumen via luminal K^+ channels. For every H^+ secreted, one HCO_3^- leaves the blood side of the cell in exchange for Cl^- (**2**). The HCO_3^- ions are obtained from the CO_2 and OH^- produced by cellular metabolism. This reaction is catalyzed by carbonic anhydrase (CA). Cl^- accumulates within the cell and then diffuses into the lumen via Cl^- channels (**3**). Thus, one Cl^- ion reaches the lumen for each H^+ ion secreted. H^+ and Cl^- combine to form HCl. (ATP, adenosine triphosphate)

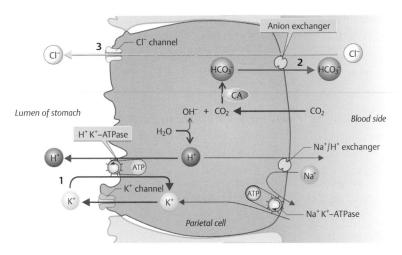

Stimulation of Acid Secretion

Vagal stimulation. The vagus nerve directly stimulates parietal cells to secrete H^+. Acetylcholine is released from the nerve, which activates muscarinic (M_3) receptors on parietal cells. The M_3 receptor is G-protein coupled (G_s), and activation causes the release of the second messenger, IP_3, and increased intracellular $[Ca^{2+}]$. These stimulate the H^+–K^+ ATPase, which actively secretes H^+ into the lumen of the stomach in exchange for K^+.

— The vagus nerve indirectly stimulates the secretion of H^+ by acting on G cells, causing the secretion of gastrin. Gastrin-releasing peptide (GRP, or bombesin) is the neurotransmitter.

Gastrin. Gastrin is released in response to amino acids, distention of the stomach, and vagal stimulation.

— Gastrin stimulates parietal cells to secrete H^+ via interaction with the cholecystokinin B (CCK_B) receptor. The CCK_B receptor is G_s coupled, and activation causes the release of IP_3 and increased intracellular $[Ca^{2+}]$. These stimulate the H^+–K^+ ATPase, which actively secretes H^+ into the lumen of the stomach in exchange for K^+.

Histamine. Histamine is released by enterochromaffin (H) cells in the fundus of the stomach.

— It acts on H_2 receptors on parietal cells, causing the release of H^+. The H_2 receptor is G_s-protein coupled, and activation causes an increase in the second messenger, cAMP. cAMP stimulates the H^+–K^+ ATPase, which actively secretes H^+ into the lumen of the stomach in exchange for K^+.

Note: There is potentiation of the effects of acetylcholine, histamine, and gastrin on gastric acid secretion, as each stimulates gastric acid production via a different mechanism. The implications of this are that low concentrations of two or more stimulants can produce a large increase in gastric acid production. One of the stimulants must be histamine for potentiation to occur (i.e., acetylcholine and gastrin do not potentiate each other's effects on gastric acid production).

■ Zollinger–Ellison syndrome

Zollinger–Ellison syndrome is a condition caused by gastrin-secreting pancreatic adenomas that lead to multiple ulcers in the stomach and duodenum. These ulcers are frequently drug resistant and are accompanied by diarrhea and steatorrhea (increased fat content of stools), as well as all of the usual peptic ulcers symptoms (e.g., burning abdominal discomfort, heartburn, nausea and vomiting, and weight loss). Tests for the condition will show raised serum gastrin and gastric acid levels. Treatment involves the use of proton pump inhibitors (e.g., omeprazole) that inhibit gastric acid production and surgical resection of the offending tumor, if this is possible. If surgery is not an option, or if full resection is not possible, then chemotherapy may be employed to slow tumor growth. The 5-year survival rate is low (19%) if there are metastases (usually to the liver).

Inhibition of Acid Secretion

Low pH. Following gastric emptying, H⁺ secretion causes the pH of the lumen of the stomach to decrease. When pH is < 2, gastrin secretion by G cells is inhibited by somatostatin released from D cells. This, in turn, causes negative feedback inhibition of further H⁺ secretion.

Somatostatin. Somatostatin inhibits the release of gastrin and histamine (from G cells and enterochromaffin cells, respectively), which indirectly decreases H⁺ secretion.

— It also activates G_i, which, in turn, inhibits adenylate cyclase and reduces cAMP levels. This antagonizes increases in cAMP produced by histamine.

Note: The effect of somatostatin on G cells is much more important than its effects on enterochromaffin cells.

Prostaglandins. Prostaglandins also inhibit gastric acid secretion by activating G_i, which, in turn, inhibits adenylate cyclase and reduces cAMP levels. This antagonizes increases in cAMP produced by histamine.

The regulation of gastric acid secretion is illustrated in **Fig. 21.6.**

Fig. 21.6 ▶ **Regulation of gastric acid secretion.**
Gastric acid secretion is stimulated in phases by neural, local gastric, and intestinal factors. Food intake leads to reflex secretion of gastric juices (**1**). The vagus nerve releases acetylcholine (ACh), which directly activates parietal cells in the fundus at M_3 cholinoceptors (**2**). Gastrin-releasing peptide (GRP) also released by the vagus nerve stimulates gastrin secretion from G cells in the antrum (**3**). Gastrin released into the systemic circulation activates parietal cells via cholecystokinin B (CCK_B) receptors. The glands in the fundus contain H cells (enterochromaffin cells), which are activated by gastrin, ACh, and β_2-adrenergic substances. H cells release histamine, which has a paracrine effect on neighboring parietal cells (via H_2 receptors). Gastric acid secretion is inhibited by a pH < 3.0. This inhibits G cells via a negative feedback mechanism and activates antral D cells, which secrete somatostatin (SIH). SIH inhibits H cells in the fundus (minor effect) and G cells in the antrum (major effect). GRP released by neurons acts on D cells in the fundus and antrum to inhibit SIH secretion. Secretin and gastrointestinal polypeptide (GIP) released from the small intestine have an inhibitory effect on gastric acid secretion (**1**). This adjusts the composition of chyme from the stomach to the needs of the small intestine. (cGRP, calcitonin gene-related peptide)

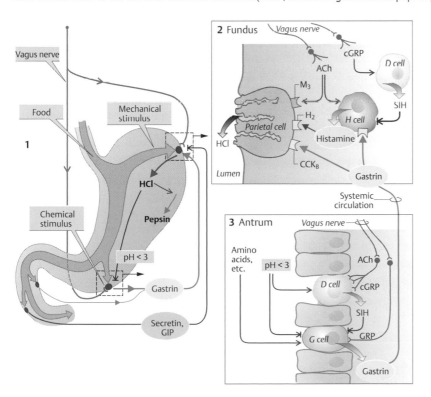

Peptic Ulcer Disease

An ulcer is a lesion extending through the mucosa and submucosa into deeper structures of the wall of the GI tract. Ulcers are the result of the breakdown of the mucosal barrier (mucus and HCO_3^-) that normally protects the lining of the GI tract and/or increased secretion of H^+ or pepsin.

Gastric Ulcers

Gastric ulcers are commonly found on the lesser curvature between the corpus and antrum of the stomach.

— They are often caused by *Helicobacter pylori,* a gram-negative spiral bacillus, which secretes cytotoxins that disrupt the mucosal barrier, causing inflammation and destruction.

— *H. pylori* secretes high levels of membrane urease, which converts urea to ammonia (NH_3). NH_3 neutralizes gastric acid around the bacterium, allowing it to survive in the acidic lumen of the stomach.

The following features are also found in gastric ulcers:

— ↓ gastric H^+: The damaged mucosa allows H^+ secreted from parietal cells into the lumen of the stomach to reenter the mucosa.

— ↑ gastrin: The decrease in H^+ stimulates gastrin secretion.

Duodenal Ulcers

Duodenal ulcers are the most common ulcers and are often associated with increased gastric H^+ secretion (but not necessarily).

— Duodenal ulcers frequently occur due to *H. pylori. H. pylori* inhibits somatostatin secretion, leading to increased gastric H^+ secretion. There is also decreased HCO_3^- secretion in the duodenum, which impedes neutralization of the excess H^+ delivered from the stomach.

21.3 Secretion by the Liver

Formation of Bile

Bile is produced continuously by hepatocytes in the liver and stored in the gallbladder. An average adult produces ~400 to 800 mL of bile each day (gallbladder stores ~ 50 mL of bile). Bile empties into the duodenum at the major duodenal papilla (of Vater) (**Fig. 21.7**).

Composition of Bile

Bile is composed of water, bile salts, lecithin, cholesterol, bile pigments, and electrolytes (**Fig. 21.8**). Unlike other secretions, it does not contain digestive enzymes, but it does facilitate the digestion of lipids by acting as an emulsifying agent to increase the surface area of the lipids.

Bile Salts

Bile salts aid in digestion and absorption of lipids and lipid-soluble vitamins in the small intestine.

— Primary bile salts are glycine or taurine conjugates of cholate and chenodeoxycholate that are synthesized from cholesterol by hepatocytes.

— Secondary bile salts—conjugated deoxycholate and lithocholate—are produced by bacterial alteration of primary salts in the intestinal lumen.

Fig. 21.7 ▶ Extrahepatic bile ducts and pancreatic ducts.
Anterior view with the gallbladder opened and the duodenum opened and windowed. The common bile duct (formed from the joining of the left and right hepatic ducts with the cystic duct) and the pancreatic duct both empty their secretions into the duodenum at the major duodenal papilla (of Vater). Superior to the major papilla is the minor duodenal papilla, which receives secretions from the accessory pancreatic duct.
From *Thieme Atlas of Anatomy, Neck and Internal Organs*, © Thieme 2006, Illustration by Markus Voll.

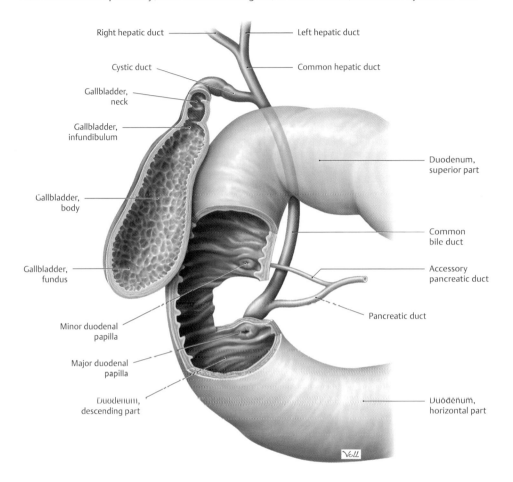

Micelles. Bile salts, along with phospholipids (e.g., lecithin), cholesterol, and ingested fats, form stable droplets called micelles. The hydrophilic portions of bile salts and phospholipids are exposed at the outer surface of the micelle, while the hydrophobic portions, lipids, and cholesterol are sequestered in the hydrophobic interior. This emulsifies lipids and cholesterol, allowing them to mix with the aqueous contents of the intestinal lumen, exposing them to digestive enzymes (e.g., pancreatic lipase, phospholipase A_2, and cholesterol esterase). These enzymes attach to the surface of the micelles, displacing bile salts so the lipids can be digested.

Bile Pigment

Bilirubin is a breakdown product of hemoglobin. It is hydrophobic and is present in plasma bound to albumin. Hepatocytes extract bilirubin from plasma and conjugate it with glucuronic acid to make it water soluble. Bilirubin diglucuronide is then secreted into bile (**Fig. 21.9**). Unlike bile salts, most bile pigment is not reabsorbed but is excreted in the feces.

Electrolytes

Bile has elevated levels of HCO_3^-, Na^+, and Cl^-. These electrolytes become concentrated in the gallbladder, where Na^+ is actively reabsorbed, and HCO_3^- and Cl^- follow passively.

— HCO_3^- in bile helps neutralize acidic chyme in the intestine.

Implications of the shared emptying of the common bile duct and pancreatic duct

The fact that the common bile duct and the pancreatic duct both empty into the major duodenal papilla has important clinical implications. A tumor at the head of the pancreas, for example, may obstruct the common bile duct, causing biliary reflux into the liver with jaundice. Similarly, a gallstone that lodges in the common bile duct may obstruct the terminal part of the pancreatic duct, causing acute pancreatitis.

Drug dosage in hepatic disease

Hepatic disease (e.g., hepatitis and cirrhosis) has the potential to affect the pharmacokinetics of many drugs; however, because hepatic reserves are large, disease has to be severe for changes in drug metabolism to occur. The mechanisms by which the pharmacokinetics may be altered include reduced hepatic blood flow, which reduces the first-pass metabolism of drugs taken orally (and rectally to a lesser extent); reduced plasma protein binding, affecting both the distribution and elimination of drugs; and reduced plasma clearance of a drug if it is eliminated by metabolism and/or into bile. A dose reduction may be necessary in hepatic disease, depending on the extent of the disease and the particular drug or drugs being administered.

Gallstones

The majority of gallstones (75%) are formed when the amount of cholesterol in bile exceeds the ability of bile salts and phospholipids to emulsify it, causing cholesterol to precipitate out of solution. Gallstones may also be caused by an increased amount of unconjugated bilirubin (often in the form of calcium bilirubinate) in the bile ("pigment stones"). Gallstones may be asymptomatic, or they can produce obstruction of a duct, causing severe pain, vomiting, and fever. Drugs (e.g., ursodiol) may be used to dissolve small cholesterol gallstones. Ursodiol decreases secretion of cholesterol into bile by reducing cholesterol absorption and suppressing liver cholesterol synthesis. This alters bile composition and allows reabsorption of cholesterol-containing gallstones. Because reabsorption is slow, therapy must continue for at least 9 months. Other treatment includes lithotripsy (shock wave obliteration of gallstones that allow the stone fragments to be excreted) and surgical removal of the gallbladder (cholecystectomy).

Jaundice

Jaundice refers to the yellow pigmentation of the skin, sclerae, and mucous membranes due to raised plasma bilirubin.

Prehepatic (or hemolytic) jaundice: Excess bilirubin (e.g., from hemolysis), or an inborn failure of bilirubin metabolism results in unconjugated bilirubin remaining in the bloodstream. Unconjugated bilirubin is water insoluble and so does not appear in urine.

Hepatocellular (or hepatic) jaundice: In hepatocellular jaundice, there is diminished hepatocyte function, leading to an increased amount of both conjugated and unconjugated bilirubin. Diminished hepatocyte function may follow cirrhosis, autoimmune diseases, drug damage (e.g., acetaminophen and barbiturates), or viral infections (e.g., hepatitis A, B, and C; Epstein−Barr virus).

Posthepatic (obstructive) jaundice: This form of jaundice usually occurs following blockage of the common bile duct by gallstones. In this case, plasma-conjugated bilirubin rises. Conjugated bilirubin is water soluble and appears in urine (making it dark). At the same time, less conjugated bilirubin passes into the gut and is converted to stercobilin, making feces appear paler.

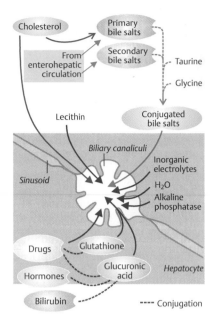

Fig. 21.8 ▶ **Bile components and hepatic secretion of bile.** Bile contains electrolytes, bile salts (bile acids), cholesterol, lecithin, bilirubin, steroid hormones, and drug metabolites, among other things. Bile salts are essential for fat digestion. Hepatocytes secrete bile into biliary canaliculi. The liver synthesizes cholate and chenodeoxycholate (primary bile salts) from cholesterol. The intestinal bacteria convert some of them into secondary bile salts, such as deoxycholate and lithicholate. Bile salts are conjugated with taurine and glycine in the liver and are secreted into bile in this form.

Fig. 21.9 ▶ **Bilirubin metabolism and excretion.**
Bilirubin is (mostly) formed from the breakdown of hemoglobin (Hb). It is conjugated by hepatocytes to form bilirubin diglucuronide and is secreted in bile into the small intestine. In the gut, anaerobic bacteria break bilirubin down into the colorless compound stercobilinogen. It is partially oxidized to stercobilin, the brown compound that colors the stools, and excreted in feces (~85%). About 15% of bilirubin is deconjugated by bacteria and returned to the liver via the enterohepatic circulation. A small portion (~1%) reaches the systemic circulation and is excreted by the kidneys as urobilinogen.

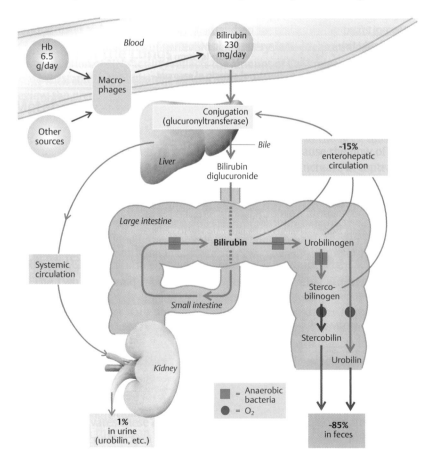

Fig. 21.10 ▶ **Mechanism for the release of bile from the gallbladder.**
When fats enter the duodenum, cholecystokinin (CCK) is released from
I cells of the duodenum. CCK binds to CCK$_A$ receptors and induces
gallbladder contraction, and thus the secretion of bile. Gallbladder
contraction is also stimulated by the neuronal plexus of the gallbladder
wall, which is innervated by preganglionic fibers of the vagus nerve.
(ACh, acetylcholine)

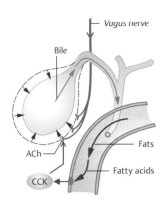

Regulation of Bile Secretion

Cholecystokinin

Entry of lipid-rich chyme into the duodenum causes increased secretion of cholecystokinin,
which causes the sphincter of Oddi to relax and the gallbladder to contract, gradually expelling
bile into the small intestine (**Fig. 21.10**).

— Cholecystokinin also stimulates pancreatic secretion of lipase (the major enzyme responsible for lipid breakdown), so both factors for lipid digestion are present in the duodenum.

Acetylcholine

Release of acetylcholine by the vagus nerve also stimulates gallbladder contraction.

Bile Salt Recirculation to the Liver

Bile salts are almost entirely reabsorbed in the ileum and recirculated to the liver (**Fig. 21.11**).
This is referred to as *enterohepatic circulation*. Most primary and secondary bile salts are reab
sorbed in the ileum by Na$^+$-dependent secondary active transport and returned to the liver via
the portal vein. The presence of bile salts throughout the length of the small intestine permits
lipids to be absorbed.

— The rate of bile secretion by the liver is determined mainly by the rate of return of bile salts
   from the ileum. Small quantities of bile salts are excreted in feces.

Fig. 21.11 ▶ **Enterohepatic circulation of bile salts.**
Bile salts are synthesized in the liver from cholesterol and
released into the small intestine via the common bile duct.
Conjugated bile salts are reabsorbed in the terminal ileum
by the Na$^+$ symport carrier ISBT (ileum sodium bile acid
cotransporter) and returned to the liver via the portal vein.
This allows the bile salts to be available for fat digestion and
absorption throughout the small intestine.

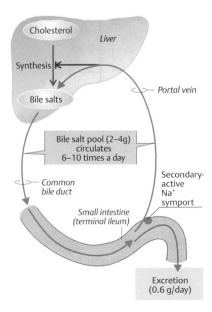

21.4 Secretion by the Exocrine Pancreas

Formation and Composition of Pancreatic Secretion

The volume of pancreatic juice secreted daily is ~500 to 1500 mL. It empties into the duodenum at the major duodenal papilla (of Vater) (see **Fig. 21.7**).

Acinar Cells

Acinar cells secrete a solution that is rich in digestive enzymes: proteases, amylase, and lipase, which break down proteins, starch, and lipids, respectively.

Ductal Cells

Ductal cells secrete an isotonic solution that contains a high concentration of HCO_3^- and lowered levels of Cl^- (compared with plasma) (**Fig. 21.12**). This alkaline secretion neutralizes acid in the small intestine and provides an alkaline environment for pancreatic enzymes to operate optimally.

Stimulation of Exocrine Pancreas Secretion

Secretion of pancreatic juice is controlled by secretin and cholecystokinin. Vagal activity potentiates the effects of these hormones.

Secretin

Secretin is released from the S cells in the duodenum in response to acidic chyme entering the duodenum.

— Secretin stimulates pancreatic ductal cells (via cAMP) to increase the volume of secretion and the concentration of HCO_3^- in that secretion (see **Fig. 21.12**). As the HCO_3^- neutralizes the acidic environment of the small intestine, secretin release is inhibited.

— Secretin also decreases gastric acid secretion.

Fig. 21.12 ▶ Secretion in pancreatic duct cells.
HCO_3^- is secreted from the luminal membrane of the ducts via an anion exchanger (AE) that simultaneously absorbs Cl^- from the lumen (**1**). This apical Cl^-–HCO_3^- exchanger relies on keeping the intracellular Cl^- concentration below a critical threshold so that the pump can continue to operate. The mechanism for achieving this is secretin-mediated Cl^- secretion via the cystic fibrosis transmembrane conduction regulator (CFTR) channel in response to elevated intracellular cyclic adenosine monophosphate (cAMP) (**2**). Movement of Cl^- by this mechanism draws water into the duct lumen. This is why in cystic fibrosis the duct contents are bicarbonate-poor and viscous because failure to secrete Cl^- inhibits the apical exchanger, and water is not moved into the duct lumen. The secreted HCO_3^- is absorbed in part from the blood via a Na^+–HCO_3^- symporter (NBC; [**3**]) and in part is the product of the CO_2 + OH^- reaction in the cytosol (**4**) catalyzed by carbonic anhydrase (CA). For each HCO_3^- secreted, one H^+ ion leaves the cell on the blood side via the Na^+–H^+ exchanger (NHF1; [**5**]). (ATP, adenosine triphosphate; PKA, protein kinase A)

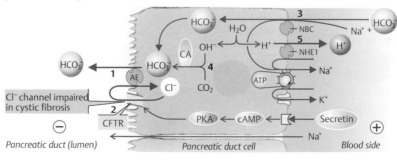

Cholecystokinin

Cholecystokinin is released from the I cells in response to amino acids, small peptides, and lipids entering the duodenum.

— Cholecystokinin stimulates pancreatic acinar cells (via IP_3 and increased intracellular $[Ca^{2+}]$) to release proteases, amylase, and lipase.

— Cholecystokinin also stimulates gallbladder contraction and ejection of bile into the duodenum. Pancreatic lipase and bile is needed to deal with the lipid load being presented to the duodenum.

Acetylcholine

— Vagovagal reflexes cause the release of acetylcholine in response to H^+, small peptides, amino acids, and fatty acids entering the duodenum.

— Acetylcholine stimulates the release of pancreatic enzymes.

Note: Cholecystokinin and acetylcholine augment the effects of secretin on pancreatic ductal cells to increase HCO_3^- secretion. This is because they stimulate their respective pancreatic secretions via different second messenger pathways.

The overall effect of secretin, cholecystokinin, and acetylcholine in pancreatic juice secretion is shown in **Fig. 21.13.**

Fig. 21.13 ▶ **Stimulation of pancreatic juice secretion.**
Pancreatic juice secretion is stimulated by vagal mechanisms (acetylcholine, ACh) and the hormone cholecystokinin (CCK, minor effect). Both cause an increase in cytostolic $[Ca^{2+}]$, which stimulates Cl^- and proenzyme secretion. Secretin increases HCO_3^- and H_2O secretion by the ducts. The effects of secretin are potentiated by ACh and CCK by increasing $[Ca^{2+}]$.

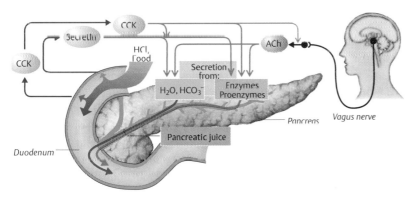

Table 21.1 summarizes GI secretion.

Table 21.1	Summary of Gastrointestinal (GI) Secretions		
Source of GI Secretion	**GI Secretion**	**Stimulated by**	**Inhibited by**
Oral cavity	Saliva	Parasympathetic nervous system (ACh at M_3 receptors) Sympathetic nervous system (norepinephrine at β_2 receptors) Muscarinic agonist drugs (e.g., pilocarpine)	Dehydration Sleep Fear Antimuscarinic drugs (e.g., atropine)
Stomach	Gastric acid (HCl)	Gastrin at CCK_B receptors Histamine at H_2 receptors ACh at M_3 receptors (direct effect) Vagal release of GRP at G cells (indirect effect)	Low pH Acidic chyme in duodenum (via secretin and GIP) Somatostatin inhibits the actions of gastrin and histamine on parietal cells Prostaglandins inhibit histamine-activated HCl secretion Drugs (e.g., omeprazole, cimetidine, and atropine)
	Intrinsic factor (IF)	ACh	Autoimmune destruction of parietal cells
	Pepsin	ACh	
	Mucus	ACh	
Liver	Bile	CCK stimulates sphincter of Oddi to relax and gallbladder to contract ACh	Ileal resection, which prevents enterohepatic circulation of bile and loss of bile into feces
Pancreas	Proteases, amylase, lipase	CCK ACh	—
	HCO_3^-	Secretin CCK (augments secretin) ACh (augments secretin)	—

Abbreviations: ACh, acetylcholine; CCK, cholecystokinin; GIP, gastrointestinal polypeptide; GRP, gastrin-releasing peptide.

22 Digestion and Absorption

Digestion is the chemical breakdown of food into its component nutrients. It occurs primarily in the stomach and small intestine and is due to the actions of gastric acid (on proteins only) and digestive enzymes.

Absorption is the movement of molecules from the lumen of the gastrointestinal (GI) tract into enterocytes (intestinal cells) and then into the bloodstream or lymph. The absorption of nutrients and water occurs primarily in the small intestine; the absorption of most of the remaining water occurs primarily in the colon.

22.1 Carbohydrate Digestion and Absorption

Carbohydrate Digestion

Salivary and pancreatic α-amylase hydrolyzes the α-1–4 glycosidic bonds in starch, forming maltose, maltotriose, and α-limit dextrins. The effects of salivary α-amylase are limited because it is inactivated by gastric acid. Isomatose breaks down the α-1–6 glycosidic bonds in α-limit dextrins to form maltose (**Fig. 22.1**).

Fig. 22.1 ▶ **Carbohydrate digestion and absorption.**
Salivary and pancreatic amylase break down ingested polysaccharides into their component disaccharides. Lactase, maltase, and sucrase break down their respective disaccharides into monosaccharides in the lumen. Monosaccharides are then able to be absorbed into the bloodstream.

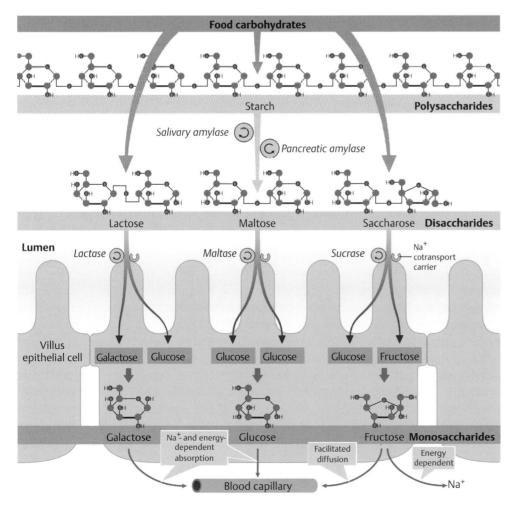

Lactose intolerance

Lactose intolerance is an inability to digest lactose in dairy products caused by a deficiency (partial or total) of lactase. Lactase is a brush border enzyme that converts lactose to the monosaccharides glucose and galactose, which are then absorbed into the bloodstream. Without lactase, lactose remains in the intestinal lumen, where it presents as osmotic load, resulting in diarrhea. Excess gas is also produced by fermentation of the luminal lactose to methane and hydrogen gas. Lactose intolerance can be diagnosed by the hydrogen breath test, which measures the amount of hydrogen produced after ingestion of a known amount of pure lactose (following an initial period of fasting). If the amount of hydrogen rises by 20 ppm, then a diagnosis of lactose intolerance can be made. The symptoms of lactose intolerance can be minimized in lactose-intolerant individuals by limiting their intake of dairy products or by oral ingestion of lactase-containing foods or drugs (e.g., Lactaid) with a meal that is high in dairy. By introducing the exogenous enzyme and the substrate at the same time, the digestion of the substrate can proceed almost normally within the GI tract.

Lactase, maltase, and sucrase in the brush border of the small intestine break down the disaccharides lactose, maltose, and sucrose, respectively, into monosaccharides:

— Lactose is broken down into galactose and glucose.

— Maltose is broken down into two glucose molecules.

— Sucrose is broken down into glucose and fructose.

Humans are unable to digest fiber, for example, cellulose (found in the cell walls of plants), because we lack the enzymes necessary to break down the β-acetyl linkages. However, dietary fiber has many effects on the GI tract (**Fig. 22.2**).

Fig. 22.2 ▶ **Fiber: Effects.**
In the stomach, fiber binds water, which enlarges the particle size so that fiber particles pass the pyloric sphincter later and therefore undergo delayed gastric emptying. In the ileum and colon the water-binding (swelling) capacity lowers transit time. Fiber may bind mineral and trace elements as well as fat-soluble vitamins, which may not allow them to be absorbed. The binding of steroids leads to an increased excretion of bile acids and cholesterol, which may be helpful in people with fat metabolism disorders. Glucose absorption is also delayed by high-fiber intake, improving glucose control in diabetics. Stool volume is increased, and the consistency of stool is softer with fiber intake. However, intestinal bacteria ferment the polysaccharides in fiber, producing methane and CO_2. During fermentation, short-chain fatty acids (FA) are produced that positively affect the composition of the intestinal flora and the intestinal pH. Finally, the binding of ammonia increases fecal nitrogen excretion, thereby unburdening the liver and kidneys.

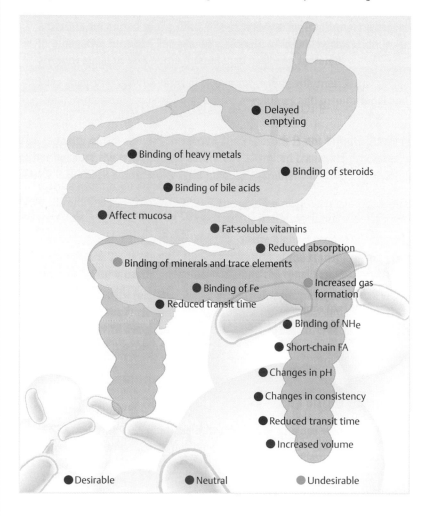

Carbohydrate Absorption

Only the monosaccharides galactose, glucose, and fructose are absorbed into the bloodstream (**Fig. 22.1**).

Active transport of monosaccharides from the lumen into the bloodstream occurs throughout the length of the small intestine. The rate of absorption is greatest in the proximal intestine and decreases distally. Under normal conditions, all carbohydrates have been absorbed by the time the chyme reaches the mid-jejunum.

Galactose and Glucose Absorption

Galactose and glucose are absorbed into enterocytes by Na^+-dependent cotransport (SGLT1). Galactose and glucose are transported "uphill," and Na^+ is transported "downhill." The Na^+ gradient is maintained by the $Na^+–K^+$ ATPase pump on the basolateral membrane of enterocytes. Galactose and glucose are then transported from enterocytes into the portal circulation by facilitated diffusion (via GLUT2 transporter).

Fructose Absorption

Fructose is absorbed by facilitated diffusion into enterocytes via GLUT5 on the luminal membrane and into blood via GLUT2 on the basolateral membrane. Enterocytes possess the appropriate biochemical pathways for the conversion of fructose to other substrates (e.g., glucose); this keeps the intracellular concentration low enough to support its continued facilitated diffusion into the cell.

22.2 Protein Digestion and Absorption

Protein Digestion

Gastric acid (HCl) denatures proteins, unfolding them to expose their bonds to pepsin, which breaks those bonds by hydrolysis (**Fig. 22.3**).

Fig. 22.3 ▶ **Protein digestion and absorption of peptides and amino acids.**
Ingested proteins are denatured by HCl. HCl also activates pepsin from pepsinogen (**1**). Pepsin becomes inactive in the small intestine (pH 7–8). Pancreatic juice also contains proenzymes of other peptidases that are activated in the duodenum (**2**). The endopeptidases trypsin, chymotrypsin, and elastase hydrolyze the protein molecules into short-chain peptides. Carboxypeptidase A and B (from the pancreas), as well as dipeptidases and aminopeptidase (brush border enzymes), also break proteins down into tripeptides, dipeptides, and amino acids. Tripeptides and dipeptides are absorbed by H^+-dependent active transport. Amino acids are transported into the portal blood by a number of specific carriers by Na^+-dependent active transport. (AA, amino acids)

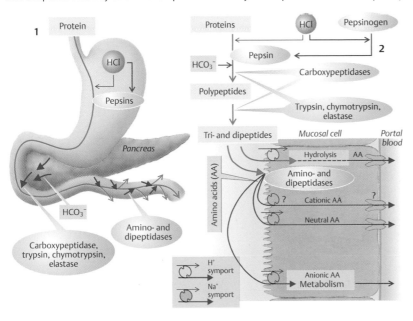

Pancreatic proteolytic enzyme precursors trypsinogen, procarboxypeptidase, chymotrypsinogen, and proelastase are secreted into the small intestine, where they are activated.

— Trypsinogen is activated to trypsin by enteropeptidase, an enzyme in the brush border.

— Trypsin activates procarboxypeptidase, chymotrypsinogen, and proelastase to their active forms (carboxypeptidase, chymotrypsin, and elastase, respectively) and stimulates production of more trypsin from trypsinogen.

Trypsin, carboxypeptidase, chymotrypsin, and elastase break down polypeptides into tri- and dipeptides.

Protein Absorption

Most amino acids and peptides are absorbed in the jejunum, but some amino acids may be absorbed in the ileum.

Free Amino Acids

Amino acids are absorbed from the lumen into enterocytes by a variety of Na^+-dependent and -independent cotransport systems in the luminal membrane. They are then transported into the portal circulation by facilitated diffusion.

Dipeptides and Tripeptides

Dipeptides and tripeptides are readily absorbed by enterocytes via H^+-dependent cotransport systems (PepT1) in the luminal membrane. Once inside enterocytes, peptides are hydrolyzed into free amino acids and pass into the portal circulation by facilitated diffusion.

22.3 Lipid Digestion and Absorption

Some digestion of lipids occurs in the stomach (10%); the remaining digestion and the entire absorption process occurs in the small intestine.

Lipid Digestion

Lipid Digestion in the Stomach

Contractions in the stomach stir and emulsify lipid droplets so lingual lipase can break them down (short-chain partially water-soluble lipids only) into monoglycerides and free fatty acids (**Fig. 22.4**).

Note: The effects of lingual lipase are functionally minor compared with pancreatic lipase.

Lipid Digestion in the Small Intestine

Emulsified lipids combine with bile salts in the duodenum to form simple micelles. Simple micelles are much smaller than emulsified lipid droplets and thus greatly increase their surface area. Bile salts prevent lipids from coalescing, but they also inhibit pancreatic lipase from binding to the droplets, limiting its effect. This is overcome by colipase, a polypeptide secreted by the pancreas. In the presence of colipase, pancreatic lipase is able to hydrolyze triglycerides into monoglycerides and free fatty acids. Similarly, cholesterol esterase hydrolyzes cholesterol; and phospholipase A_2 hydrolyzes phospholipids. These lipid digestion products and bile salts then combine to form mixed micelles, which have their hydrophilic polar heads on the outside and hydrophobic nonpolar tails on the inside. Mixed micelles are then able to penetrate through the unstirred water layer adjacent to the luminal surface of the brush border membrane, thereby facilitating lipid absorption.

Fig. 22.4 ► **Lipid digestion.**
After ingestion, first lingual lipase and then gastric lipase are secreted and mixed with the food particles. Both enzymes are active at low pH values. Gastric motility ensures thorough mixing with the enzymes and the breakdown of fat into smaller particles. The emulsified lipids are then released into the duodenum. Bile acids attach to the fat particles (termed *simple micelles*), causing their surface to become negatively charged, which allows colipase to attach to the triglycerides. Pancreatic lipase, which is inhibited by bile acids, now binds to colipase and hydrolyzes the triglycerides. Similarly, cholesterol esterase hydrolyzes cholesterol, and phospholipase A_2 hydrolyzes phospholipids. Together with bile acids, the resulting products of fat digestion assemble spontaneously into mixed micelles, with polar heads facing out and nonpolar tails inside the micelle. Mixed micelles are able to traverse the unstirred water layer above the microvilli, and in the slightly acidic environment found here, they spontaneously break apart, allowing their lipid content to be released and absorbed by enterocytes, largely by passive diffusion. The bile salt component of the mixed micelle is then recycled via the enterohepatic circulation.

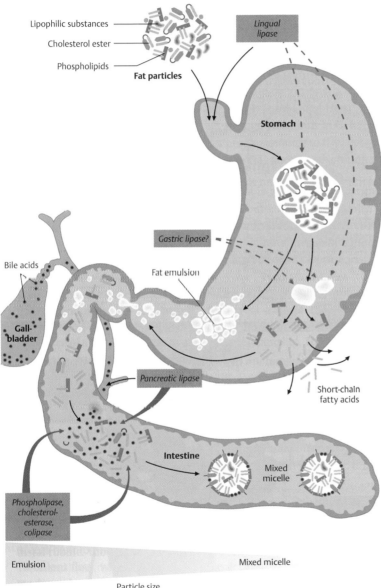

Lipophilic substances
Cholesterol ester
Phospholipids
Fat particles

Lingual lipase

Stomach

Gastric lipase?

Bile acids

Fat emulsion

Gall-bladder

Pancreatic lipase

Short-chain fatty acids

Phospholipase, cholesterol-esterase, colipase

Intestine

Mixed micelle

Emulsion

Mixed micelle

Particle size

Lipid Absorption

Short-chain fatty acids diffuse passively across the enterocyte membrane, but carrier-mediated absorption is now the proposed mechanism for the uptake of long-chain fatty acids (aliphatic chain > 12 carbons) and sterically large lipids (**Fig. 22.5**). The lipid transporters identified in the small intestine are the plasma membrane fatty acid–binding protein (FABPpm), the fatty acid transport protein 4 (FATP4), and the fatty acid translocase (FAT/CD36).

Intestinal cholesterol absorption is also reported to be facilitated by the scavenger receptor SR-BI and by Niemann–Pick C-1-like 1 (NPC1L1). This is the reason that the absorption of cholesterol can now be inhibited by drugs. Lipid transport may be bidirectional, and some lipids may be secreted by intestinal cells.

Once monoglycerides and free fatty acids are absorbed into enterocytes, they are reesterified into triglycerides and form droplets. A lipoprotein layer coats the droplet, forming a chylomicron. These leave the cell by exocytosis and enter the bloodstream via the lymphatic circulation because they are too large to enter the bloodstream directly.

Note: All ingested fat is digested and absorbed. The 2 to 4% of stool that is fat usually comes from exfoliated cells and colonic bacteria.

Fig. 22.5 ▶ **Lipid absorption.**
Dietary triglycerides (TG) are broken down into free fatty acids (FFA) and monoglycerides (MG) in the gastrointestinal (GI) tract. Short-chain FFA (e.g., acetic and butyric acid) do not need to be ferried across the unstirred water layer in mixed micelles because they are sufficiently water soluble to do this by themselves. In addition, they are also sufficiently lipid soluble to diffuse out of the enterocyte at the basolateral membrane and go directly to the bloodstream (portal venous blood), without the necessity of being incorporated into chylomicrons. Long-chain FFA and MG are not soluble in water. They are resynthesized to TG in the enterocytes. Because TG are not soluble in water, they are subsequently loaded into chylomicrons, which are exocytosed into the extracellular fluid, then passed on to the intestinal lymph (thereby bypassing the liver), from which they finally reach the greater circulation.

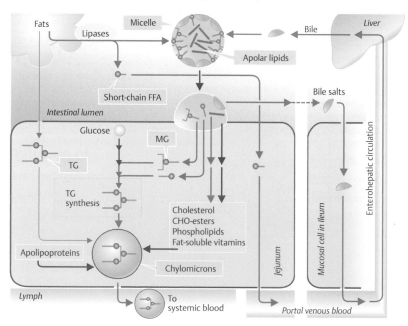

22.4 Absorption of Vitamins, Water, and Electrolytes

Absorption of Vitamins

Fat-soluble Vitamins

Fat-soluble vitamins (A, D, E, and K) require bile salts to facilitate micelle formation for efficient absorption. They are then incorporated into chylomicrons and enter the bloodstream via the lymphatic system.

Although diffuse intestinal absorption remains the only pathway described for vitamins D and K, the uptake of vitamins A and E may also be mediated by lipid transporters. SR-B1 is involved in the uptake of vitamin E, and NPC1L1 and ABCA1 may also play a role in the facilitated transport of vitamin A.

Water-soluble Vitamins

Water-soluble vitamins are (mostly) absorbed by Na^+-dependent cotransport systems (**Fig. 22.6**).

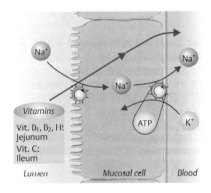

Fig. 22.6 ▶ **Absorption of water-soluble vitamins.** Water-soluble vitamins (except vitamin B_{12}) are absorbed by Na^+-dependent cotransport in the small intestine. (ATP, adenosine triphosphate)

Vitamin B_{12} absorption, however, is more complicated and involves several steps (**Fig. 22.7**): B_{12} is released from dietary proteins by gastric acid. It then binds to R proteins, which are secreted in saliva. In the duodenum, trypsin digests the R protein, liberating B_{12}, which then forms a complex with intrinsic factor (IF), a glycoprotein secreted by gastric parietal cells. This B_{12}–IF complex is resistant to the effects of trypsin and travels to the terminal ileum, where it binds to specific receptors and is absorbed.

Vitamin K

Vitamin K is absorbed in the small intestine during lipid digestion and transported to the liver in chylomicrons. It is required for blood coagulation. Inactive precursors of the coagulation factors are synthesized in the liver and activated by γ-glutanyl carboxylase. Vitamin K is a cofactor for this reaction. Vitamin K results in the activation of factors II, VII, IX, and X, which are then released into the bloodstream.

Vitamin C (ascorbic acid)

The majority of vitamin C is absorbed in the proximal small intestine and is excreted by the kidneys. However, increasingly large amounts are excreted in feces following megadoses. Vitamin C is a reductant substance, and this property allows it to act as a H^+ donor in hydroxylation reactions (e.g., during the biosynthesis of catecholamines when it is a cofactor for dopamine-β-hydroxylase). Other biological effects are based on different mechanisms, some of which are unknown (e.g., it is involved in collagen biosynthesis). It is also involved in the synthesis of bile acids from cholesterol, the synthesis of hormones (e.g., gastrin, gastrin-releasing peptide [GRP, bombesin], corticotropin-releasing hormone [CRH], and thyrotropin-releasing hormone [TRH]), the synthesis of cytochrome P-450 enzymes in the liver. It enhances iron absorption (by reducing Fe^{3+} to Fe^{2+}) and it has well known antioxidant properties.

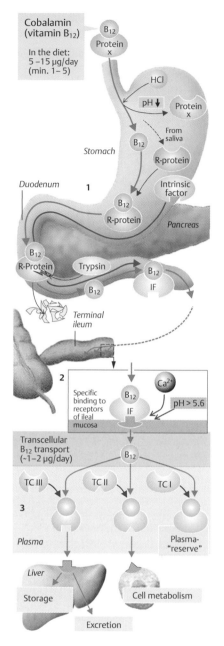

Fig. 22.7 ▸ Vitamin B₁₂ absorption.

Vitamin B_{12} is a relatively large and lipophobic molecule that requires transport proteins. During passage through the GI tract, plasma, and other compartments, vitamin B_{12} binds to intrinsic factor (IF), which is secreted by gastric parietal cells; transcobalamin II (TC II) in plasma; and R proteins in plasma (TC I), granulocytes (TC III), saliva, bile, milk, and other agents. Gastric acid releases vitamin B_{12} from dietary proteins. In most cases, the B_{12} then binds to R protein from saliva or (if the pH is high) IF (**1**). The R protein is digested by trypsin in the duodenum, which liberates B_{12}, which is then bound by IF (trypsin resistant). The mucosa of the terminal ileum has highly specific receptors for the B_{12}–IF complex. It binds to these receptors, and B_{12} is absorbed by receptor-mediated endocytosis, provided the pH is > 5.6 and Ca^{2+} ions are available (**2**). B_{12} then binds to TC I, II, and III in plasma (**3**). TC II mainly distributes B_{12} to all cells undergoing division. TC III (from granulocytes) transports excess B_{12} to the liver, where it is either stored or excreted as bile. TC I has a half-life of ~10 days and serves as a short-term reservoir of B_{12} in plasma.

Absorption and Secretion of Water and Electrolytes

An average adult ingests 1 to 2 L of water per day, but the fluid load to the small intestine is 9 to 10 L, 8 to 9 L being added by the secretions of the GI system. Most absorption of water and electrolytes occurs in the small intestine, with some water absorbed in the colon as well.

Absorption of NaCl (Fig. 22.8)

Na^+ is absorbed from the lumen into enterocytes by

— passive diffusion through Na^+ channels

— cotransport with glucose and amino acids

— Na^+–H^+ exchange

— neutral Na^+–Cl^- cotransport

Cl^- is absorbed via

— cotransport with Na^+

— $Cl^--HCO_3^-$ exchange

Aldosterone stimulates Na^+ (and Cl^-) absorption by stimulating the Na^+-K^+ ATPase pump.

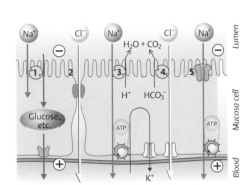

Fig. 22.8 ▸ **Na^+ and Cl^- absorption.**
Na^+ is absorbed by various mechanisms and the Na^+-K^+ ATPase in the basolateral membrane is the primary driving mechanism for all of them. Na^+ passively diffuses into cells in the duodenum and jejunum via symport carriers, which actively cotransport glucose, amino acids, and other compounds (**1**). A lumen-negative potential forms that drives Cl^- out of the lumen (**2**). Na^+ ions in the lumen of the ileum are exchanged for H^+ ions (**3**), while Cl^- is exchanged for HCO_3^- at the same time (**4**). The H^+ and HCO_3^- combine to form H_2O and CO_2, which diffuse out of the lumen (not shown in figure). Na^+, Cl^-, and H_2O (by osmosis) are absorbed by this electroneutral transport mechanism. Na^+ in the colon is mainly absorbed through luminal Na^+ channels (**5**). This type of transport is electrogenic and aldosterone-dependent. The lumen-negative potential either leads to K+ secretion or drives Cl- out of the lumen (**2**).

Absorption of Water

The absorption of solutes (e.g., Na^+ and Cl^-) makes the lumen of the intestine slightly hypotonic, which then creates the osmotic gradient for water to move paracellularly. This keeps the lumen close to isotonic.

Absorption and Secretion of K^+

K^+ is absorbed passively in the small intestine. Most of its movement is paracellular. As Na^+ and H_2O are absorbed, the volume of the luminal contents decreases, resulting in an increase in $[K^+]$, permitting passive diffusion.

K^+ is secreted in the colon (mechanism is the same as that for K^+ secretion in the distal tubule of the kidney). This is stimulated by aldosterone via stimulation of the Na^+-K^+ ATPase pump.

Note: K^+ secretion is a function of the luminal concentration. The colon can absorb or secrete K^+ depending on the local conditions.

Absorption of Ca^{2+}

Calcium ions are actively absorbed in the proximal small intestine and passively by paracellular diffusion in the rest of the small intestine. The active absorption is dependent upon 1,25-dihydroxycholecalciferol ($1, 25-(OH)_2-D_3$), the active form of vitamin D, which is synthesized in the kidney.

Active transcellular transport of Ca^{2+} occurs across the apical membrane via channel-like Ca^{2+} transporter (CaT1), through the cytoplasm via calbindin 9k (CaBP9k), and across the basolateral membrane via the Na^+-Ca^{2+} exchanger (NCX1), along with the plasma membrane Ca^{2+} ATPase (PMCA1).

Cholera

Cholera is an infectious disease caused by *Vibrio cholerae*, a gram-negative bacteria. It is spread via the fecal-oral route. *V. cholerae* toxin increases cyclic adenosine monophosphate (cAMP) concentrations in intestinal mucosal cells, causing the opening of Cl^- channels and massive secretion of Cl^-. This results in the production of a profuse amount of watery diarrhea, which, in turn, causes severe dehydration. This can lead to kidney failure, shock, coma, and death. Treatment requires rapid replacement of lost body fluids with oral or intravenous solutions containing salts and sugar. Tetracycline reduces fluid loss and diminishes transmission of the bacteria.

Diarrhea

Exudative diarrhea is a consequence of gross structural damage to the intestinal or colonic mucosa. It is characterized by the presence of blood in feces and frequent passage of a small volume of feces. Examples of causes of exudative diarrhea are inflammatory bowel diseases such as ulcerative colitis and Crohn disease, mucosal invasion by a protozoan (e.g., *Entamoeba histolytica*), and certain bacterial infections (e.g., *Shigella* and *Salmonella*).

Secretory diarrhea occurs in disorders in which there is increased secretion of fluids and electrolytes (e.g., cholera) or inhibited reabsorption of fluids and electrolytes.

Osmotic diarrhea occurs when ingested solutes that are osmotically active are retained in the lumen of the GI system. Because water must also be retained to maintain isotonicity, the volume of fluid delivered to the colon may exceed its absorptive capacity. Osmotic diarrhea may result from ingestion of substances that are difficult to absorb (e.g., $MgSO_4$, a common ingredient in laxatives), limited absorption of solutes, or resection of the jejunum or ileum, which decreases the surface area for absorption of digested food. It may also occur if bile salts enter the colon, both because they are irritants and because they present an osmotic load.

Iron deficiency anemia

Iron deficiency anemia is a condition in which there is insufficient hemoglobin in red blood cells due to a lack of iron (which is an essential component of heme). Approximately two thirds of the iron content of the body is bound to hemoglobin. It is usually caused by blood loss from menses in premenopausal women, but it can also be due to inadequate dietary intake of iron, GI bleeding (e.g., from peptic ulcers, long-term nonsteroidal antiinflammatory drug use, or certain GI cancers), from GI conditions that decrease the absorption of iron (e.g., Crohn disease), or by pregnancy (in which the need for iron is increased due to the increase in maternal blood volume and for fetal hemoglobin synthesis). Treatment involves replenishment of iron in the form of ferrous sulfate tablets.

Hemochromatosis

Hemochromatosis is a condition in which there is failure of regulation of iron absorption in the bowel. This leads to excessive iron in the body, which then gets deposited in organs such as the liver, heart, pancreas, and the pituitary gland. Iron toxicity causes lipid peroxidation of cell membranes, DNA damage, and increased collagen formation. Signs include fatigue, arthralgia, skin pigmentation, liver disease (cirrhosis and carcinoma), cardiomyopathy, heart failure, arrhythmias, diabetes, and infertility. Management is by the drainage of venous blood until the patient is iron deficient.

Antibiotic-associated pseudomembranous colitis

Pseudomembranous colitis is inflammation of the colon due to superinfection with *Clostridium difficile*, a gram-positive bacillus. It typically occurs following a course of antibiotic treatment in which the normal gut commensal bacteria are eradicated, allowing *C. difficile* to colonize the gut unimpeded. The most common antibiotics that cause this condition are the penicillins, cephalosporins, fluoroquinolone, and clindamycin. Symptoms of pseudomembranous colitis include diarrhea, fever, and abdominal pain. It is treated with metronidazole or vancomycin.

Absorption of Iron

Iron ions are mainly absorbed in the proximal small intestine.

Ingested ferric ions (Fe^{3+}) are reduced to ferrous ions (Fe^{2+}) by acid in the stomach and by the enzyme ferric reductase, which is present on the brush border of the proximal small intestine. Gastric acid also allows iron to form complexes with substances such as ascorbic acid that reduce it to the Fe^{2+} form.

Free Fe^{2+} is transported by the divalent metal ion transporter (DMT1) into enterocytes in association with H^+. Fe^{2+} is then transported by ferroportin 1, a transporter on the basolateral membrane, into the blood. This process is facilitated by a protein called hephaestin, which is associated with ferroportin 1. Fe^{2+} is transported in blood bound to transferrin. It is stored in the liver and transported to the bone marrow for the synthesis of hemoglobin.

The overall absorption and secretion of water and electrolytes in the gut are shown in **Fig. 22.9.**

Fig. 22.9 ► **Water and electrolyte absorption and secretion.**
H_2O is primarily absorbed in the jejunum and the ileum, with smaller amounts being absorbed in the colon. When solutes (e.g., Na^+ and Cl^-) are absorbed, H_2O follows by osmosis. Na^+ reabsorption in the colon is dependent on aldosterone. This also causes the secretion of K^+. HCO_3^- is secreted throughout the GI system except the jejunum, where it is absorbed. Ca^{2+} and Fe^{2+} are absorbed in the duodenum. Active Ca^{2+} absorption requires calcitriol.

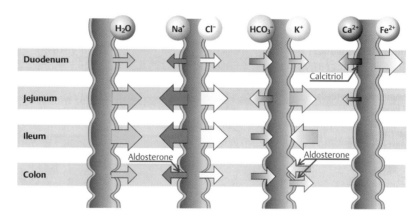

Review Questions

1. Most vagal preganglionic efferent nerves to the gastrointestinal (GI) tract synapse with
   A. smooth muscle cells.
   B. endocrine cells.
   C. neurons in the myenteric plexus.
   D. ganglia outside the GI tract.
   E. exocrine glands.

2. Peristalsis is characterized by
   A. intestinal smooth muscle contractions that are controlled by only extrinsic neural input.
   B. contractions that proceed in an aboral direction down the gut for a distance of 1 to 2 ft.
   C. contractions that can start only in the duodenum.
   D. contractions that are prominent during fasting.
   E. the gastrocolic reflex.

Questions **3** to **6** refer to the clinical scenario that follows.
A 50-year-old female patient has symptoms that include heartburn, reflux of blood, and waking up in the night coughing and choking with a mouthful of bitter-tasting fluid. She reports that this problem has been worsening over the past several months. Other symptoms noted by the patient are regurgitation of carbonated drinks, a persistent cough that is worse during the night, and recurrent substernal discomfort that can sometimes be relieved by over-the-counter (OTC) antacids. When asked, the patient reports being in generally good health for the past year, with no loss of appetite; her weight is unchanged relative to her last two annual checkups at her physician's office.

3. What is the most likely cause of the patient's symptoms?
   A. Weakness of the cricothyroid muscle
   B. Diffuse esophageal spasm
   C. An incompetent lower esophageal sphincter (LES)
   D. Achalasia
   E. Nonulcer dyspepsia

4. Which procedure would be most helpful in determining the extent of reflux-induced damage to this patient's esophageal mucosa?
   A. Endoscopy
   B. Radiologic examination
   C. Esophageal pressure measurement
   D. Nuclear magnetic resonance (NMR) visualization of the lower esophagus
   E. Ultrasonic visualization of the lower esophagus

5. Which treatment would be most helpful to the patient?
   A. Vagotomy
   B. Reducing the intake of fatty foods
   C. Administration of atropine
   D. Administration of gastrin
   E. A proton pump inhibitor

6. Under which of the following circumstances would this patient's symptoms worsen?
   A. During increased intragastric pressure
   B. Following administration of secretin
   C. Following administration of somatostatin
   D. If the patient takes OTC antacids
   E. If the patient is given metoclopramide to augment cholinergic neuroeffector transmission

7. Which condition would you expect to find in a patient after resection of the fundus and upper part of the corpus of the stomach?
   A. Increase in the rate of slow-wave depolarizations
   B. Increased secretion of pepsinogen
   C. Decreased gastrin production
   D. Failure of the stomach to empty
   E. Loss of receptive relaxation of the stomach

8. You would expect gastric emptying to slow in response to
   A. a high-fat meal.
   B. elevated levels of motilin.
   C. a high pH in the duodenal lumen.
   D. an isotonic NaCl solution.
   E. vagotomy.

9. Stimulation of the parasympathetic nerves innervating the parotid glands causes
   A. decreased concentration of HCO_3^- in saliva.
   B. decreased O_2 consumption.
   C. increased volume of secretion.
   D. vasoconstriction in the parotid gland.
   E. nausea.

10. A patient with a duodenal ulcer is given a test meal, and the serum levels of several hormones are measured. The hormonal response to the test meal is repeated 24 hours after administration of the H^+-K^+ ATPase inhibitor omeprazole. The serum level of which of the following would be higher following drug treatment?
    A. Gastrin
    B. Cholecystokinin
    C. Secretin
    D. Somatostatin
    E. Histamine

11. Which of the following substances when present in the duodenal lumen would produce the greatest stimulus for pancreatic exocrine secretion?
    A. Lipid and glucose
    B. K^+ and HCO_3^-
    C. Na^+ and H^+
    D. Amino acids and Fe^{2+}
    E. H^+ and lipid

12. A 36-year-old male patient is about to be treated for Crohn disease by surgical removal of his terminal ileum. You are charged with explaining to the patient the nature of the postoperative complications he is likely to experience. Which of the following would you tell him to anticipate?
    A. Reduced Ca^{2+} and Fe^{2+} absorption
    B. Lowered serum cholesterol and pernicious anemia
    C. Decreased hepatic bile salt synthesis and an increased risk of pancreatitis
    D. Peripheral edema linked to protein malabsorption
    E. Night blindness as a consequence of fat-soluble vitamin deficiency

13. In humans, which of the following would be an effective source of energy in the absence of exocrine pancreatic secretion?
    A. Glycogen
    B. Sucrose
    C. Cellulose
    D. Starch
    E. Amylose

14. When chyme first enters the duodenum, the direction of net flux of water is primarily determined by the
    A. rate of Na^+ absorption.
    B. rate of organic solute absorption.
    C. rate of Cl^- absorption.
    D. osmolarity of the chyme.
    E. serum secretin concentration.

15. A patient had portions of his ileum and ascending colon removed after complications to repair an inguinal hernia. His subsequent steatorrhea and weight loss were treated with a low-fat, low-oxalate, and reduced lactose diet. What is the cause of his low serum Ca^{2+}?
    A. Increased Ca^{2+} secretion by the pancreas
    B. Increased binding of Ca^{2+} to the pancreatic enzymes
    C. Increased binding of Ca^{2+} to unabsorbed fatty acids
    D. Increased binding of Ca^{2+} to oxalate
    E. Increased binding of Ca^{2+} to unabsorbed proteins

Questions **16** and **17** refer the clinical scenario that follows.
A patient who has recently returned from an extended stay in a developing country develops a diarrhea that is characterized by the passage of a large volume of fluid, and he rapidly becomes dehydrated. Although the patient reports not eating for almost 48 hours, his diarrhea shows no signs of stopping, and although he has tried to increase his fluid intake, he continues to be severely dehydrated.

16. Which of the following is the most likely cause of this patient's symptoms?
    A. Celiac disease (sprue)
    B. Irritable bowel syndrome (IBS)
    C. Cholera enterotoxin
    D. Ulcerative colitis
    E. Crohn disease

17. Which of the following would be the most appropriate solution for oral rehydration of this patient?
    A. Distilled water
    B. Distilled water with glucose
    C. Isotonic NaCl
    D. A slightly hypotonic NaCl and glucose solution
    E. An isotonic NaCl and glucose solution

Answers and Explanations

1. **C.** Most extrinsic nerves innervating the GI tract exert their effects via the myenteric plexus (p. 202).
   A,B,D,E Very few efferents synapse directly on smooth muscle cells, endocrine cells, ganglia outside the GI tract, or exocrine glands.

2. **B.** Peristalsis moves a bolus a short distance along the gut in an aboral direction (p. 206).
   A Peristalsis involves intrinsic but not extrinsic innervation of the gut.
   C Peristalsis can start anywhere along the length of the small or large intestine.
   D Migrating myoelectric complex (MMC) refers to those GI contractions that take place during fasting. Peristalsis are those contractions that occur in the GI in response to stretching of the gut wall, and is most important following gastric emptying.
   E The gastrocolic reflex is characterized by increased colonic motor activity shortly after ingestion of a meal.

3. **C.** This patient is suffering from gastroesophageal reflux disease (GERD), and she is refluxing acidic gastric contents into her esophagus through an incompetent (partially open) LES. The "bitter taste" indicates the gastric origin of the refluxate because this shows it to be acidic. At this point, she is exhibiting symptoms consistent with erosive esophagitis (bleeding from the inflamed esophagus), suggesting that this is a long-standing problem (p. 207).
   A The cricothyroid muscle supports the larynx, not the esophagus.
   B Patients with diffuse esophageal spasm report excruciating pain during swallowing, this is a motility disorder that is not characterized by mucosal damage or regurgitation of the gastric contents into the mouth.
   D Based on this patient's general health, appetite, and weight, she is not nutritionally compromised, which indicates that food is passing from her esophagus to her stomach without difficulty. For this reason, and in the absence of any other likely related problems (e.g., aspiration pneumonia), a diagnosis of achalasia (failure of the LES to open) does not match her presenting symptoms. Although achalasia patients may report regurgitation, this refluxate does not have a "bitter taste" because it originates in the distal esophagus, not the acidic gastric lumen.
   E Symptoms of nonulcer dyspepsia include abdominal discomfort, bloating, nausea, and early satiety. This is a functional bowel disorder that is not associated with pathophysiological damage to any structures in the upper GI tract. The presence of blood in this patient's reflux precludes a diagnosis that is characterized by a lack of structural damage.

4. **A.** Endoscopy provides direct visualization of the mucosal surface.
   B,D,E Radiology, NMR, and ultrasound all show the structure of the esophagus as a whole, but not the integrity of the mucosal epithelium.
   C Pressure measurements may give an indication of altered motility linked to esophageal damage, but this tells us nothing quantitative about the mucosal surface itself.

5. **E.** A proton pump inhibitor will reduce the acidity of the gastric contents refluxing into the esophagus and allow mucosal healing to take place; these drugs are pharmacologically selective, essentially acting at only one anatomical location (parietal cells), and so would be the treatment of choice in this case (p. 207).
   A Vagotomy will help to reduce acid secretion, but it is an invasive surgical procedure, and its effect on motility due to loss of receptive relaxation likely outweighs any potential benefits under these circumstances.
   B Low-fat meals empty more rapidly from the stomach and provide a lesser stimulus for acid secretion, but this alone will not significantly alter the reflux of acid into the distal esophagus.
   C Muscarinic cholinergic antagonists (atropine-like drugs) can also reduce acid secretion, but they are so nonselective in terms of where they act that the side effects would outweigh any benefits for this patient.
   D Gastrin stimulates gastric acid secretion, which will make the patient's condition worse. Note that this patient may benefit from a referral to have an assessment for a fundoplication to increase LES pressure.

6. **A.** Increased intragastric pressure increases reflux and would worsen the patient's symptoms (p. 207).
   B,C Secretin and somatostatin both reduce acid secretion and would lessen the patient's symptoms.
   D Antacids can provide symptomatic relief because they neutralize gastric acid in the distal esophagus and reduce the sensation of discomfort that causes.
   E Metoclopramide is one of a class of drugs known as "prokinetic agents" because they directly or indirectly stimulate GI smooth muscle. In this case, the drug would augment LES tone and, along with an appropriate acid-reducing drug, help to reduce the reflux of acidic gastric contents into the distal esophagus, thus reducing the patient's symptoms.

7. **E.** Receptive relaxation occurs in the proximal stomach (fundus and corpus), so this would be lost following resection (p. 208).
   A Slow-wave depolarizations are not affected by the presence of the proximal stomach.
   B Some pepsinogen is secreted by the proximal stomach, so secretion decreases.
   C Gastrin is secreted in the distal stomach, so it is unaffected by this surgery.
   D Stomach motility is reduced by this surgery, but the stomach still empties.

8. **A.** Lipid in the duodenal lumen inhibits gastric emptying via the enterogastric reflex (p. 209).

   B Motilin has a promotility action on the stomach; for example, it plays an important controlling role in migrating myoelectric complex (MMC), powerful peristaltic contractions that remove non-digestible solids left behind in the stomach and small intestine.

   C Gastric emptying is slowed by low pH in the duodenal lumen due to the presence of acidic chyme. The latter stimulates the release of secretin which slows gastric emptying by decreasing gastric motility and by increasing the tone of the pyloric sphincter. This is part of the hormonal mediation of the enterogastric reflex.

   D As a consequence of the enterogastric reflex, isotonic NaCl empties more rapidly than solutions that are hyper- or hypotonic.

   E Vagotomy causes the stomach to lose the ability to expand without contracting as the volume of the luminal contents increases (receptive relaxation); in the absence of this vagally mediated inhibition, gastric motility and rate of emptying are increased.

9. **C.** Parasympathetic efferent fibers stimulate salivation via muscarinic M_3 receptors, resulting in increased volume of secretion (p. 215).

   A The concentration of HCO_3^- in saliva increases as the rate of salivary secretion is increased.

   B O_2 consumption is increased by parasympathetic stimulation of the parotid gland.

   D Parasympathetic stimulation causes the release of kallikrein, which activates bradykinin. Bradykinin is a potent vasodilator substance.

   E Stimulation of the salivary glands via their parasympathetic innervation does not cause nausea, but nausea, as a prelude to vomiting, can be associated with increased salivary secretion.

10. **A.** Because accumulation of gastric acid in the antrum inhibits gastrin secretion via the release of somatostatin, gastrin levels increase when less acid is "pumped" into the stomach. This effect, however, will take time to manifest (p. 219).

   B Secretion of cholecystokinin is not affected by changes in gastric acid production.

   C Secretin release is stimulated by acid exiting the stomach and entering the duodenum, so its plasma concentration would be reduced in a patient taking a proton pump inhibitor such as omeprazole.

   D Somatostatin is a paracrine substance, and although its release would be reduced if gastric acid production falls, it does not travel in the systemic circulation.

   E Histamine secretion is stimulated by gastrin, so its levels increase, but it acts locally and does not change its concentration in blood.

11. **E.** Lipid in the duodenum stimulates secretion of the hormone cholecystokinin, and H^+ stimulates the release of the hormone secretin. In combination, these hormones provide the largest stimulus for pancreatic exocrine secretion (pp. 224, 225).

   A–C Glucose, K^+, Na^+, Fe^{2+}, and HCO_3^- in the duodenal lumen do not affect the volume of pancreatic exocrine secretion either directly or indirectly.

   D Amino acids only stimulate secretion of cholecystokinin, and Fe^{2+} does not affect the volume of pancreatic exocrine secretion.

12. **B.** Serum cholesterol decreases because it is used for hepatic synthesis of bile salts, and removal of the terminal ileum would interrupt the enterohepatic circulation by which this pool is returned to the liver from the small intestine. Vitamin B_{12} (necessary for red blood cell formation) can only be absorbed in the terminal ileum. In its absence, absorption ceases, hepatic stores of this vitamin become depleted, and pernicious anemia follows (pp. 223, 253).

   A Ca^{2+} and Fe^{2+} absorption occurs in the most proximal regions of the small intestine and would be unaffected by this surgical procedure.

   C The reduction of the circulating bile salt pool would cause increased hepatic synthesis of bile salts, but there is no relationship between this phenomenon and the pancreas.

   D Loss of the terminal ileum would not significantly change protein digestion or amino acid absorption, so a negative nitrogen balance and peripheral edema will not be a concern for this patient.

   E Although this patient may initially experience diarrhea, lipid and therefore fat-soluble vitamin (A, D, K, and E) absorption is largely complete before the luminal contents reach the terminal ileum, so the patient is unlikely to experience night blindness as a consequence of long-term vitamin A malabsorption.

13. **B.** Sucrose can be digested by the brush border enzyme sucrase to produce one molecule of glucose and one molecule of fructose, each of which can be absorbed by the enterocyte and used for adenosine triphosphate (ATP) formation (p. 228).

   A,D,E The other carbohydrates listed require a contribution from pancreatic amylase for digestion to occur.

   C Cellulose is not digested to produce tri-, di-, or monosaccharides in the human small intestine, although this does occur in herbivores.

14. **D.** Chyme is initially adjusted to isotonicity by secretion (hypertonic chyme) or absorption (hypotonic chyme) of water. This process is facilitated by the "leaky" duodenal mucosa, which results from the large intercellular spaces in the epithelium (p. 235).

   A–C After adjustment to isotonicity, absorption of water is dependent upon solute absorption.

   E Secretin does not affect absorption of water.

15. **C.** Ca^{2+} binds to free fatty acids more readily than to pancreatic enzymes, oxalate, or proteins (B,D,E).

   A The pancreas does not secrete Ca^{2+}.

16. C. The enterotoxin released by the bacterium *Vibrio cholerae* elicits uncontrolled intestinal secretion of Cl^- via the cystic fibrosis transmembrane conduction regulator (CFTR), and this draws massive quantities of water into the gut lumen. The fluid accumulating in the lumen by this mechanism stimulates motility, which in turn results in a decreased reabsorption time and extensive fluid loss from the body. Given the patient's recent travel history, this option is the best match for his symptoms (p. 234).

A In celiac disease, the patient exhibits symptoms that are most likely to occur shortly after eating any gluten-containing food; this can be alleviated by diet modification. The continued presence of symptoms after a 48-hour fast should exclude this condition as a diagnosis.

B Patients with IBS may report, among other things, abdominal discomfort, bloating, gas, diarrhea and/or constipation, and/or a sensation of incomplete evacuation, but this condition is not associated with the severe dehydration and volume loss this patient is reporting.

D,E Ulcerative colitis and Crohn disease, collectively known as inflammatory bowel disease, are associated with abdominal pain and bloody diarrhea, but not normally extensive fluid loss and extreme dehydration, particularly the first time the patient presents, so these options do not correlate well with this patient's symptoms.

17. D. Oral replacement therapy includes NaCl plus glucose. This works because substrate-coupled cotransport of glucose and sodium into the enterocyte via the Na^+-dependent cotransport (SGLT1) is not disrupted by cholera toxin or the elevated intracellular cyclic adenosine monophosphate (cAMP) that it causes in intestinal epithelial cells; the inward movement of Na^+ by this mechanism draws water out of the intestinal lumen and opposes the fluid loss associated with cholera toxin–induced hypersecretion of Cl^- (pp. 233–235).

A Distilled water alone does not work because there is no net driving force that would take it out of the intestinal lumen and into the body, so the dehydration would just continue even if it were to be given.

B,C Na^+ facilitates glucose absorption which serves as an energy (adenosine triphosphate [ATP]) source. NaCl is required to replace losses from the diarrhea.

E Slightly hypotonic is a better treatment than isotonic because this favors water absorption from the gut lumen.

23 General Principles of Endocrine Physiology

The endocrine and nervous systems are the major control systems of the body. The endocrine system regulates functions by releasing hormones, whereas the nervous system regulates functions by releasing neurotransmitters. There is some overlap between hormones and neurotransmitters; for example, epinephrine acts as both.

23.1 Hormones

Hormone Synthesis

Protein and Peptide Hormone Synthesis

Protein and peptide hormones are synthesized in the endoplasmic reticulum, where they are assembled into chains of amino acid residues and then folded.

— These hormones are hydrophilic and most bind to plasma proteins in the blood. They require membrane receptors to act on cells.

Tyrosine-derived Hormone Synthesis

Tyrosine-derived hormones include norepinephrine, epinephrine, dopamine, and the thyroid hormones triiodothyronine (T_3) and thyroxine (T_4).

— Norepinephrine and dopamine are synthesized in nerve terminals. Epinephrine (and some norepinephrine) is synthesized in the adrenal medulla (see pages 276–277). These hormones are hydrophilic and act on membrane receptors.

— Thyroid hormones are synthesized in thyroid follicular cells (see pages 251–254). They are lipophilic and so can diffuse into cells and act on cytoplasmic or nuclear receptors.

Steroid Hormone Synthesis

Steroid hormones are synthesized from cholesterol in the adrenal cortex, testes, and ovaries (see page 270).

— Steroid hormones are lipophilic and therefore diffuse into cells and act on cytoplasmic receptors.

— Eicosanoids are derived from fatty acids, mainly arachidonic acid. They interact with cell membrane receptors.

Table 23.1 provides examples of each of the hormone types discussed.

Table 23.1 ▶ Examples of Hormone Classes			
Class	**Type**	**Receptor Location**	**Examples**
Protein	Protein	Cell membrane	Insulin, parathyroid hormone, calcitonin, and glucagon
	Peptide	Cell membrane	Oxytocin and antidiuretic hormone
Tyrosine	Hydrophilic	Cell membrane	Epinepherine, norepinephrine, dopamine
	Lipophilic	Cytoplasm	T_3 and T_4
Steroid	Steroid	Cytoplasm	Progesterone, cortisol, testosterone, aldosterone, and estradiol
	Eicosanoid	Cell membrane	Prostaglandins, prostacyclin, thromboxane, and leukotrienes

---

◼ Eicosanoid synthesis

Eicosanoids (*eicosa* = Greek for 20) are a group of autocoids (biological factors synthesized and released locally that play a role in vasoconstriction, vasodilation, and/or inflammation) that are derived from the 20-carbon cell membrane fatty acid arachidonic acid. Arachidonic acid forms arachidonate when acted upon by phospholipase A_2. Phospholipase A_2 is activated by hormones or other stimuli and is inactivated by steroids. Arachidonate is then further metabolized by cyclooxygenases (COX-1, COX-2, and COX-3) to form prostaglandin H_2. PGH_2 is the parent substance for prostacyclins, prostaglandins, and thromboxanes. In a different pathway, PGH_2 may be acted upon by lipooxygenase, forming hydro- and hydroperoxy fatty acids, which then form leukotrienes. Acetylsalicylic acid and related nonsteroidal antiinflammatory drugs (NSAIDs) inhibit the cyclooxygenases, thereby blocking the formation of most eicosanoids. These drugs have analgesic, antipyretic, and antirheumatic effects.

Hormone Transport

Catecholamines, peptides, and small proteins generally circulate in free form.

Steroid hormones, and thyroid hormones, and some protein hormones are bound to transport proteins. There are nonselective transporters (e.g., albumin and prealbumin) and specific transporters (e.g., thyroxine-binding globulin [TBG] and corticosteroid-binding globulin [CBG]). These are produced in the liver. The function of the transporters is to keep hormones in an inactive state until the target is reached, to act as a reservoir, to prevent small hormone molecules from passing into the urine, to slow the actions of the liver to transform a hormone into an inactive form, and to regulate a hormone's half-life.

Hormone Release

Hormone release may be initiated by neuronal stimulation, by the action of releasing hormones, or as a direct response to fluctuating plasma levels. **Table 23.2** provides examples of hormones released by these mechanisms.

Table 23.2 ▶ General Stimuli for Hormone Release

Stimulus for Release	Examples
Direct neural innervation	Hypothalamic hormones Posterior pituitary hormones Catecholamines (from the adrenal medulla)
Releasing hormones (from the hypothalamus)	All anterior pituitary hormones
Plasma levels	Ca^{2+} and glucose

Hormone Regulation

The release of hormones is tightly regulated because very low levels of hormones in the blood can act quickly and have powerful effects on the body. Regulation of secretion occurs by feedback control. Regulation of the duration and magnitude of response occurs via the rate of degradation in the bloodstream (half-life) and upregulation or downregulation of receptors.

Feedback Control

Negative feedback. Negative feedback occurs when hormones feedback at any preceding point in the cascade to directly or indirectly inhibit its own secretion or that of another hormone.
- This is the most common mechanism of hormonal regulation and is a self-limiting process.

Note: In pathological hormone derangement, negative feedback can steer hormone production toward normal, but is generally ineffective in overcoming the original defect.

Positive feedback. Positive feedback is when a hormone directly or indirectly stimulates its own secretion or that of another hormone. This system is designed to be self-perpetuating and to increase the magnitude of the response in target tissues.
- In hormonal terms, this is a rare form of regulation.

Regulation of Receptors

Upregulation of receptors. Upregulation of receptors is an increase in the number of receptors or their affinity for a hormone.

Downregulation of receptors. Downregulation of receptors is a decrease in the number of receptors or their binding affinity for a hormone.

■ Half-life of hormones

The half-life of any substance in the bloodstream is the duration of time until its concentration decreases by one-half of its original concentration. Steroid hormones are largely bound to plasma proteins, so they have a half-life of 1 or 2 hours. Thyroxine is bound and has the longest half-life of 5 to 7 days. Catecholamines have the shortest half-life of < 1 minute.

— This occurs if hormone secretion is abnormally high for an extended period.
Table 23.3 provides examples of each of the mechanisms of hormonal regulation.

Table 23.3 ▸ **Examples of Mechanisms of Hormonal Regulation**	
Mechanism of Hormonal Regulation	**Example**
Negative feedback	If plasma cortisol levels are elevated, cortisol causes feedback inhibition of CRH release from the hypothalamus and ACTH release from the anterior pituitary.
Positive feedback	The LH surge that initiates ovulation occurs due to the positive feedback of estrogen on the anterior pituitary. LH then acts on the cells in the ovaries to secrete more estrogen.
Upregulation of receptors	The presence of estrogen upregulates the receptors for estrogen and for LH in the cells of the ovary.
Downregulation of receptors	High levels of insulin cause target cells to remove insulin receptors from their cell membranes.

Abbreviations: ACTH, adrenocorticotropic hormone; CRH, corticotropin-releasing hormone; LH, luteinizing hormone.

Hormone–Cell Interactions

Hormones interact with receptors on the cell membrane, cytoplasm, or nucleus. The number of available receptors and the binding affinity between receptor and hormone are important for a hormonal effect. See pages 7 to 10 for a discussion of receptors and the mechanisms of signal transduction. **Table 23.4** lists the mechanisms of signal transduction of hormones.

Table 23.4 ▸ **Signal Transduction of Hormones**	
Mechanism	**Examples**
Second messengers	
↑ Cyclic AMP	CRH, ACTH, LH, FSH, TSH, PTH, calcitonin, glucagon, ADH (V_2 receptors), hCG, EPI, and NE (at β receptors)
↓ Cyclic AMP	Prolactin
DAG and IP_3	GHRH, GnRH, TRH, ADH (V_1 receptors), oxytocin, EPI, and NE (acts via IP_3 at α_1 receptors)
Cyclic GMP	NO
Tyrosine kinase	Insulin, IGF-1
Steroid mechanism	**Cortisol, aldosterone, testosterone, estrogen, progesterone, and calcitriol (vitamin D)**
Nuclear receptors	Thyroid hormones (T_3 and T_4)

Abbreviations: ACTH, adrenocorticotropic hormone; ADH, antidiuretic hormone; AMP, adenosine monophosphate; CRH, corticotropin-releasing hormone; DAG, diacylglycerol; EPI, epinephrine; FSH, follicle-stimulating hormone; GHRH, growth hormone–releasing hormone; GMP, guanosine monophosphate; GnRH, gonadotropin-releasing hormone; hCG, human chorionic gonadotropin; IGF-1, insulin-like growth factor; IP, inositol 1,4,5-triphosphate; LH, luteinizing hormone; NO, nitric oxide; NE, norepinephrine; PTH, parathyroid hormone; T_3, triiodothyronine; T_4, thyroxine; TRH, thyrotropin-releasing hormone; TSH, thyroid-stimulating hormone.

24 The Pituitary Gland

24.1 Interaction between the Hypothalamus and Pituitary

The hypothalamus and pituitary constitute a functional unit that represents the greatest integration of central nervous system control of endocrine systems.

Hypothalamic–Anterior Pituitary Axis

Hypothalamic neuropeptide hormones regulate the biosynthesis and release of hormones from the anterior pituitary. They are synthesized in hypothalamic neurons whose axons terminate at a primary capillary plexus. Upon stimulation, these neuropeptide hormones are released into the capillary bed, where they travel quickly to the anterior pituitary via the hypothalamic-hypophysial portal system. This terminates at a second, highly fenestrated capillary bed in the anterior pituitary, where they are free to diffuse into the extracellular fluid bathing the anterior pituitary cells. This bed drains into the cavernous sinus and jugular vein, allowing for the peripheral transport of any released anterior pituitary hormones (**Fig. 24.1**).

Hypothalamic–Posterior Pituitary Axis

The posterior pituitary contains axons whose cell bodies are located in the hypothalamic nuclei (paraventricular nuclei and supraoptic nuclei). These neurons are much larger than those regulating the anterior pituitary. Posterior pituitary hormones are synthesized in the cell bodies in the hypothalamus and travel down the axons to be released by the posterior pituitary upon stimulation (**Fig. 24.1**). The posterior pituitary hormones also reach the peripheral circulation and target organs via the cavernous sinus and jugular vein.

Table 24.1 provides a summary of hypothalamic and pituitary hormones.

> **Embryology of the hypothalamus and pituitary glands**
>
> The anterior pituitary arises from an invagination of oral ectoderm (Rathke pouch). The hypothalamus and posterior pituitary are derived from neuroectoderm.

Table 24.1 ▶ Summary of Hypothalamic and Pituitary Hormones and Their Target Organs		
Hypothalamic Releasing or Inhibiting Hormone	**Anterior Pituitary Hormone**	**Target Organ(s)**
GHRH	↑ GH	Liver, skeletal muscle, bone, and adipose tissue
Somatostatin (or GHIH)	↓ GH	Liver, GI tract, and pancreas
CRH	ACTH	Adrenal cortex
GnRH	LH and FSH	Gonads
TRH	TSH	Thyroid gland
PIH (dopamine)	Prolactin	Mammary glands and gonads
Posterior Pituitary Hormones		**Target Organ**
Oxytocin		Uterine and other smooth muscle
ADH		Kidney tubules (mainly), also vascular smooth muscle, liver, and anterior pituitary gland

Abbreviations: ACTH, adrenocorticotropic hormone; ADH, antidiuretic hormone; CRH, corticotropin-releasing hormone; FSH, follicle-stimulating hormone; GH, growth hormone; GHRH, growth hormone–releasing hormone; GHIH, growth hormone–inhibiting hormone; GI, gastrointestinal; GnRH, gonadotropin-releasing hormone; LH, luteinizing hormone; PIH, prolactin-inhibiting hormone; TRH, thyroid-releasing hormone; TSH, thyroid-stimulating hormone.

Fig. 24.1 ▶ **Hypothalamic–pituitary hormone secretion.**
The hypothalamus responds to impulses from the central nervous system by releasing stimulating and/
or inhibiting hormones that are carried to the pituitary via axoplasmic transport (movement through the
cytoplasm of an axon). Hypothalamic hormones induce hormone release in the anterior and posterior
pituitary. Hypothalamic releasing or inhibiting hormones are released into a capillary bed in the anterior
pituitary. Axons from the hypothalamus terminate in the posterior pituitary to release hormones. (ACTH,
adrenocorticotropic hormone; ADH, antidiuretic hormone; FSH, follicle-stimulating hormone; GH, growth
hormone; IH, inhibiting hormone; LH, luteinizing hormone; MSH, melanocyte-stimulating hormone; NE,
norepinephrine; NPY, neuropeptide Y; PRL, prolactin; RH, releasing hormone; TSH, thyroid-stimulating
hormone)

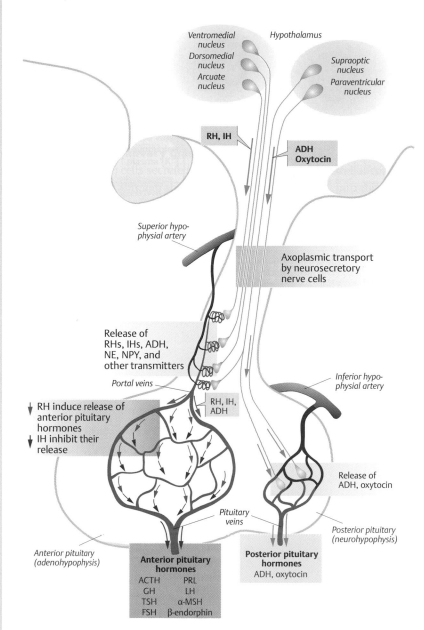

Melanocyte-stimulating hormone and melanin

Melanocyte-stimulating hormone (MSH) is produced by cells in the intermediary lobe of the pituitary gland. Both MSH and adrenocorticotropic hormone (ACTH) are derived from the precursor peptide pro-opiomelanocortin (POMC). MSH stimulates the synthesis and secretion of melanin by melanocytes in skin and hair. It also plays a role in appetite and sexual arousal. MSH is increased in pregnancy and Cushing disease (due to increased levels of ACTH). Melanin is responsible for giving color to the hair, skin, and iris of the eyes. It also helps protect skin from sun damage by collecting in vesicles called melanosomes, which migrate to the epidermis and cover the nucleus, thus protecting DNA from damage from the sun's ionizing radiation.

24.2 Anterior Pituitary Hormones

Growth hormone (GH) and prolactin have several target organs and are discussed in the following section. All remaining anterior pituitary hormones are discussed in relation to their target gland in the following chapters (**Fig. 24.2**).

Fig. 24.2 ▸ **Effects of pituitary hormones on target tissues.**
(ACTH, adrenocorticotropic hormone; ADH, antidiuretic hormone; FSH, follicle-stimulating hormone; GH, growth hormone; IGFs, insulinlike growth factors; LH, luteinizing hormone; TSH, thyroid-stimulating hormone)

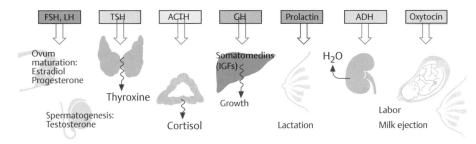

Growth Hormone

Regulation of Secretion

Hypothalamic control. Growth hormone–releasing hormone (GHRH) causes increased GH synthesis and secretion.

Somatostatin, released from specific hypothalamic neurons, decreases the response of the anterior pituitary to GHRH, thus decreasing GH levels. It is identical to somatostatin secreted by the endocrine pancreas and intestinal mucosa. Somatostatin also inhibits the secretion of thyroid-stimulating hormone (TSH), ACTH, and prolactin.

Negative feedback control

— GH inhibits its own secretion by inhibiting GHRH secretion in the hypothalamus and by desensitizing the pituitary to GHRH. It also increases the secretion of somatostatin from the hypothalamus.
— Somatomedins (insulinlike growth factors [IGFs]) inhibit the secretion of GH by stimulating the secretion of somatostatin from the hypothalamus and by directly inhibiting the cells of the anterior pituitary.

Other factors. **Table 24.2** lists factors that stimulate or inhibit GH secretion.

Table 24.2 ▸ Other Factors Affecting Growth Hormone Secretion	
Stimuli	**Inhibitors**
Major stimuli	Somatostatin
GHRH	Hyperglycemia
Deep sleep (slow-wave)	Cortisol**
Hypoglycemia*	
Stress	
Metabolites (e.g., amino acids and free fatty acids after eating a high-protein meal)	
Secondary stimuli	
Exercise	
Pyrogens (substances that induce a fever)	
Glucagon	
Vasopressin (ADH)	
Opioids	

*Hypoglycemia may stimulate a-adrenergic neurons in the hypothalamus, causing stimulation of growth hormone–releasing hormone (GHRH) production by nearby neurons.
** Cortisol inhibits growth hormone secretion, as it also causes hyperglycemia (which lowers GHRH secretion).
Abbreviation: ADH, antidiuretic hormone.

Growth hormone and nitrogen balance

GH creates a positive nitrogen balance in the body. This is mainly due to an increased rate of lipolysis, which provides the energy the body needs while sparing proteins and glucose. Diseases and conditions in which there is a negative nitrogen balance, such as acquired immunodeficiency syndrome (AIDS), cachexia (loss of lean body mass that cannot be corrected with increased calorific intake), trauma, and severe burns, can be treated with GH to improve lean body mass and improve wound healing.

Growth at epiphyseal plates

The epiphyseal plate consists of hyaline cartilage at the end of long bones. Chondrocytes in the epiphyseal plates are constantly undergoing mitosis. The older cells (at the diaphysis end) are then ossified by osteoblasts. This progressive laying down of bone leads to longitudinal growth. Growth of long bones continues throughout childhood and adolescence but ceases in adulthood. GH acts to increase the mitosis of chondrocytes in the epiphyseal place and thus is ineffective for bone growth in adults.

Actions

GH is essential for the growth of infants to maturity. **Table 24.3** summarizes the effects of GH and somatomedins (IGFs).

Table 24.3 ▶ Effects of Growth Hormone and Insulinlike Growth Factors	
Effects Mediated by GH**	**Effects Mediated by Somatomedins (IGFs)**

↑ Somatomedin synthesis*
↑ gluconeogenesis
↑ lipolysis
↑ protein synthesis
↑ amino acid uptake in the gut
↓ insulin (causing ↓ glucose uptake into cells)
	↑ protein synthesis, resulting in the following effects:
	    ↑ muscle mass
	    ↑ cartilage growth (this causes linear bone growth)
	    ↑ size of most organs

* Somatomedins (IGFs) are intermediaries for some growth hormone (GH) actions.
** Many of the actions of GH occur in association with cortisol.
Abbreviations: GH, growth hormone; IGF, insulinlike growth factor.

Pathophysiology

Growth hormone deficiency.   **Table 24.4** summarizes some of the more common causes of GH deficiency, the symptoms produced, and treatment.

Table 24.4 ▶ Growth Hormone Deficiency			
Causes	**Effects**	**Symptoms**	**Treatment**
Genetic syndromes with deficient GH secretion Tumors of the hypothalamus or anterior pituitary Physical trauma (cranial injury with involvement of the sella turcica) Liver unable to produce somatomedin C (IGF I) Defective GH receptors	↓ GH	Short stature with normal body proportions Delayed puberty	Replacement with recombinant GH

Abbreviations: GH, growth hormone; IGF, insulinlike growth factor.

Growth hormone excess.   GH excess causes gigantism in children and acromegaly in adults. **Table 24.5** summarizes the causes, symptoms, and treatment of these conditions.

Table 24.5 ▶ Growth Hormone Excess			
Causes	**Effects**	**Symptoms**	**Treatment**
Pituitary adenoma GHRH-producing pancreatic tumor Ectopic GH-producing tumor	↑ GH	Gigantism (in children) — Tall stature Acromegaly (in adults) — Coarsened features of the face caused by exaggerated growth of the mandible and frontal and nasal bones. The hands and feet are similarly coarsened. — Hypertrophy of the connective tissue of the heart, liver, and kidneys — Reduced glucose tolerance, which can lead to cardiomyopathy, hypertension, and diabetes mellitus	Surgical resection of the causative tumor Octreotide therapy (somatostatin analogue) Pegvisomant (GH receptor antagonist)

Abbreviations: GH, growth hormone; GHRH, growth hormone–releasing hormone.

Prolactin

The chemical structure of prolactin is closely related to that of GH.

Regulation of Secretion

Hypothalamic control.   Dopamine (prolactin-inhibiting hormone [PIH]), secreted by the hypothalamus, tonically inhibits prolactin secretion. Thyrotropin-releasing hormone (TRH) increases prolactin secretion.

Negative feedback control.   Prolactin inhibits its own secretion by stimulating the hypothalamus to release dopamine (via neuroendocrine reflex from the breast). It also inhibits the secretion of gonadotropin-releasing hormone (GnRH) in high concentrations.

Actions

- ↑ breast development
- ↑ milk production in the breast
- ↑ the synthesis of vitamin D_3 for Ca^{2+} absorption
- ↓ ovulation by decreasing the synthesis and release of GnRH
- ↓ spermatogenesis by decreasing GnRH

Pathophysiology

Prolactin deficiency.   **Table 24.6** summarizes the causes, symptoms, and treatment of prolactin deficiency.

Table 24.6 ▶ Prolactin Deficiency

Causes	Effects	Symptoms	Treatment
Damage to the anterior pituitary Dopamine agonist drugs (e.g., bromocriptine)	↓ serum [prolactin]	Inadequate lactation/ failure to lactate Menstrual disorders Delayed puberty Subfertile states/ infertility	Inadequate lactation may respond to metoclopramide (a dopamine antagonist) Subfertile states may be overcome with clomophene citrate or gonadotropin therapy

Prolactin excess.   **Table 24.7** summarizes the causes, symptoms, and treatment of prolactin excess (hyperprolactinemia).

Table 24.7 ▶ Prolactin Excess (Hyperprolactinemia)*

Causes**	Effects	Symptoms	Treatment
Hypothalamic disorders causing ↓ dopamine secretion or ↑ TRH secretion Pituitary tumors Hypothyroidism (↑ TRH) Drugs: estrogens, dopamine antagonists (e.g., metoclopramide and methyldopa), and phenothiazines	↑ serum [prolactin]	Galactorrhea (↑milk flow) Amenorrhea, loss of libido, and impotence (in men) due to ↓ LH and FSH secretion Tendency toward hyperglycemia	Bromocriptine therapy (dopamine agonist) Surgical resection for tumors (transsphenoidal approach)

* Hyperprolactinemia is the most common hormonal disturbance of the pituitary gland.
** Hyperprolactinemia may be physiological during pregnancy, breast-feeding, sleep, or stress.
Abbreviations: FSH, follicle-stimulating hormone; LH, luteinizing hormone; TRH, thyrotropin-releasing hormone.

Transsphenoidal pituitary surgery

Most pituitary tumors are removed endoscopically through the sphenoid sinus, which can be accessed through the nostril. This approach provides relatively easy surgical access to the pituitary and leaves no visible scarring.

24.3 Posterior Pituitary Hormones

Oxytocin

Regulation of Secretion

Oxytocin secretion is stimulated by neural reflexes in response to

— breast stimulation (suckling)
— uterocervical stimulation
— emotional stimulation (e.g., sight of a neonate)

Actions

— ↑ uterine contractility involved in parturition (oxytocin receptors are upregulated during pregnancy)
— ↑ contraction of mammary myoepithelium, which results in the ejection of milk ("let-down")

Antidiuretic Hormone (Vasopressin)

Regulation of Secretion

Regulation of ADH secretion occurs at both the neuronal soma and axonal levels. Its secretion is stimulated by the following:

— ↑ plasma osmolality (e.g., dehydration)
— ↓ plasma volume (e.g., hemorrhage, or hypovolemia)
— ↓ blood pressure (via direct effects on the baroreceptors and by decreased plasma volume)

Note: ADH release is most sensitive to plasma osmolality, yet larger quantities of ADH are released in response to major changes in blood pressure and blood volume.

Actions

Table 24.8 summarizes the effects of ADH (vasopressin) at V_1 and V_2 receptors.

Table 24.8 ▶ **Action of Antidiuretic Hormone at Its Receptors**		
Receptor	**Second Messenger System**	**Effect**
V_1	DAG and IP_3	Contraction of vascular smooth muscle (as vasopressin)
V_2	Cyclic AMP	Acts in the collecting ducts of the kidneys to increase aquaporin permeability (↑ water reabsorption)
Abbreviations: AMP, adenosine monophosphate; DAG, diacylglycerol; IP, inositol 1,4,5-triphosphate.		

Pathophysiology

ADH deficiency.   ADH deficiency causes diabetes insipidus (summarized in **Table 24.9**).

Table 24.9 ▶ **Diabetes Insipidus**			
Causes	**Effects**	**Symptoms**	**Treatment**
Damage to the hypothalamus or pituitary (trauma/surgery/autoimmune disease) Tumors	↓ ADH	Polyuria Polydipsia Dehydration	Dextrose solution Desmopressin (synthetic ADH analogue)
Abbreviation: ADH, antidiuretic hormone.			

ADH excess.   Excess ADH release causes the syndrome of inappropriate ADH secretion (SIADH; Schwartz–Bartter syndrome). **Table 24.10** summarizes this condition.

Table 24.10 ► **Syndrome of Inappropriate Antidiuretic Hormone Secretion**			
Causes	**Effects**	**Symptoms**	**Treatment**
ADH-secreting pituitary tumor Cirrhosis Nephrotic syndrome Congestive heart failure	↑ ADH ↓ serum [Na$^+$]	Water retention Hyponatremia causes nausea, lethargy, anorexia, headaches, cramps, weakness, and, if severe, seizures and coma Kidney stones may occur	Hypertonic saline Conivaptan, tolvaptan, or demeclocycline (ADH antagonists) Furosemide (loop diuretic) Symptomatic control
Abbreviation: ADH, antidiuretic hormone.			

25   Thyroid Hormones

25.1 Triiodothyronine and Thyroxine

The primary function of the thyroid gland is to synthesize and secrete two hormones into the circulation: triiodothyronine (T_3) and thyroxine (T_4). T_3 and T_4 control essential functions, such as the regulation of energy metabolism and basal metabolic rate (BMR) and the promotion of protein synthesis and growth (**Fig. 25.1**).

Fig. 25.1 ▶ Thyroid hormones (overview).
Thyrotropin-releasing hormone (TRH) from the hypothalamus stimulates the anterior pituitary to secrete thyroid-stimulating hormone (TSH). TSH is involved in regulating each step in thyroid hormone synthesis. Thyroid follicular cells preferentially synthesize T_4, which is then converted to T_3 (a more physiologically active form) in the periphery. T_3 causes feedback inhibition of TSH to maintain optimal thyroid hormone levels. (SIH, somatostatin)

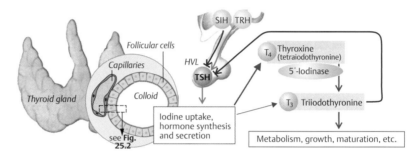

Synthesis

T_3 and T_4 are iodine-containing derivatives of tyrosine that are synthesized in thyroid follicular cells (**Figs. 25.2** and **25.3**). The steps involved in thyroid hormone synthesis are as follows:

— Thyroid follicular cells synthesize and store thyroglobulin (TG), a glycoprotein with tyrosine residues.
— The Na^+-I^- pump on the follicular cell membrane actively transports iodide (I^-) into the colloid.
— Iodide is oxidized to iodine (I_2) by thyroid peroxidase as it moves through the membrane into the colloid.
— Thyroid peroxidase also conjugates the iodine onto the tyrosine residues of TG, forming monoiodotyrosine (MIT) and diiodotyrosine (DIT). Thus, iodine found in the follicular colloid is bound to TG and is not free.
— MIT and DIT undergo coupling reactions while bound to TG. T_4 is produced when two molecules of DIT combine, and T_3 is formed when one molecule of MIT and one molecule of DIT combine.
— Iodinated TG in colloid is endocytosed into follicular cells under stimulation of thyroid-stimulating hormone (TSH).
— Endocytotic vesicles fuse with lysosomes, and proteolysis of TG occurs. This releases T_4, T_3, MIT, and DIT.
— T_4 and T_3 are released into the circulation. MIT and DIT are deiodinated by iodotyrosine dehalogenase, and the iodine is recycled.

The synthesis of T_4 is generally favored over that of T_3 (10–20:1).

▬ Na^+-I^- pump

Iodine (from ingested food) is necessary for thyroid hormone synthesis. Because dietary intake inevitably varies, the thyroid gland must sequester iodine so that adequate amounts are always available for thyroid hormone synthesis. It does this via the Na^+-I^- pump on the cell membrane. This pump binds iodide with high affinity and specificity. The transport of iodide into the cytoplasm requires oxidative phosphorylation and a membrane Na^+-K^+ ATPase. TSH is the major physiological regulator of the iodide pump, but high intracellular levels of iodide inhibit the activity of the pump.

Fig. 25.2 ▶ **Thyroid hormone synthesis and secretion.**
Thyroglobulin (TG) is synthesized from tyrosine residues and carbohydrates in thyroid follicular cells. TG is packaged in vesicles that are exocytosed in colloid. Iodide is actively transported into the follicular cells by the Na^+-I^- symporter. Iodide is oxidized to iodine as it moves through the follicular membrane into colloid. Thyroid peroxidase also causes iodine to attach to TG, forming monoiodotyrosine (MIT) and diiodotyrosine (DIT). MIT and DIT combine to form T_3 and T_4 in the colloid. T_3 and T_4 are then endocytosed into the follicular cell, where they fuse with lysozymes, whose proteolytic enzymes cleave off TG. T_3 and T_4 are released into the circulation. The iodine is removed from MIT and DIT, and it pools in the follicular cell for reuse. Each of these steps is regulated by TSH.

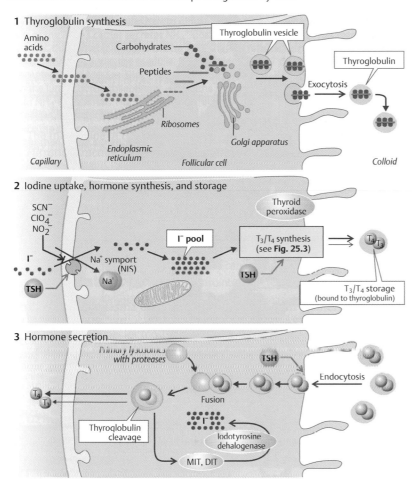

Regulation of Secretion

Hypothalamic–anterior pituitary control. Thyrotropin-releasing hormone (TRH) from the hypothalamus stimulates the anterior pituitary to release TSH. TSH then acts on the thyroid gland to increase the synthesis and release of T_3 and T_4 (**Table 25.1**).

Table 25.1 ▶ Mechanisms by Which TSH Increases the Release of Thyroid Hormones
Upregulation and ↑ Sensitivity of TSH Receptors to TSH
↑ TG synthesis
↑ thyroid peroxidase and glucose oxidase levels, which increase the iodination of TG
↑ activity of the iodide pump
↑ Na^+-K^+ ATPase activity, which increases the capacity for iodide intake
TSH favors a shift to more T_3 formed relative to T_4 under acute increases in metabolic demand.
Abbreviations: T_3, triiodothyronine; T_4, thyroxine; TG, thyroglobulin; TSH, thyroid-stimulating hormone.

Fig. 25.3 ► Synthesis, storage, and mobilization of the thyroid hormones.

TG is exocytosed from the follicular cell into colloid. Iodide also moves into colloid in exchange for Cl⁻, and as it does so, it is oxidized to iodine. TG and iodine combine to form MIT or DIT. One molecule of MIT and one of DIT form T_3, and two molecules of DIT combine to form T_4. TG is then endocytosed back into the follicular cell where it is cleaved to release T_3 and T_4 into the circulation. (TPO, thyroid peroxidase)

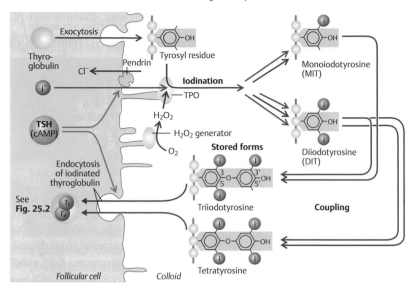

Negative feedback control.

Both TRH and TSH are controlled by the level of circulating thyroid hormone (T_4, which is converted to T_3 in pituitary cells) in a classic negative feedback loop, but TRH release also increases in response to increased metabolic demand.

Iodine regulation of T_3 and T_4.

Increased levels of iodine prevent thyroid hormone production in the following ways:

— Acute increases in circulating iodine inhibit iodine incorporation into TG.
— Chronic increases in iodine reduce the number of iodide pump molecules.
— High intracellular levels of iodine decrease iodide pump activity. The iodine molecules also compete with thyroid peroxidase conjugation of TG and iodine.

Metabolism

Plasma-binding proteins.

T_3 and T_4 mainly circulate bound to thyroid-binding globulin (TBG). Thyroid hormones that are bound to plasma proteins are biologically inactive but provide a reservoir to buffer against changes in thyroid gland function or metabolic demand, which increases their half-life (decreased renal clearance).

T_4 conversion to T_3.

T_4 is converted to T_3 in peripheral tissues by 5'– monodeiodination. This accounts for 80% of T_3.

Reverse T_3 (rT_3) is an inactive metabolite of T_4 that is formed during this deiodination.

Actions

Development.

T_3 is essential for the development of the central nervous system (CNS). Fetal deficiencies in thyroid function lead to a slowing of all intellectual functions due to irreversible CNS lesions (cretinism). If hypothyroidism develops from ~2 years of age onward, the effects on mental development are minor and reversible.

▮ Wolff–Chaikoff effect

The Wolff–Chaikoff effect is a reduction in the synthesis and release of thyroid hormones caused by a large amount of iodine. This effect lasts ~10 days, after which iodine incorporation into TG and thyroid peroxidase function returns to normal. It is widely believed that the resumption of normal functioning is due to downregulation of the iodide pump on the follicular cell membrane. The Wolff–Chaikoff effect is the principle behind the use of iodine for the treatment of hyperthyroidism.

Growth.   Growth depends on the normal function of the thyroid in postnatal life. Thyroid hormones promote growth by

— ↑ protein synthesis
— ↑ bone formation with growth hormone (GH) and insulinlike growth factor (IGF)
— ↑ ossification and fusion of growth plates

Basal metabolic rate. T_3 and T_4 increase BMR to provide an increased amount of substrate for adenosine triphosphate (ATP) production in the face of higher oxygen consumption throughout the body.

T_3 stimulates Na^+–K^+ ATPase activity, thus increasing the use of ATP by the Na^+–K^+ pump. This requires higher oxygen consumption.

The increase in BMR and ATP hydrolysis combines to cause an increase in heat production.

Metabolic.   Thyroid hormones are catabolic and stimulate the following:

— ↑ intestinal absorption of glucose
— ↑ gluconeogenesis, glycogenolysis, and glucose oxidation
— ↑ lipolysis
— ↑ cholesterol turnover and plasma clearance of cholesterol
— T_3 potentiates the "hypoglycemic" actions of insulin by increasing glucose uptake into muscle and adipose tissue.

Systemic. T_3 upregulates adrenergic β-receptors, leading to increased heart rate, cardiac output, and ventilation and decreased peripheral vascular resistance. These actions support increased oxygen demand in tissues.

T_3 also potentiates epinephrine, enhancing adrenergic-mediated glycogenolysis.

T_3 facilitates the actions of cortisol, glucagon, and GH.

Pathophysiology

The causes of thyroid hormone derangement are illustrated in **Fig. 25.4**.

Thyroid function tests

There are several tests for measuring the function of the thyroid gland. The most common is a simple blood test for TSH, free T_4 (FT$_4$), and T_3 levels. Usually TSH and FT$_4$ are combined to identify thyroid dysfunction. A high TSH level combined with a low FT$_4$ is indicative of primary hypothyroidism (the thyroid gland itself is failing). A low TSH combined with a low FT$_4$ is indicative of secondary hypothyroidism (pituitary gland dysfunction), and low TSH/high FT$_4$ is indicative of hyperthyroidism. T_3 will also be high in hyperthyroidism, but it is not useful for the diagnosis of hypothyroidism, as T_3 is the last to become abnormal. Thyroid antibody titers for autoimmune thyroid disorders (Hashimoto thyroiditis and Graves disease) can also be measured in a blood test. Additional tests include radioactive iodine uptake (RAIU) and thyroid scans. These tests can determine whether the thyroid is overactive or underactive.

Recurrent laryngeal nerve paralysis with thyroidectomy

The recurrent laryngeal nerve (RLN) is a branch of the vagus nerve (cranial nerve X). It supplies visceromotor innervation to the posterior cricothyroid, the muscle that abducts the vocal cords. It also supplies sensation to the laryngeal mucosa. The recurrent laryngeal nerves, especially the right RLN, is vulnerable to damage during thyroidectomy, as they run immediately posterior to the thyroid gland. Unilateral RLN damage results in hoarseness; bilateral damage causes respiratory distress (dyspnea) and aphonia. Aspiration pneumonia may occur as a complication. The presence of damage may be confirmed by laryngoscopy showing the absence of movement of the vocal cord on the affected side. There is no cure for this condition, but there are surgical options that may help prevent aspiration.

Fig. 25.4 ▶ Causes of hypothyroidism, hyperthyroidism, and goiter.
Hypothyroidism is usually caused by dysfunction of the thyroid gland itself. Defects may occur at any step of thyroid hormone synthesis or at the target organ receptors. It is very commonly caused by inflammatory damage to the gland or due to thyroidectomy (for cancer). It is rarely due to defects of the hypothalamus or anterior pituitary. Hyperthyroidism is commonly due to autoimmune disease where thyroid-stimulating immunoglobulin (TSI) mimics TSH at its receptors (Graves disease). Other causes are tumors, thyroiditis, excess TSH, and excessive supply of thyroid hormones. Goiter is caused by uncontrolled thyroid cell proliferation due to a tumor or overstimulation by TSH or TSI. (DIT, diiodotyrosine, MIT, monoiodotyrosine)

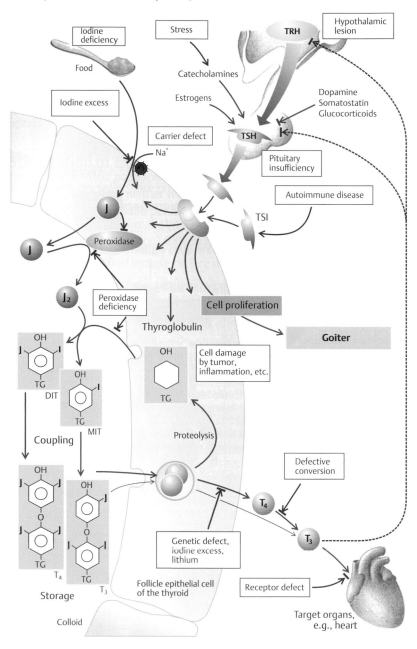

Hypothyroidism. **Table 25.2** summarizes the causes, effects, symptoms, and treatment of hypothyroidism. The pathophysiological mechanism of hypothyroidism is illustrated in **Fig. 25.5**.

Table 25.2 ▶ Hypothyroidism			
Causes	**Effects**	**Symptoms**	**Treatment**
Hypothalamic or pituitary tumor* Congenital absence or disappearance of thyroid tissue A block in thyroid hormone synthesis (e.g., due to deficiency in iodide trap, peroxidase function, or iodothyrodine dehalogenase) Destruction of thyroid tissue by autoimmune antibodies against TG (Hashimoto thyroiditis)	↓ TRH and TSH if hypothyroidism due to pathology of the hypothalamus or anterior pituitary ↑ TRH and TSH if cause of hypothyroidism is unrelated to the hypothalamus or anterior pituitary (due to ↓ feedback inhibition by T_3 and T_4) ↓ T_3 and T_4	Weight gain without increase of caloric intake Low body temperature and intolerance to a cold environment Decreased sweating Dry skin and hair Hoarse voice Depression Lethargy May or may not be associated with goiter (enlarged thyroid gland)	Replacement of thyroid hormones (levothyroxine is the drug of choice)

* Myxedema is hypothyroidism secondary to a hypothalamic or pituitary tumor.
Abbreviations: TG, thyroglobulin; TRH, thryotropin-releasing hormone; TSH, thyroid-stimulating hormone.

Hyperthyroidism. **Table 25.3** summarizes hyperthyroidism. The pathophysiological mechanism of hyperthyroidism is illustrated in **Fig. 25.6**.

Table 25.3 ▶ Hyperthyroidism			
Causes	**Effects**	**Symptoms**	**Treatment**
Thyroid adenoma	↑ T_3 and T_4 ↓ TSH	Weight loss Increased appetite Hypersensitivity to heat and increased sweating Nervousness Tremor Palpitations and tachycardia Lid lag Goiter	Propylthiouracil or methimazole Partial thyroidectomy Thyroid destruction by radioactive iodine (^{131}I) β-blockers for symptomatic relief of hyperthyroidism or for the treatment of hyperthyroid crisis (thyrotoxic storm)
Graves disease (autoimmune disease associated with stimulatory antibodies mimicking TSH actions)	↑ T_3 and T_4 ↓ TSH	Symptoms as above plus exophthalmos (bulging eyes) and other opthalmopathy Graves disease is associated with IDDM and pernicious anemia	As above plus steroid therapy (if required) for edema of the eye

Abbreviations: IDDM, insulin-dependent diabetes mellitus; TSH, thyroid-stimulating hormone.

Fig. 25.5 ► **Pathophysiology of hypothyroidism.**
Each of the signs and symptoms of hypothyroidism can be related to hyposecretion of thyroid hormones and the effect this has on development, growth, and metabolism. (CO, cardiac output; VLDL, very-low-density lipoprotein)

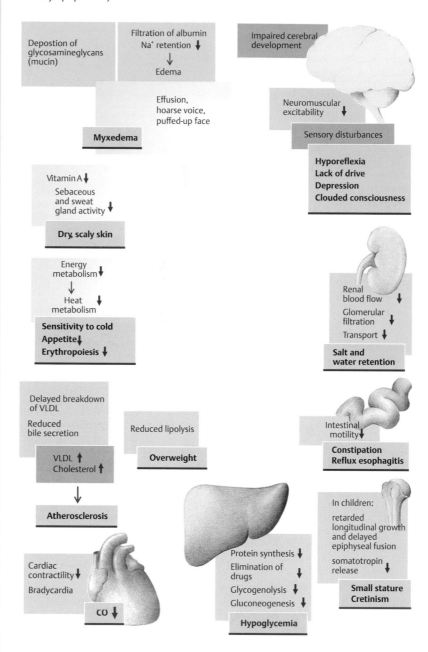

Fig. 25.6 ► **Pathophysiology of hyperthyroidism.**
Each of the symptoms of hyperthyroidism can be related to hypersecretion of thyroid hormones and the effect this has on development, growth, and metabolism. (GFR, glomerular filtration rate; LDL, low-density lipoprotein; RPF, renal plasma flow; VLDL, very-low-density lipoprotein)

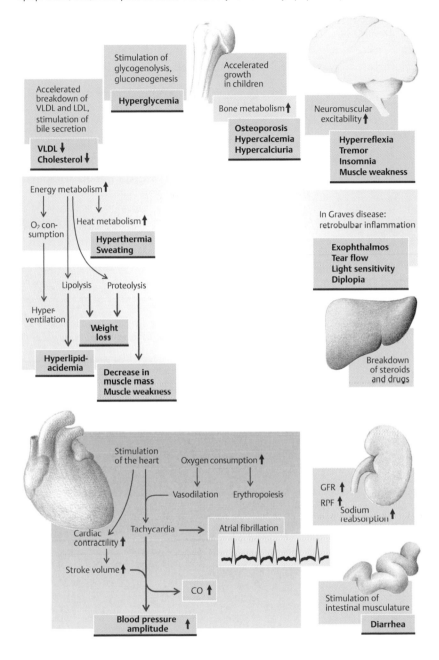

Thyrotoxic storm

Thyrotoxic storm is a life-threatening condition in which there is severe hyperthyroidism, fever, sweating, restlessness, shaking, tachycardia, diarrhea, and vomiting. It may precipitate heart failure, pulmonary edema, and coma and may be fatal. It usually occurs in people with untreated hyperthyroidism who encounter a stressor, such as infection or trauma. Treatment involves reducing thyroid hormone synthesis and secretion with propylthiouracil and Lugol solution (iodine solution), systemic steroids (dexamethasone) for insufficiency and to stabilize the integrity of blood vessels, β-blockers to control the β-mediated effects, antibiotics (if infection is causative), and fluid replacement.

Drugs for hyperthyroidism

Propylthiouracil and methimazole act by inhibiting the iodination of tyrosyl residues in TG, and the coupling of iodotyrosines, by inhibiting the peroxidase enzyme. Effects are not apparent until the thyroid reserve is depleted. It is common for patients taking these drugs to develop a rash.

Thyroid opthalmopathy

Thyroid opthalmopathy occurs in Graves disease, in which autoantibodies cause inflammation and lymphocytic infiltration around the orbit. This condition tends to affect men more than women and may occur before other signs of hyperthyroidism are evident. Upper lid retraction is the most common sign, but others include decreased visual acuity, proptosis, lid lag, diplopia (double vision), edema, erythema, conjunctivitis, incapacity to fully close the eyelids (lagophthalmos), dysfunction of the lacrimal gland, and limitation of eye movement (especially upward gaze, due to tethering and fibrosis of the inferior rectus muscle). The patient may complain of eyes feeling dry and gritty. In severe cases, the optic nerve may be compressed, which, if untreated, could lead to blindness. Treatment involves normalizing thyroid hormone levels (opthalmopathy may not respond to this), smoking cessation, artificial tears, tarsorrhaphy (eyelids are partially sutured together), systemic steroids to reduce inflammation, surgical decompression of the optic nerve, and other surgeries to improve the patient's comfort level (e.g., lid lengthening).

26 Calcium Metabolism

Calcium and phosphate are fundamental intracellular components that participate in the functions of all cells. Half of circulating calcium is bound to serum albumin; the other half is composed of free calcium, calcium phosphates, citrates, and salts of other organic acids. Free, ionized calcium is physiologically active.

Plasma calcium levels are maintained within strict limits and are under the endocrine control of three hormones:

— Parathyroid hormone (PTH)
— Calcitriol (1,25-dihydroxycholecalciferol, or 1,25-$(OH)_2$vitamin D_3)
— Calcitonin

Note: PTH and calcitriol are the main hormones that control calcium levels. Calcitonin is not involved in maintaining calcium homeostasis on an acute basis. Its secretion is only of importance in hypercalcemic states. **Fig. 26.1** provides an overview of calcium homeostasis.

Phosphate $(PO_4{}^{3-})$ binds to Ca^{2+} and therefore decreases the amount of free, ionized Ca^{2+} in the blood. Phosphate levels are not regulated directly but rather occur as a secondary effect of calcium regulation (**Fig. 26.2**).

26.1 Hormones that Regulate Calcium Metabolism

Parathyroid Hormone

PTH is the major regulator of serum $[Ca^{2+}]$.

Synthesis

PTH is an 84 amino acid peptide synthesized and secreted by chief cells of the parathyroid gland.

Regulation of Secretion

PTH secretion is controlled by serum $[Ca^{2+}]$ via negative feedback. Secretion of PTH is stimulated by the following:

— ↓ serum $[Ca^{2+}]$
— ↑ serum $[PO_4{}^{3-}]$
— ↓ serum [calcitriol]

Note: The effects of phosphate and calcitriol are indirect; they are due to the subsequent decrease in circulating free calcium levels caused by the increased phosphate levels.

Secretion of PTH is inhibited by decreases in serum $[Mg^{2+}]$.

Receptors and Cell Mechanisms

The actions of PTH on the kidney are via specific membrane receptors coupled to adenylate cyclase. Most of the cyclic adenosine monophosphate (cAMP) found in urine is generated by activated PTH receptors.

PTH activates specific membrane receptors on osteoblasts. PTH-activated osteoblasts stimulate osteoclasts via paracrine mechanisms, as there are no PTH receptors on osteoclasts. Activated osteoclasts demineralize bone via collagenases, hyaluronic acid, and acid phosphatases.

Actions

PTH targets the kidneys and bone to increase serum $[Ca^{2+}]$ and decrease serum $[PO_4{}^{3-}]$. Its effects are first seen in the kidneys, as they are direct and occur at lower PTH concentrations. Conversely, the effects of PTH on bone are indirect and require a higher concentration of PTH.

Fig. 26.1 ▶ **Calcium homeostasis.**
Calcium homeostasis is achieved by three main hormones, parathyroid hormone (from parathyroid gland), calcitonin (from parafollicular cells of the thyroid gland), and calcitriol (mainly produced in the kidneys). In low serum Ca^{2+} states, the actions of parathyroid hormone (PTH) and calcitriol predominate, causing increased Ca^{2+} uptake from the gut and bone and decreased renal Ca^{2+} excretion. In high serum Ca^{2+} states, the action of calcitonin predominates, causing decreased Ca^{2+} uptake from the gut, increased renal excretion, and storage of excess Ca^{2+} in bone. (UV, ultraviolet)

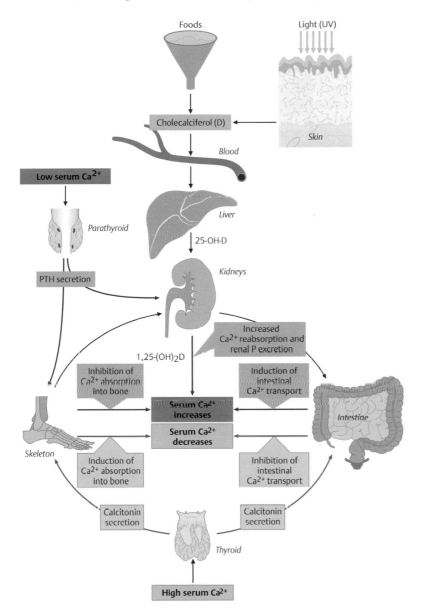

Kidneys

— PTH increases the reabsorption of Ca^{2+} in the distal tubule (primary effect), thereby reducing its excretion.

— PTH rapidly increases phosphodiuresis (urinary excretion of phosphate) by inhibiting phosphate reabsorption in the proximal tubule. This property prevents hyperphosphatemia secondary to the phosphate-liberating effect of PTH on bone. Enhanced clearance of phosphates increases Ca^{2+} ionization because of less formation of calcium phosphate.

— PTH increases the synthesis and activity of 1-α hydroxylase in the proximal tubule. This enzyme is needed for calcitriol synthesis.

Fig. 26.2 ▶ **Regulation of phosphate levels.**
Regulatory mechanisms of phosphate are closely tied to calcium homeostasis. High plasma phosphate levels reduce the level of free, unionized calcium, thus increasing PTH secretion. This causes increased renal phosphate excretion. High plasma phosphate levels also cause a reduction in 25-(OH)D hydroxylase activity, which lowers the formation of 1,25-(OH)$_2$D$_3$, causing decreased intestinal phosphate absorption and decreased phosphate release from bones. These homeostatic mechanisms allow plasma phosphate levels to be reduced without severely affecting calcium metabolism.

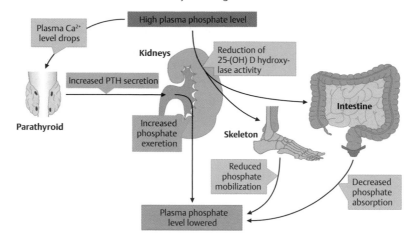

— PTH increases urinary cAMP (generated by activated renal PTH receptors).
— PTH increases the excretion of hydroxyproline (from the resorption of bone organic matrix).

Bone

— PTH liberates Ca^{2+} and PO$_4^{3-}$ from bone.
— PTH also releases hydroxyproline (from the organic matrix), magnesium, citrate, and osteocalcin from bone.
— PTH indirectly increases the resorption of both calcium and phosphate from bone by increasing calcitriol synthesis.

Intestines

— PTH indirectly increases intestinal Ca^{2+} absorption by increasing calcitriol synthesis.

Pathophysiology

Hypoparathyroidism. Table 26.1 summarizes hypoparathyroidism.

Table 26.1 ▶ **Hypoparathyroidism**			
Causes	**Effects**	**Symptoms**[*]	**Treatment**
Total thyroidectomy	Low serum [Ca^{2+}]	Convulsions	Oral calcium
Destruction of the para-	High serum [PO$_4^{3-}$]	Psychotic states	and calcitriol
thyroid gland after	Usually low calcitriol	Tetany (repetitive spontaneous	
neck injury	Decreased renal	muscle contractions)	
Thyroid carcinoma	phosphate excretion	Paresthesias (tingling sensations)	
Autoimmune destruction	Low urinary cAMP[**]	Numbness	
of PTH synthesizing		Laryngospasm	
cells		Degeneration of teeth and nails	
		Cataracts (lack of transparency of	
		lenses)	

[*] The symptoms of hypoparathyroidism are due to hypocalcemia (extracellular).
[**] cAMP levels in urine are an indication of PTH levels in blood.
Abbreviation: PTH, parathyroid hormone.

Hyperparathyroidism. **Tables 26.2** and **26.3** summarize primary hyperparathyroidism and secondary hyperparathyroidism, respectively.

Table 26.2 ▸ Primary Hyperparathyroidism

Primary hyperparathyroidism is due to a defect of the parathyroid glands, which causes excessive PTH production relative to serum calcium levels.

Causes	Effects	Symptoms*	Treatment
Parathyroid adenoma Parathyroid hyperplasia	High serum [Ca^{2+}] Low serum [PO_4^{3-}] Increased calcitriol High serum [alkaline phosphatase] Increased renal phosphate excretion Increased renal Ca^{2+} excretion (due to increased amount of Ca^{2+} filtered) High urinary cAMP**	Mental confusion and headaches Depression Bradycardia Muscle fatigue Polyuria and polydipsia Calcified corneas, kidney stones, and gallstones Pancreatitis associated with gallstones Bone resorption (evidence seen radiographically)	Surgical resection of adenoma, or most of gland if there is hyperplasia High fluid intake to prevent stone formation

* The symptoms of primary hyperparathyroidism are due to hypercalcemia.
** cAMP levels in urine are an indication of PTH levels in blood.

Table 26.3 ▸ Secondary Hyperparathyroidism

Secondary hyperparathyroidism is increased PTH secretion caused by a disease process occurring outside the parathyroid glands, which causes low calcium.

Causes	Effects	Symptoms of Secondary Hyperparathyroidism*	Treatment
Calcitriol deficiency due to - Chronic renal failure - Insufficient dietary intake of vitamin D	Low serum [Ca^{2+}] Low calcitriol	Convulsions Psychotic states Tetany (repetitive spontaneous muscle contractions) Paresthesias (tingling sensations) Numbness Laryngospasm Degeneration of teeth and nails Cataracts (lack of transparency of lenses)	Calcium and calcitriol

* The symptoms of secondary hyperparathyroidism are due to hypocalcemia, so they are the same as for hypoparathyroidism.

Calcitriol

Calcitriol (1,25-dihydroxycholecalciferol [1,25-$(OH)_2D_3$]) is the most active metabolite of vitamin D.

Synthesis

Calcitriol (1,25-$(OH)_2D_3$) is synthesized in the kidney from 25-hydroxycholecalciferol (25-(OH) D_3 under the enzymatic control of 1-α hydroxylase (**Fig. 26.3**).

Regulation of secretion

Modulation of 1-α hydroxylase activity in the kidneys controls calcitriol synthesis and secretion. The factors that stimulate and inhibit 1-α hydroxylase activity are shown in **Table 26.4**.

Table 26.4 ▸ Regulators of 1-α Hydroxylase Activity

Stimulatory Factors	Inhibitory Factors
Low serum [Ca^{2+}] Low serum [PO_4^{3-}] High PTH	High serum [Ca^{2+}] (by decreasing PTH levels) High serum [PO_4^{3-}] High serum [calcitriol] (by negative feedback) Calcitonin

Multiple endocrine neoplasia type 1

Multiple endocrine neoplasia type 1 (MEN1) is a rare, inherited disorder that causes multiple tumors (usually benign) in the endocrine glands and duodenum. It affects men and women equally and is usually not detected until adulthood, when tumors start growing. The parathyroid glands are most commonly affected. Tumors here cause hyperparathyroidism, which leads to hypercalcemia and its symptoms. It also commonly affects the pancreas, causing gastrinomas (from excess gastrin secretion), which, in turn, cause ulcers. These ulcers are more sinister than normal gastric ulcers and are highly prone to perforate. Multiple gastrinomas causing ulcers is referred to as Zollinger–Ellison syndrome. Pancreatic tumors may also cause insulinomas (leading to hypoglycemia), glucagon excess (leading to diabetes), or vasoactive intestinal peptide (leading to watery diarrhea). Pituitary tumors may also occur, leading to derangement of its hormones. People with MEN1 are more likely to develop cancerous tumors in later life. MEN1 can be detected early by gene testing, and people affected have a 50% chance of passing the disease to their children. There is no cure for MEN1, but there are various drugs and surgical options available to treat the effects.

Alkaline phosphatase

Alkaline phosphatase (ALP) is an enzyme that is predominantly made in the liver and bone, but it may also be made in the kidneys and the intestines, as well as in the placenta during pregnancy. It acts to cleave phosphate groups from molecules, such as nucleotides and proteins. ALP is a by-product of osteoblastic activity, so levels of ALP are raised during periods of rapid bone growth (puberty), in bone diseases that cause bone turnover (e.g., Paget disease and osteomalacia), and during calcium derangement (e.g., hyperparathyroidism). Elevated ALP levels are also a sign of damage to liver cells. Tests that measure ALP are therefore used to detect liver and bone disease, but they lack specificity.

Chronic renal failure

In chronic renal failure, the kidneys are unable to perform the necessary $α_1$-hydroxylation reactions to produce calcitriol, and they have a reduced capacity to excrete phosphate. This leads to secondary hyperparathyroidism due to hypocalcemia and hyperphosphatemia. Derangement of bone remodeling occurs, which is referred to as *renal osteodystrophy*. The symptoms of renal osteodystrophy include bone and joint pain, bone deformation, and increased likelihood of bone fractures. Chronic renal failure requires hemodialysis several times per week until renal transplantation can occur. Renal osteodystrophy is treated by calcium and calcitriol, restricting dietary intake of phosphate, and by the administration of medication that binds phosphate, such as calcium carbonate or calcium acetate.

Pseudohyperparathyroidism type 1a

Pseudohyperparathyroidism type 1a, or Albright osteodystrophy, is a disease in which there is resistance to PTH in the kidneys and bone due to a defective G_s protein. This causes hypocalcemia, hyperphosphatemia, and increased PTH levels. People with this condition have short fourth and fifth metacarpals, rounded facies, and the symptoms of hypoparathyroidism (convulsions, psychosis, tetany, paresthesias, numbness, laryngospasm, degeneration of teeth and nails, and cataracts). Unlike hypoparathyroidism, however, pseudohyperparathyroidism type 1a cannot be corrected by exogenous PTH. Treatment for pseudohyperparathyroidism type 1a involves correcting the hypocalcemia with oral calcium and calcitriol, or intravenous (IV) calcium if severe. This should suppress PTH secretion and avoid increased bone remodeling.

Fig. 26.3 ▶ Vitamin D metabolism.
Ultraviolet B (UVB) light converts 7-dehydrocholesterol to cholecalciferol (vitamin D_3). Ingested vitamin D is fat soluble and is transported to the liver in chylomicrons. All free vitamin D is transported in the blood and liver by a specific vitamin D–binding protein (DBP). DBPs are needed because steroid derivatives are hydrophobic. The liver converts vitamin D to 25-hydroxycholecalciferol [25-(OH)D], which is then transported to the kidneys, where it is converted to its active form, calcitriol, or 1,25-(OH)$_2$D, under the influence of parathyroid hormone (via the enzyme 1-α-hydroxylase). The effects of this are increased mineralization of bone, increased calcium and phosphate reabsorption in the kidneys, and increased calcium absorption in the gut. Excess vitamin D is excreted into bile.

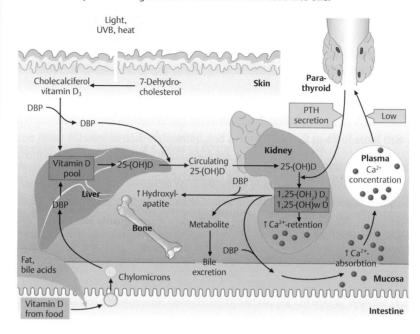

Actions

The principal role of calcitriol is to make calcium and phosphate available to extracellular fluid so that new bone can be mineralized.

Intestine

— Calcitriol increases synthesis of the protein calbindin. This is a specific Ca^{2+}-binding protein that transports Ca^{2+} through cellular cytosols. It acts as a "shuttle" for Ca^{2+} during the absorptive process (**Fig. 26.4**).
— Calcitriol increases the permeability of the entire small intestine to Ca^{2+}. This action facilitates passive absorption of Ca^{2+}.
— Calcitriol increases the intestinal absorption of phosphate.

Bone

— Calcitriol mobilizes Ca^{2+} and phosphate from mature bone for the mineralization of new bone and osteoid tissue. The Ca^{2+} that is mobilized is also available to the extracellular space.

Kidney

— Calcitriol promotes the reabsorption of Ca^{2+} and PO_4^{3-} at physiological concentrations.

Fig. 26.4 ► **Induction of calcium absorption.**

1,25-(OH)$_2$D$_3$ is necessary for the active intestinal absorption of calcium. In the cytosol, 1,25-(OH)$_2$D is thought to bind to a cytoplasmic receptor before being transferred to a DNA-associated nuclear receptor. This process induces the formation of several proteins, including calcium-binding protein (CaBP), an ATPase, alkaline phosphatase, and phytase. At the same time, lipid synthesis is increased, altering membrane lipids. The exact mechanism by which calcium is transported from the brush border of the intestines to the basal membrane and its absorption into the bloodstream is unknown.

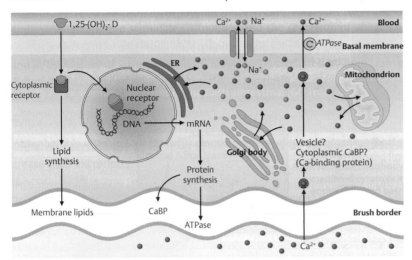

Pathophysiology

Reduced calcitriol production or resistance to calcitriol produces rickets in infants/children (**Table 26.5**) and osteomalacia in adults (**Table 26.6**).

Table 26.5 ► **Rickets**

Causes	Effects	Symptoms*	Treatment
Insufficient exposure to sunlight, especially in breast-fed infants	Low serum [Ca^{2+}] and [PO$_4^{3-}$] High PTH	Skeletal deformities: - Bowed legs - Pelvic abnormalities - Spinal curvature Fragile bones that may fracture easily Bone pain Impaired growth Muscle weakness Tetany and seizures may occur	Calcium, cholecalciferol, and phosphate replacement Skeletal deformities may require bracing to allow realignment; or surgical correction.

* Severe calcitriol deficiencies are manifested as hypocalcemia.
Note: Cholecalciferol (a calcitriol precursor) is given so that the body can synthesize calcitriol as needed.

Table 26.6 ► **Osteomalacia**

Causes	Effects	Symptoms*	Treatment
Insufficient sunlight Insufficient vitamin D intake in the diet Malabsorption (e.g., due to celiac disease, gastrectomy, or biliary disease) Hyperthyroidism Liver or kidney disease Long-term use of anticonvulsant medication	Low serum [Ca^{2+}] and [phosphate] High PTH	It typically presents with bone pain, which often starts in the lumbar spine, but may also affect the pelvis and legs Increased susceptibility of the bones to fractures Occasional skeletal deformities, such as bowed legs and compressed vertebrae Muscle weakness Pseudofractures (Looser zones) can be seen on radiographs)	Manage the underlying condition (if applicable) Cholecalciferol and calcium replacement

* Severe calcitriol deficiencies are manifested as hypocalcemia.
Note: Cholecalciferol (a calcitriol precursor) is given so that the body can synthesize calcitriol as needed.

Sarcoidosis and vitamin D hypersensitivity

Sarcoidosis is a multisystem granulomatous inflammatory disease that mainly affects the lungs, lymph nodes, eyes, and skin. It is thought to be autoimmune in origin (delayed-type hypersensitivity reaction [type IV]). Sarcoidosis may be associated with vitamin D hypersensitivity. There is increased 1-α hydroxylase activity in macrophages within the granulomas, causing calcitriol activation. This extrarenal production of calcitriol causes toxicity, producing symptoms such as lethargy and fatigue, irritability, cognitive disturbances, memory loss, and a metallic taste. Hypercalcemia and its accompanying symptoms may also be present, but the compensatory reduction in PTH hormone may prevent hypercalcemia. This vitamin D hypersensitivity reaction may also occur in some cancers, including lymphomas and sarcomas.

Calcitonin

Synthesis

Calcitonin is a 32 amino acid peptide that is synthesized and secreted by the parafollicular cells (or C cells) of the thyroid gland.

Regulation of Secretion

Secretion of calcitonin is stimulated by high serum $[Ca^{2+}]$ and is inhibited by normal or low serum $[Ca^{2+}]$.

Actions

— The main role of calcitonin is to inhibit bone resorption, which reduces liberation of the calcium and phosphate, thereby indirectly lowering serum Ca^{2+} levels. It also prevents acute hypercalcemia due to dietary intake postprandially (after eating).

Bone

— Calcitonin limits the resorption of Ca^{2+} from bone and thereby protects bone mass. It has no effect on bone formation.

Kidneys

— Calcitonin increases the urinary excretion of Ca^{2+}, PO_4^{3-}, Na^+, and K^+ and reduces the excretion of Mg^{2+}.

Therapeutic Uses

— Calcitonin is used clinically in cases of hypercalcemia and hyperphosphatemia because it decreases blood levels of these minerals while protecting bone mass from breakdown.
— Calcitonin also inhibits the loss of bone mass in aging, so it is used to treat postmenopausal women for osteoporosis (bone thinning).

Table 26.7 provides a summary of the effects of PTH, calcitriol, and calcitonin in calcium homeostasis.

Table 26.7 ▶ Summary of Main Regulatory Hormones of Calcium (and Phosphate) Homeostasis			
	PTH	**Calcitriol**	**Calcitonin**
Stimulus for secretion	↓ serum $[Ca^{2+}]$ ↑ serum $[PO_4^{3-}]$ ↓ serum [calcitriol]	↓ serum $[Ca^{2+}]$ ↓ serum $[PO_4^{3-}]$ ↑ PTH	↑ serum $[Ca^{2+}]$

Actions:
   Bone
   Kidney
 	
   Intestines
	↑ resorption of bone
	 ↑ Ca^{2+} reabsorption
	 ↓ phosphate reabsorption
	 ↑ Ca^{2+} absorption
	 (via activation of calcitriol)
		↑ resorption of bone
		 ↑ Ca^{2+} and phosphate reabsorption
		 ↑ Ca^{2+} and phosphate absorption
			↓ resorption of bone
			 ↑ excretion of Ca^{2+} and phosphate
			 None

Net effect	↑ serum $[Ca^{2+}]$ ↓ serum $[PO_4^{3-}]$	↑ serum $[Ca^{2+}]$ ↑ serum $[PO_4^{3-}]$	↓ serum $[Ca^{2+}]$

26.2 Integrated Regulation of Calcium and Phosphate in the Body

The main hormonal events and effects of low serum Ca^{2+} and high serum Ca^{2+} are summarized in **Tables 26.8** and **26.9,** respectively. The actions of other hormones on calcium homeostasis are summarized in **Table 26.10.**

Table 26.8 ▶ Coordinated Hormonal Events in Hypocalcemia that Return Ca²⁺ Levels to Normal			
Hormone Levels	Effects	Overall Result	Feedback Control
↑PTH (within ~15 min)	↑ Ca^{2+} and PO_4^{3-} from bone ↑ Ca^{2+} reabsorption and ↓ Ca^{2+} excretion in the kidney ↑ PO_4^{3-} excretion in the kidney ↑ 1-α hydroxylase activity in the kidney, leading to calcitriol synthesis (with its accompanying effects)	Rapid increase in serum [Ca^{2+}]	An increase of Ca^{2+} levels to normal removes the stimulus for PTH release Calcitriol also suppresses PTH secretion
↑ calcitriol	↑ Ca^{2+} and PO_4^{3-} absorption from the small intestine ↑ Ca^{2+} and PO_4^{3-} from bone ↑ reabsorption of Ca^{2+} and PO_4^{3-} in the kidney		
Calcitonin	Inactive in hypocalcemia		

Table 26.9 ▶ Coordinated Hormonal Events in Hypercalcemia that Return Ca²⁺ Levels to Normal			
Hormone Levels	Effects	Overall Result	Feedback Control
↓PTH*	↓ Ca^{2+} and PO_4^{3-} from bone ↓ Ca^{2+} reabsorption and Ca^{2+} excretion in the kidney ↓ PO_4^{3-} excretion in the kidney ↓ 1-α–hydroxylase activity in the kidney, leading to ↓ calcitriol synthesis (with its accompanying effects)	Decrease in serum [Ca^{2+}] Prevention of bone resorption	A decrease of Ca^{2+} levels to normal allows PTH secretion to be restored.
↓ calcitriol	↓ Ca^{2+} and PO_4^{3-} absorption from the small intestine ↓ Ca^{2+} and PO_4^{3-} from bone ↓ reabsorption of Ca^{2+} and PO_4^{3-} in the kidney		
↑ calcitonin**	↓ Ca^{2+} from bone ↑ Ca^{2+} and PO_4^{3-} excretion in the kidney		

* Removal of the action of PTH on the kidney is the major regulatory factor in returning Ca^{2+} to normal levels.
** Calcitonin plays a secondary role in returning plasma Ca^{2+} to normal.

Clinical signs of hypocalcemia

There are two specific clinical signs of hypocalcemia, Chvostek sign and Trousseau sign. Chvostek sign is twitching of the facial muscles after tapping on the facial nerve. Trousseau sign is carpopedal spasm elicited by inflating a blood pressure cuff above systolic pressure for 3 minutes. The carpopedal spasm includes flexion of the wrist, flexion of the metacarpophalangeal joints, extension of the interphalangeal joints, and adduction of the thumbs and fingers. It may be accompanied by paresthesias and fasciculations. Trousseau sign is an earlier and more sensitive sign of hypocalcemia than Chvostek sign.

Calcium balance in pregnancy

Pregnancy and lactation are two periods of potential high calcium loss. To maintain calcium balance, there is increased Ca^{2+} absorption from the intestine during pregnancy due to enhanced levels of calcitriol and plasma-binding proteins. PTH is also elevated during pregnancy, but increased calcitonin levels ensure that bone mass is protected.

Table 26.10 ▸ Effects of Other Hormones on Calcium Metabolism and Bone Turnover

Hormone	Effects	Overall Result
Thyroid hormones (T$_3$ and T$_4$)	↑ serum [Ca^{2+}] and [PO$_4$$^{3-}$] ↑ urinary [Ca^{2+}] ↓ urinary PO$_4$$^{3-}$ ↓ intestinal absorption of Ca^{2+}	T$_3$ and T$_4$ stimulate the renewal of bone, but with a greater effect of resorption
Glucocorticoids	↓ intestinal absorption of Ca^{2+} ↑ PTH ↓ number of osteoblasts	Osteopenia by promoting the breakdown of bone matrix
Estrogens	↓ serum [Ca^{2+}] and [PO$_4$$^{3-}$] ↑ urinary [Ca^{2+}] ↓ urinary hydroxyproline May ↑ 1-α hydroxylase activity, thus increasing the synthesis of calcitriol	Reduces the resorptive surface of bone ↑ retention of Ca^{2+} in bone
Growth hormone	↑ 1-α hydroxylase activity (via IGF), thus increasing the synthesis of calcitriol	Ca^{2+} and PO$_4$$^{3-}$ liberated from bone by calcitriol is available for new bone formation (primarily), and to ECF

Abbreviations: ECF, extracellular fluid; ICF, intracellular fluid.

27 Adrenal Hormones

The two adrenals (suprarenal glands) are located on top of each kidney. Each gland can be divided histologically and biochemically into two major subglands:

— The adrenal cortex
— The adrenal medulla

The adrenal medulla is a modified autonomic ganglion synthesizing mainly epinephrine and some norepinephrine. The adrenal cortex has three zones, each of which synthesizes a specific steroid hormone or hormones (**Fig. 27.1**).

Fig. 27.1 ► **Adrenal gland.**
Secretion of hormones from the adrenal cortex is under hypothalamic and pituitary control. The adrenal medulla is stimulated to release catecholamines by activation of the sympathetic nervous system. Both controlling mechanisms are activated by stress. Hyperkalemia and angiotensin II stimulate the release of aldosterone. Adrenocortical hormones cause feedback inhibition of their upstream regulatory hormones. Cortisol also acts in a paracrine fashion to stimulate catecholamine release. Epinephrine and norepinephrine act on the pituitary to stimulate the release of adrenocorticotropic hormone (ACTH). (CRH, corticotropin-releasing hormone)

Embryology of the adrenal gland

The adrenal cortex, which comprises 80% of the gland, is derived from mesothelium; whereas the adrenal medulla (20%) is derived from neural crest cells. The sympathetic nervous system is also derived from neural crest cells, which explains the link between this system and the adrenal medulla and their common functionality.

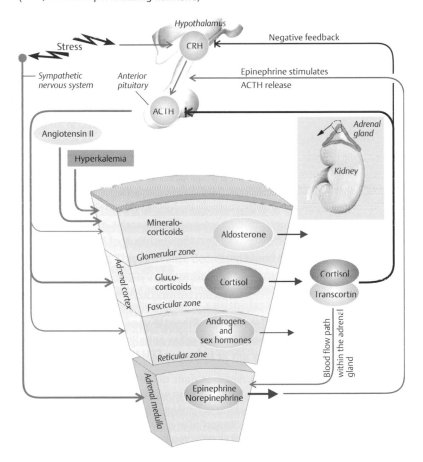

Hypothalamic-Anterior Pituitary Control of Adrenocortical Hormones

Corticotropin-releasing hormone (CRH) secreted from the hypothalamus acts on the anterior pituitary to synthesize pro-opiomelanocortin (POMC), which is the precursor of adrenocorticotropic hormone (ACTH). ACTH secreted from the anterior pituitary activates cholesterol desmolase in the adrenal cortex. This stimulates the conversion of cholesterol to pregnenolone, which is the precursor of all adrenocortical hormones.

The cascade showing the synthesis of all of the adrenocortical and androgen hormones from cholesterol is shown in **Fig. 27.2**.

Fig. 27.2 ► **Synthesis of mineralocorticoids, glucocorticoids, and androgens.**
(ACTH, adrenocorticotropic hormone)

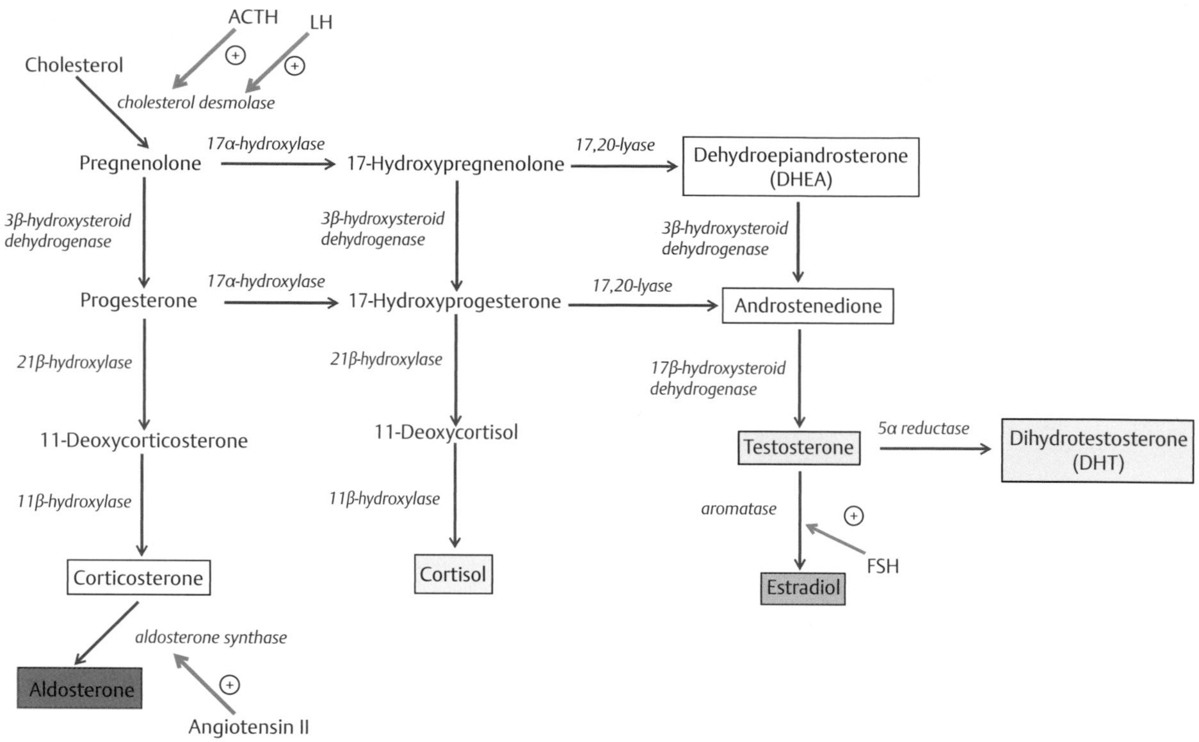

27.1 Adrenocortical Hormones

Aldosterone

Synthesis

The synthesis of aldosterone is shown in **Fig. 27.2** (red box).

Regulation of Secretion

Aldosterone secretion is mainly regulated by the renin–angiotensin system (RAS) and by serum [K$^+$]. ACTH maintains tonic control.

— Renin is an enzyme produced by juxtaglomerular (JG) cells of the afferent arteriole of the kidney. It is released when blood pressure (BP) is reduced in the renal artery, usually due to a decrease in circulating blood volume (hypovolemia). It is also released in response to sympathetic nervous system activation or decreased NaCl load at the macula densa. Renin cleaves angiotensinogen to angiotensin I; angiotensin I is acted upon by angiotensin-converting enzyme (ACE) to form angiotensin II. Angiotensin II increases the conversion of corticosterone to aldosterone, which increases the renal reabsorption of Na$^+$ and water, thus restoring blood volume to normal (**Fig. 27.3**).

■ **α$_1$ Blockers and renin release**

Alpha$_1$ blockers (e.g., prazosin, terazosin, and doxazosin) act selectively on the postsynaptic α$_1$ receptor of vascular smooth muscle, causing vasodilation and lowering total peripheral resistance. They are used in the treatment of hypertension or heart failure. However, their effects will cause counter-regulatory activation of the renin–angiotensin–aldosterone system (unwanted); thus, they are often combined with ACE inhibitors or β blockers. Orthostatic (postural) hypotension is common with the first dose.

Fig. 27.3 ▶ **Renin–angiotensin–aldosterone system.**
The renin–angiotensin–aldosterone system regulates sodium balance, fluid volume, and blood pressure (BP). Renin is released by the kidneys in response to reduced perfusion (due to decreased plasma volume and BP). Renin then stimulates angiotensinogen to convert angiotensin I to angiotensin II in the lungs. Angiotensin II causes vasoconstriction and stimulates the secretion of aldosterone from the adrenal cortex. Aldosterone causes sodium (and water) reabsorption, thus increasing fluid volume, BP, and renal perfusion. Angiotensin II and aldosterone cause feedback inhibition of renin secretion. Renin is also inhibited by normalization of renal perfusion. (GFR, glomerular filtration rate; RBF, renal blood flow)

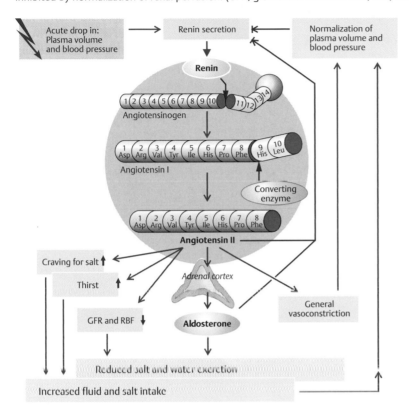

- Hyperkalemia causes increased aldosterone production, which increases the renal excretion of K⁺.

Table 27.1 summarizes the factors that stimulate and inhibit aldosterone secretion.

Table 27.1 ▶ **Regulation of Aldosterone Secretion**	
Stimulates Aldosterone	**Inhibits Aldosterone**
Activation of RAS	Atrial natriuretic peptide
Hyperkalemia	Hypokalemia
Abbreviation: RAS, renin–angiotensin-aldosterone system.	

Actions

Aldosterone acts on the distal tubule and collecting duct of the kidney to regulate BP and blood volume via the following mechanisms:

- ↑Na⁺ (and water) reabsorption
- ↑K⁺ secretion
- ↑H⁺ secretion

Fig. 27.4 summarizes the secretion, effects, and degradation of aldosterone.

Angiotensin-converting enzyme inhibitors

ACE inhibitors (e.g., captopril, enalapril, and lisinopril) prevent the activation of angiotensin I to angiotensin II in the lung by inhibiting ACE. In doing so, they prevent the direct vasoconstrictive effects of angiotensin II on blood vessels, causing decreased peripheral vascular resistance and decreased sympathetic tone. They also prevent aldosterone release from the adrenal cortex, which prevents sodium and water reabsorption, thus decreasing preload. ACE inhibitors prevent the degradation of bradykinin (ACE also acts as a kininase). Bradykinin is a vasodilator, and increasing its level may contribute to the effect of ACE inhibitors. These drugs provide symptomatic improvement and reduce mortality in hypertension and heart failure. Common side effects include a persistent cough, hyperkalemia, first-dose hypotension, and taste disturbances.

Atrial natriuretic peptide

Atrial natriuretic peptide (ANP) is a 27 amino acid peptide secreted by atrial myocytes. It is involved in the long-term homeostasis of sodium and water, BP, and arterial pressure. It is secreted in response to atrial distention, sympathetic stimulation of β receptors, hypernatremia (indirect effect), angiotensin II, and endothelin secretion (a vasoconstrictor). It acts by two mechanisms. First, it causes direct vasodilation of vascular smooth muscle. This decreases cardiac output by increasing venous capacitance, which decreases preload, and by decreasing peripheral resistance, which decreases afterload. Second, ANP acts on the kidney to inhibit renin secretion and the reabsorption of sodium. Overall, ANP provides counter-regulation to the renin–angiotensin–aldosterone system.

Fig. 27.4 ▶ **Secretion, effects, and degradation of aldosterone.**
Aldosterone is secreted by the zona glomerulosa of the adrenal cortex partly in response to stimulation by ACTH from the anterior pituitary. Aldosterone secretion increases mainly in response to drops in BP and blood volume (mediated by angiotensin II) and by hyperkalemia. Aldosterone leads to retention of Na^+, a moderate increase in H^+ secretion, and increased K^+ excretion. It also increases the number of Na^+–K^+–ATPase molecules in the target cells. Aldosterone is degraded by glucuronidation in the liver and excreted into bile and feces. (GFR, glomerular filtration rate; RBF, renal blood flow)

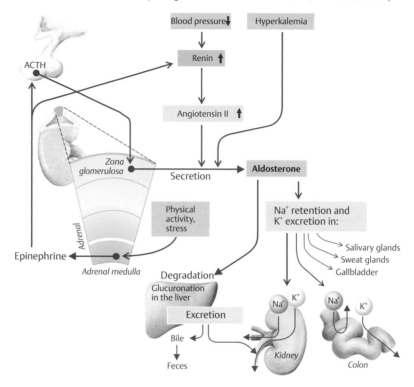

Pathophysiology

Aldosterone deficiency.   Aldosterone deficiency may be a part of overall adrenocortical insufficiency (Addison disease), which is discussed on page 275. It is a major problem due to loss of Na^+ and retention of K^+.

Aldosterone excess.   Primary aldosteronism and secondary aldosteronism are summarized in **Tables 27.2** and **27.3.** respectively.

Table 27.2 ▶ **Primary Aldosteronism***			
Causes	**Effects**	**Symptoms**	**Treatment**

Adrenocortical adenoma**
 Bilateral adrenocortical
  hyperplasia
 Adrenal carcinoma
	Hypertension
	 ↓ serum [K^+]
	 ↑ serum [Na^+]
	  (usually mild)
	 Alkalosis
		Edema	Conn syndrome:
			 - Spironolactone treatment for
			   4 weeks, then surgical resection
			 Hyperplasia:
			 - Dexamethasone

* With primary aldosteronism, excess aldosterone production is independent of the renin–angiotensin-aldosterone system. ** Primary aldosteronism due to adrenocortical adenoma is Conn syndrome.			

Table 27.3 ► **Secondary Aldosteronism***

Causes	Effects	Symptoms	Treatment
High renin due to: Renal artery stenosis Hypertension Heart failure Hepatic failure Diuretics	Hypertension ↓ serum [K⁺] ↑ serum [Na⁺] (usually mild) Alkalosis	Edema	Restrict sodium intake Treat underlying cause

* Secondary aldosteronism is due to excess renin secretion secondary to a disease process.

Cortisol

Synthesis

The synthesis of cortisol is shown in **Fig. 27.2** (yellow box).

Regulation of secretion

Negative-feedback control: When cortisol decreases, CRH and ACTH increase, producing cortisol release. An increase in cortisol decreases CRH and ACTH concentrations (**Figs. 27.1** and **27.5**).

Cortisol levels follow a circadian rhythm of secretion, with a peak in its concentration occurring just after awakening in the morning (**Fig. 27.6**).

Fig. 27.5 ► **Regulation of cortisol and epinephrine concentrations in plasma.**
The hypothalamus is stimulated to release corticotropin-releasing hormone (CRH) by the limbic system. This then causes the downstream release of adrenocorticotropic hormone (ACTH) from the anterior pituitary and cortisol release from the adrenal cortex. Cortisol may, in turn, stimulate the release of epinephrine and norepinephrine from the adrenal medulla. Autonomic centers in the medulla may also directly stimulate the adrenal medulla to produce epinephrine and norepinephrine via postganglionic sympathetic nerves. Epinephrine and norepinephrine may positively feed back on the anterior pituitary to stimulate the release of ACTH and thus amplify the original signal. Cortisol negatively feeds back on the limbic system, hypothalamus and anterior pituitary. ACTH may also negatively feed back on the hypothalamus. This allows the original signal to be inhibited when plasma levels of ACTH or cortisol become too high. (ADH, antidiuretic hormone)

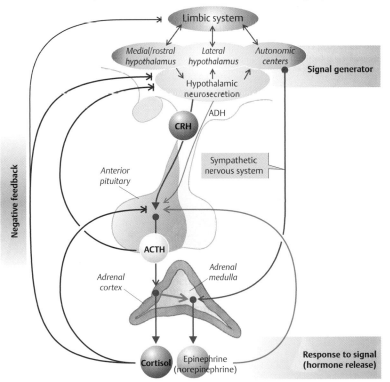

Dosing regimen of glucocorticoid drugs

The dosing regimen of glucocorticoid drugs can minimize adrenocortical atrophy, e.g., if they are given when normal cortical secretion is high, and feedback inhibition is low (late morning), the glucocorticoid drug is eliminated during daytime, and normal endogenous cortisol production starts early the following morning.

The respiratory burst

The respiratory burst is the rapid release of reactive oxygen species (superoxide radical and hydrogen peroxide) from phagocytes (neutrophils and monocytes) when a microbe is encountered and phagocytosed. It is one of the mechanisms by which phagocytes exert their microbicidal effects and is an important immune defense. The reactive oxygen species are generated by the partial reduction of oxygen in the respiratory chain (electron transport chain). They combine with Cl^- to form hypochlorous acid, which dissociates to form hypochlorite ions, which kill the microbes. Cortisol (and exogenous corticosteroids) inhibits the respiratory burst and so may predispose an individual to infection.

Stress and the hypothalamic-pituitary-adrenocortical axis

The hypothalamic-pituitary-adrenocortical (HPA) axis and the sympathetic nervous system are involved in maintaining homeostasis in response to physiological or psychological stress. Sympathetic (catecholaminergic) mechanisms provide short-term, rapid changes in response to perturbations in homeostasis, whereas the HPA axis (via elevations in cortisol) provides a longer-term mechanism that allows continued resistance to homeostatic perturbations. The major physiological stressors are hypoglycemia, hypotension, hypovolemia, hypo- or hyperthermia, and pain. Psychological stressors are more difficult to identify but may include novel environments (change), unpredictability of situation, conflict, sleep deprivation, and psychological "pressure." The magnitude of the HPA response is determined by past experience or exposure (coping or adapting) to these stressors. Many of the actions of glucocorticoids (especially at high levels) are damaging, so prolonged exposure to stressors can cause an emergence of stress-related disease. Under "normal" regulation, cortisol provides negative feedback, which rapidly terminates HPA axis activation following exposure to a stressor, but chronic or repeated exposure to a stressor dampens cortisol negative feedback, allowing "hypersecretion" of cortisol at basal levels and following acute stress. Activation of the HPA axis in stress may also alter the pulsatile and episodic release of other hormones (via disruption of circadian rhythms) and may disrupt feedback inhibition.

Fig. 27.6 ▶ Circadian rhythm of adrenocorticotropic hormone (ACTH) and cortisol secretion. Peak plasma cortisol concentration is observed at ~8:00 A.M., and the lowest concentration is at ~8:00 P.M. There are fluctuations around the mean concentration, with some secondary peaks occurring throughout the day. The rhythm for release of ACTH is similar to that for cortisol. (CRH, corticotropin-releasing hormone; ADH, antidiuretic hormone)

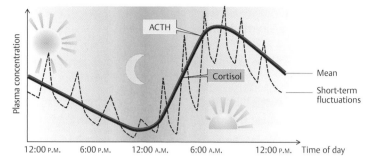

Actions

Cortisol coordinates the body's response to stress and is essential for life. It regulates the expression of a large number of genes, enzymes, and proteins and is generally considered "permissive," allowing the actions of other hormones.

The actions of cortisol are summarized in **Table 27.4.** These actions may also be elicited by synthetic glucocorticoids.

Table 27.4 ▶ Summary of the Actions of Cortisol

Tissues	Organs	Inflammatory Response	Immune Response
Muscle: ↑ proteolysis ↓ protein synthesis **Fat:** ↑ lipolysis* ↑ lipogenesis in the abdomen, trunk, and face **Bone:** ↓ bone formation** ↑ bone resorption **Arteries:** ↑ vascular sensitivity to epinephrine (due to upregulation of α_1 receptors) and angiotensin II ↓ vascular permeability **Tissue and organ maturation in the fetus**	**Liver:** ↑ gluconeogenesis (via amino acids from protolysis, and glycerol from lipolysis) ↑ glycolysis ↓ glycogenolysis **Kidney:** ↑ free water clearance ↑ renin production and release ↓ 1-α hydroxylase levels **Parathyroid gland:** ↑ PTH **Pancreas:** ↑ glucagon production*** **Anterior pituitary:** ↓ growth hormone **Brain:** ↑ glucose uptake	↓ production of phospholipase A, which inhibits the production of prostaglandins, leukotrienes, and thromboxanes ↓ production of leukocytes ↓ recruitment of leukocytes to the site of trauma or infection ↓ leukotriene-activated respiratory burst from phagocytes, which decreases their bacteriocidal activity ↓ fibroblast activity and connective tissue proliferation ↑ stabilization of mast cells	↓ T cells, especially Th cells, by inducing apoptosis ↓ T-cell recruitment to the site of an antigen ↑ T-cell redistribution to lymph nodes B cells are not modulated by cortisol

* Cortisol alone is only weakly lipolytic; however, it is permissive for maximal stimulation of fat metabolism by epinephrine and growth hormone.
** Excessive cortisol inhibits bone formation by decreasing type I collagen synthesis (the fundamental component of bone matrix), by suppressing the differentiation of osteoprogenitor cells to active osteoblasts, and by antagonizing the action of calcitriol.
*** Cortisol induces enzymes that activate glucagon and epinephrine, which are the primary hormones that counter the actions of insulin (except in the central nervous system).
Abbreviation: PTH, parathyroid hormone.

Inactivation

Cortisol can be inactivated by conversion to cortisone, an inactive metabolite of cortisol, by the enzyme 11-β-hydroxysteroid dehydrogenase (11-β-HSD). This enzyme is expressed by mineralocorticoid receptor (MR)–responsive cells in the kidney.

Uses of Exogenous Corticosteroids (e.g., hydrocortisone)

— *Endocrine uses*:

- — Adrenocortical insufficiency (Addison disease)
- — Congenital adrenal hyperplasia (to suppress ACTH release)

— *Nonendocrine uses*:

- — Rheumatoid arthritis
- — Leukemia
- — Lymphoma
- — Allergic reactions
- — Asthma
- — Inflammatory and autoimmune disorders
- — Immunosuppression for transplantation
- — Collagen disorders
- — Cerebral edema
- — Bacterial meningitis

Pathophysiology

Adrenocortical insufficiency. Adrenocortical insufficiency, or Addison disease, is summarized in **Table 27.5.**

Table 27.5 ► Addison Disease

Causes	Effects	Symptoms	Treatment
Hypothalamic dysfunction: - CRH neuronal dysfunction Pituitary dysfunction: - ACTH hyposecretion Adrenal dysfunction*: - Defects in key enzymes in the cortisol biosynthetic pathway - ACTH receptor hypofunction Idiopathic (probably autoimmune)	↓ CRH ↓ ACTH ↓ cortisol (↓ aldosterone) (↓ androgens) ↓ serum [Na⁺] Hypoglycemia ↑ sensitivity to insulin ↑ melanocyte-stimulating hormone ↑ serum [K⁺] and [Ca²⁺]	Excessive fatigue Abdominal pain Infrequent periods Weight loss Anorexia (loss of appetite) Vomiting and diarrhea Concentrated urine, inability to clear free water Depression Occasional increase in skin pigmentation	Replacement of endogenous steroids by synthetic steroids, e.g., hydrocortisone given orally or intramuscularly

* Adrenal dysfunction may be specific to the fasciculata zone, or it may affect adrenocortical steroidogenesis in general. If generalized, then symptoms of aldosterone and androgen hypofunction will also be present.
Abbreviations: ACTH, adrenocorticotropic hormone; CRH, corticotropin-releasing hormone.

Perinatal role of cortisol

The most important perinatal role of cortisol is its effects on lung maturation. It is essential for production of lung surfactant, development of the alveoli, flattening of the lining cells, and thinning of the lung septa. Cortisol is also essential for the maturation of digestive enzyme capacity of the intestinal mucosa. This permits the neonate to utilize disaccharides present in milk.

Acute adrenal crisis (Addisonian crisis)

Acute adrenal crisis (addisonian crisis) is due to acute insufficiency of adrenal steroids, mainly cortisol. It usually occurs in people with known Addison disease who undergo some form of stress, such as surgery, trauma, or infection, but it may also occur when someone on long-term steroids abruptly stops them or forgets to take their medication. The main sign of acute adrenal crisis is shock (hypotension, tachycardia, and oliguria), but there may also be acute abdominal pain, diarrhea, vomiting, hypoglycemia, fever, weakness, and confusion. It may progress to seizures, coma, and death if untreated. If there is a high index of suspicion for acute adrenal crisis, treatment should begin before any laboratory results are in. Treatment involves giving intravenous (IV) fluids, IV hydrocortisone, antibiotics, and IV glucose if necessary. In the longer term, the patient can be switched to oral steroids, and the precipitating factor should be treated.

Perioperative steroid coverage

Patients who have been on long-term steroids or who have stopped steroids in the last few months will have some adrenal suppression (due to feedback inhibition of ACTH secretion by exogenous steroids). Consequently, the perioperative administration of steroids prior to undergoing the stress of surgery is necessary to prevent adrenal crisis (addisonian crisis), the major effect of which is shock.

Congenital virilizing adrenal hyperplasia

Congenital virilizing adrenal hyperplasia (adrenogenital syndrome) is a deficiency of both aldosterone and cortisol biosynthesis. It is associated with a defect in adrenal steroidogenesis, most commonly with 21-hydroxylase, but defects with 3-β-hydroxysteroid dehydrogenase (3-β-HSD) may also occur. When these enzymes are not expressed or nonfunctional, then steroidogenesis proceeds through androgen formation in all zones of the cortex, but there is little or no synthesis of cortisol and aldosterone. Because cortisol is not being synthesized, there is no feedback on the HPA axis, leading to hypersecretion of ACTH. The high levels of ACTH induce adrenocortical hyperplasia, which augments the already elevated androgen production. The individual appears "masculinized" or virilized (male secondary features), allowing the disease to be diagnosed more easily in female neonates than male neonates, in whom it is typically missed. Treatment involves replacement of the deficient hormones.

Role of cortisol in fasting states

Cortisol is essential for survival during fasting states because of its proteolytic effects. During fasting, liver glycogen stores quickly become depleted. Gluconeogenesis continues using amino acids (from protein catabolism) in the presence of cortisol. However, if cortisol is deficient, death from hypoglycemia occurs. Plasma levels of cortisol during fasting are only slightly to moderately elevated (usually relative to the degree of initial hypoglycemia), but this elevation of cortisol is sufficient because cortisol has a permissive effect on key metabolic enzymes. During fasting, the body also breaks down adipose tissue into glycerol and fatty acids. Glycerol can then be converted to glucose in the liver. Cortisol enhances this lipolysis.

Adrenocortical excess.   Adrenocortical excess, or Cushing syndrome, is summarized in **Table 27.6.** Note that Cushing disease is adrenocortical excess that is caused by an ACTH-secreting tumor.

Table 27.6 ▶ Cushing Syndrome

Causes	Effects	Symptoms	Treatment
Pituitary tumors secreting ACTH Adrenal adenomas or carcinomas ACTH-secreting tumors of ovary, lung (oat cell carcinoma), thymus, and pancreas Corticosteroid administration	↑ cortisol ↑ insulin ↓ ACTH (due to feedback inhibition from cortisol) ↑ ACTH (if there is a pituitary tumor, or ACTH-secreting tumor) ↓ CRH (due to feedback inhibition) ↑ serum [K⁺] ↓ serum [Na⁺]	Redistribution of body fat: - Obesity that affects the trunk with relative sparing of the limbs - Moon face - "Buffalo hump" on the back Purple striae on abdomen Hypertension Muscle wasting and weakness Osteoporosis Virilization Amenorrhea Steroid diabetes: - Hyperglycemia - Polyuria and polydipsia - Delayed wound healing	Treat the cause

Abbreviations: ACTH, adrenocorticotropic hormone; CRH, corticotropin-releasing hormone.

Androgens

Synthesis

The synthesis of androgens is shown in **Fig. 27.2** (blue and purple boxes).

27.2 Adrenal Medullary Hormones

Epinephrine (and norepinephrine)

The adrenal medulla is a modified sympathetic ganglion. It is innervated by preganglionic sympathetic axons that stimulate the release of two catecholamines:

— Epinephrine (80%)
— Norepinephrine (20%)

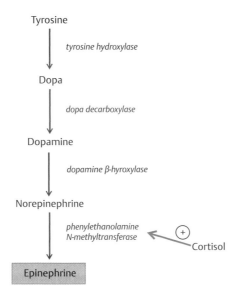

Fig. 27.7 ▶ **Biosynthesis of catecholamines.**

Note: Most norepinephrine released in the body comes from sympathetic ganglia rather than the chromaffin cells of the adrenal medulla.

Synthesis

The synthesis of epinephrine and norepinephrine is shown in **Fig. 27.7**. The conversion of tyrosine to dihydroxyphenylalanine (Dopa) by the enzyme tyrosine hydroxylase is the rate-limiting step and is well regulated.

Mechanism of Norepinephrine Secretion

Acetylcholine, released from preganglionic sympathetic neurons, reacts with nicotinic receptors on chromaffin cells in the adrenal medulla, causing increased Na^+ and Ca^{2+} permeability. Increased cytosolic Ca^{2+} moves granules to the membrane and causes exocytosis of norepinephrine.

Regulation of Secretion

The secretion of epinephrine and norepinephrine is controlled by the sympathetic nervous system, the HPA axis (via cortisol), and feedback inhibition of enzymes caused by high levels of norepinephrine.

— Acutely high levels of norepinephrine stimulate the release of tyrosine hydroxylase.
— Chronically elevated norepinephrine stimulates the release of tyrosine hydroxylase and dopamine hydroxylase, as well as PNMT (via cortisol).

Refer to **Figs. 27.1** and **27.5**.

Actions

Epinephrine mobilizes energy for "fight or flight." Its principal actions are summarized in **Table 27.7**, but these are covered in detail in Chapter 4 on the autonomic nervous system.

Table 27.7 ▶ Summary of the Main Actions of Epinephrine	
Organ/Organ System	**Effect(s)**
Heart	↑ Heart rate and contractility*
CNS	↑ Alertness
GI tract	↓ Gut motility
Liver	↑ Glycogenolysis ↑ Gluconeogenesis
Pancreas	↑ Glucagon secretion ↓ Insulin secretion
Arteries	↑ Vasodilation
Fat	↑ Lipolysis
Muscle	↑ Glycogenolysis

* These effects are enhanced in the presence of thyroid hormone and cortisol.
Abbreviations: CNS, central nervous system; GI, gastrointestinal.

Inactivation

The half-life of catecholamines is only 1 to 2 minutes in the circulation. Catecholamines are largely inactivated by catechol-*O*-methyltransferase (COMT) in the liver, or they may be deaminated by monoamine oxidase (MAO) within nerve terminals.

Pathophysiology

Epinephrine and norepinephrine.   An excess of epinephrine and norepinephrine results from an adrenal tumor known as a pheochromocytoma (see **Table 27.8**).

Table 27.8 ▶ Pheochromocytoma			
Cause	**Effects**	**Symptoms**	**Treatment**
Pheochromocytoma is a rare, benign tumor of adrenal chromaffin tissue - 90% are unilateral	↑ Epinephrine ↑ Norepinephrine ↑ VMA in urine Hypertension Cardiomyopathy Hyperglycemia	Headaches (associated with arterial hypertension) Weight loss Periods of crisis lasting ~15 minutes, characterized by fear, tachycardia, palpitations, tremors, sweating, nausea, and pallor	Stabilize blood pressure with phenyoxybenzamine (an α-blocker) and propranolol a β-blocker) Surgery to remove the tumor

Abbreviations: VMA, 3-methoxy-4-hydroxymandelic acid, a metabolite produced from the degradation of norepinephrine, epinephrine, and dopamine.

28 Endocrine Pancreas

The endocrine part of the pancreas is the islets of Langerhans. The islets contain four major cell types, each of which secrete a specific hormone, as shown in **Table 28.1**. These hormones regulate energy metabolism by coordinating the secretion and storage of endogenous glucose, free fatty acids, amino acids, and ketone bodies. This ensures that energy needs are met during basal and active states.

Table 28.1 ▶ Hormones Secreted by Pancreatic Islet Cells	
Cell Type	**Hormone Secreted**
α cells	Glucagon
β cells	Insulin
δ cells	Somatostatin
PP cells	Pancreatic polypeptide

Gap junctions connect α cells and β cells to themselves and each other, permitting rapid communication. Beta cells inhibit the activity of neighboring α cells and thereby suppress the release of glucagon.

28.1 Insulin, Glucagon, and Energy Metabolism

Insulin

Synthesis

Insulin is synthesized in pancreatic islet β cells in the form of a prohormone. This is a single polypeptide chain of 86 amino acids. Active insulin is generated by the proteolysis of C peptide by a specific endopeptidase, leaving two polypeptide chains joined by disulfide bonds. The A chain has 21 residues, and the B chain has 30 residues. C peptide, which is more stable than insulin, is cosecreted. Measurement of C peptide gives a reliable estimate of endogenous insulin production (**Fig. 28.1**).

Mechanism of Secretion

Glucose binds to a GLUT-2 transporter located in the plasma membrane of β cells and enters the cell by facilitated diffusion. Once inside, it is rapidly metabolized, generating adenosine triphosphate (ATP) and the reduced form of nicotinamide adenine dinucleotide (NADH). This rise in ATP levels causes K^+ channels to close, resulting in depolarization of the β cell. Depolarization opens voltage-gated Ca^{2+} channels, and the subsequent Ca^{2+} influx stimulates release of insulin from stores (rapid release). This influx of Ca^{2+} also activates calmodulin and Ca^{2+}/calmodulin-dependent protein kinase (CamK). This activation increases insulin synthesis (and slows the release of insulin).

Table 28.2 lists other glucose transporters, the tissues that express them, and their actions.

Table 28.2 ▶ **Glucose Transporters**			
Transporter	**Tissues Expressing Transporters**	**Regulated by Insulin**	**Actions**
Sodium glucose transporter 1 (SGLT-1)	Intestinal epithelium Renal tubular cells	No	Mediates Na⁺-dependent secondary active transport of glucose into intestinal epithelium and renal tubular cells. This enables the body to recover glucose from the glomerular filtrate.
GLUT-1	All cells*	?	Mediates basal glucose transport into cells unidirectionally
GLUT-2	Liver Pancreas	No	Bidirectional transport of glucose, depending on glucose concentration on either side of the plasma membrane. It maintains intracellular levels of glucose equal to extracellular levels.
GLUT-3	Neural tissue	No	
GLUT-4	Skeletal muscle Cardiac muscle Adipose tissue	Yes Insulin stimulates the synthesis of GLUT-4 transporters and their insertion into plasma membranes.	Mediates glucose transport into muscle and adipose cells unidirectionally
GLUT-5	?	?	Transports fructose

* Skeletal muscle, cardiac muscle, and adipose tissue express GLUT-1. These tissues are responsible for 40% of the body's glucose metabolism. GLUT-1 concentration in these tissues is inadequate to maintain normal metabolism in the absence of insulin.

Regulation of Secretion

Table 28.3 lists the factors that regulate insulin secretion. Serum [glucose] is the main determinant.

Table 28.3 ▶ **Factors that Regulate Insulin Secretion**	
Stimulatory Factors	**Inhibitory Factors**
High serum [glucose] High levels of proteins (amino acids), ketoacids, fatty acids, and triglycerides Gastrointestinal polypeptide (GIP) Glucagon Glucagon-like polypeptide (GLP) Incretins Activity of the parasympathetic nervous system (acetylcholine)	Low serum [glucose] Low levels of proteins (amino acids), ketoacids, fatty acids, and triglycerides Somatostatin Activity of the sympathetic nervous system (α-adrenergic effect of norepinephrine and epinephrine)

Receptors and Cell Mechanisms

The insulin receptor is a tyrosine kinase receptor with extracellular α subunits and β subunits that span the cell membrane. When insulin binds to this receptor, it leads to intracellular autophosphorylation and the activation of signaling molecules (**Fig. 28.1**).

Insulin downregulates its own receptors, so the number of receptors is decreased in obesity and increased in fasting.

Central nervous system and hypoglycemia

The brain and spinal cord require glucose for energy and are therefore sensitive to large decreases in plasma glucose levels. If there is a gradual decrease in blood glucose, the CNS adapts by using ketone bodies, but this adaptation requires several weeks. Hypoglycemia inhibits insulin synthesis and secretion, causing a decrease of GLUT-4 transporters. This causes decreased glucose uptake in muscle and adipose tissue, thereby sparing glucose for the brain to use. The GLUT-3 transporter in neural tissue is not affected by insulin; hence, the brain can still use glucose in the absence of insulin.

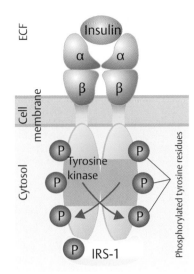

Fig. 28.1 ▶ Tyrosine kinase receptors.
The insulin receptor is a glycoprotein consisting of two α chains, located on the outside of the cell, and two β chains that are membrane bound and reach into the cytosol. Insulin binds to the exterior α chains, which phosphorylates part of the internal β chains. This activates tyrosine kinase, which then phosphorylates tyrosine R groups of a peptide (insulin receptor substrate, IRS-1), which, in turn, triggers further phosphorylation and dephosphorylation reactions, leading to the physiological effects of insulin. (ECF, extracellular fluid)

Actions

Insulin causes a reduction in serum [glucose] (principally). It does this by stimulating the uptake of glucose in tissues by the insertion of GLUT transporters into cell membranes. Insulin also increases the utilization of glucose as an energy source and promotes the storage of energy (as glycogen, protein, and fat).

Liver

— Insulin increases glycogen synthesis (and decreased glycolysis).
— Insulin decreases gluconeogenesis by shunting substrate away from glucose formation toward the production of fructose 2,6-bisphosphate. It does this by activating phosphofructokinase.
— Insulin decreases ketone formation.

Muscle

— Insulin increases the uptake of amino acids into skeletal muscle cells.
— Insulin increases protein synthesis and decreases protein degradation.
— Insulin increases glycogen synthesis (and decreases glycolysis).

Fat

— Insulin increases uptake of free fatty acids into adipose tissue.
— Insulin increases fat deposition and decreases fat degradation.

Pancreas

— Insulin decreases the secretion of glucagon.

Miscellaneous

— Insulin increases Na$^+$–K$^+$ ATPase activity, which promotes K$^+$ uptake into cells (in exchange for Na$^+$); thus, serum [K$^+$] levels are reduced.
— Insulin increases cholesterol biosynthesis.
— Insulin also increases wound healing, microvascular integrity, neural integrity, and growth factors.

Insulin and potassium balance
Potassium is an important ion in the body, with 98% being intracellular. The ratio of intracellular to extracellular potassium determines cell membrane potential. Immediate potassium balance is controlled by intracellular and extracellular potassium exchange driven by the Na$^+$–K$^+$ ATPase pump. This is controlled by insulin and β$_2$ receptors. Long-term potassium balance is controlled by renal excretion. In hyperkalemic (high potassium) states, glucose and insulin are used to drive potassium into cells by increasing the activity of the Na$^+$–K$^+$ ATPase pump.

Diagnosis of type I diabetes

The diagnosis of type I diabetes is made by conducting various blood tests:

1. *Glycated hemoglobin (A1c) test.* This test measures the percentage of blood sugar attached to hemoglobin and is an indication of average blood sugar levels over the past 2 to 3 months. An A1c > 6.5% is diagnostic of diabetes. This test has yet to be adopted by the World Health Organization as the gold standard test for diabetes.

2. *Random blood sugar level.* A random blood test is taken and blood sugar measured. A value > 200 mg/dL (11.1 mmol/L) is suggestive of diabetes, especially if the patient has associated symptoms.

3. *Fasting blood sugar test.* The patient fasts overnight, and a blood sample is taken in the morning. A value > 126 mg/dL (> 7 mmol/L) on two separate occasions allows for a diagnosis of diabetes.

4. *Oral glucose tolerance test.*

If type I diabetes is diagnosed, the patient will also be screened for autoantibodies, which are commonly associated with diabetes.

Diabetic ketoacidosis

Diabetic ketoacidosis (DKA) is a life-threatening complication of diabetes (usually type I diabetes). In DKA, there is a severe shortage of insulin, so the body ceases to use carbohydrate as an energy source and starts using fatty acids, which produce ketones (acetoacetate and β hydroxybutyrate). Dehydration (due to osmotic effect of glucose) occurs followed by acidosis (from ketones) and coma. Signs may include hyperventilation (to compensate for acidosis) and the breath smelling of ketones (sweet smell). DKA usually occurs in known diabetics and can be triggered by illness or inadequate/inappropriate insulin therapy. Treatment involves giving insulin, potassium, bicarbonate, and replacement fluids.

Glucose tolerance test

The glucose tolerance test is used to test for type II diabetes. It involves giving the patient a known oral dose of glucose, following an 8- to 12-hour fasting period, then measuring plasma glucose levels at intervals thereafter to determine how quickly plasma glucose levels fall and homeostasis is regained. Normal fasting plasma glucose levels are < 6.1 mmol/L. Glucose levels of 6.1 to 7.0 mmol/L are considered borderline and are indicative of impaired fasting glycemia. Measurements of plasma glucose taken after 2 hours should be < 7.8 mmol/L. Glucose levels of 7.8 to 11.0 mmol/L indicate impaired glucose tolerance; levels ≥ 11.1 mmol/L allow the diagnosis of diabetes.

Pathophysiology

Insulin deficiency. Insulin deficiency, or diminished effectiveness of endogenous insulin, leads to diabetes mellitus. There are two types of diabetes mellitus: type I and type II. These two types of diabetes are summarized in **Tables 28.4** and **28.5**.

Table 28.4 ► Type I Diabetes mellitus

Type I (insulin-dependent diabetes mellitus [IDDM]) comprises 10% of cases and is usually juvenile onset.

Causes	Effects	Symptoms	Complications	Treatment
β-cell destruction by autoantibodies leading to insulin deficiency Hypersecretion of glucagon	Very high serum [glucose] ↑ urinary [glucose] ↑ glucagon ↑ serum [K⁺] ↓ serum [Na⁺] ↓ insulin Metabolic acidosis Hyperlipidemia	Cardiac arrhythmias (due to ↑ serum [K⁺]) Hypovolemia Polyuria (due to osmotic diuresis caused by ↑ urinary [glucose]) Polydipsia (due to hyperosmolality of plasma) Increased appetite (due to hyperosmolality of plasma) Weight loss Dehydration Fatigue Nausea and vomiting	Microvascular disease: - Retinopathy (can cause blindness) - Nephropathy - Neuropathy Accelerated atherosclerosis causing strokes, coronary heart disease, hypertension Gangrene Diabetic ketoacidosis Poor wound healing Susceptibility to infection	Insulin replacement Regulation of carbohydrate intake

Table 28.5 ► Type II Diabetes mellitus

Type II (non–insulin-dependent diabetes mellitus [NIDDM]) comprises 90% of cases. It usually occurs in middle-aged patients and is associated with obesity.

Causes	Effects	Symptoms	Complications	Treatment
Impaired insulin secretion (β cells fail to respond to increases in serum [glucose], especially after a meal) Reduced number of insulin receptors (leading to insulin resistance)	High serum [glucose] ↑ urinary [glucose] Normal or elevated insulin	Delayed response to ingested glucose load Hypotension due to ECF volume contraction	Milder aspects of complications of type I	Caloric regulation Weight reduction Exercise Drugs including sulfonylureas, biguanides, thiazolidinediones, and α-glycosidase inhibitors May require insulin therapy

Abbreviation: ECF, extracellular fluid.

Insulin excess. Insulin excess is summarized in **Table 28.6.**

Table 28.6 ► Insulin Excess			
Causes	**Effects**	**Symptoms**[*]	**Treatment**
Inappropriate insulin intake/glucose intake in diet β-cell tumors (insulinomas)	↑ serum [insulin] Hypoglycemia The low glucose level will be more potent than the high insulin level, so glucagon will be secreted.	Irritability Confusion Seizures Drowsiness Motor incoordination Headache Paresthesias (tingling sensations) Sympathetic activation: - Rapid heart rate - Sweating - Hunger Weight gain Loss of consciousness, coma, and death (if severe)	Oral glucose (if possible) Parenteral glucose (if patient is unconscious) Glucagon Additional treatment of insulinomas: - Surgical resection is the treatment of choice - CT-guided radiofrequency ablation (if surgery is inappropriate) - Diazoxide to reduce the secretion of insulin (by activating K⁺ channels) - Hydrochlorothiazide (a diuretic) to counteract the hyperkalemia caused by diazoxide use - Octreotide (a somatostatin analogue) to prevent hypoglycemia

[*] Symptoms are due to hypoglycemia.
Abbreviation: CT, computed tomography.

Glucagon

Synthesis

Glucagon is synthesized in pancreatic islet α cells in the form of a prohormone, which is a single polypeptide chain. Active glucagon is a 29 amino acid peptide that is generated by limited proteolysis by a specific endopeptidase.

Regulation of Secretion

Table 28.7 lists all of the factors that modulate glucagon secretion.

Table 28.7 ► Regulation of Glucagon Secretion	
Stimulatory Factors	**Inhibitory Factors**
Low plasma [glucose] Increased levels of some glucogenic amino acids Decreased levels of proteins (amino acids arginine and alanine), fatty acids, ketones, and triglycerides Increased activity of the parasympathetic nervous system (acetylcholine) Activity of the sympathetic nervous system Stress Exercise	High plasma [glucose] Increased proteins, fatty acids, ketones, and triglycerides Increased plasma [insulin] Somatostatin

Actions

Glucagon increases serum [glucose] by promoting the utilization of energy stores.

Liver

— Glucagon increases glycogen breakdown to glucose.
— Glucagon increases gluconeogenesis by decreasing the activity of phosphofructokinase. This reduces the formation of fructose 2,6-bisphosphate, and substrate is shunted toward glucose production.
— Glucagon increases ketone formation.

Ketone formation

Ketones provide an alternative source of energy during periods of low glucose and after glycogen stores have been consumed. They are formed by the β oxidation of acetyl coenzyme A (acetyl-CoA), which is derived from fatty acids, in liver mitochondrial cells. Beta oxidation also yields NADH and the reduced form of flavin adenine dinucleotide (FADH₂), which then undergoes oxidative phosphorylation, producing ATP. Acetyl-CoA from fatty acid catabolism would normally enter the citric acid cycle, producing energy (via the oxidative phosphorylation of NADH and FADH₂). However, in periods of low glucose, oxaloacetate (a citric acid cycle intermediate) is used for gluconeogenesis, diverting acetyl-CoA for ketone formation. There are three ketones: acetoacetic acid, β-hydroxybutyric acid, and acetone. Acetone is produced from the spontaneous decarboxylation of acetoacetic acid, so it is produced in the least quantity of all of the ketones. Furthermore, acetone cannot be converted back to acetyl-CoA and is excreted in urine and exhaled (giving the breath a characteristic "fruity" smell in ketotic states).

Oral antidiabetic drugs

Oral antidiabetic drugs are useful in the treatment of type II diabetes. Sulfonylurea drugs (e.g., tolbutamide and glimerpiridine) directly close ATP-sensitive K⁺ channels on the surface of pancreatic β cells, causing membrane depolarization and increased insulin secretion. Biguanides (e.g., metformin) decrease hepatic gluconeogenesis, and the absorption of glucose from the gastrointestinal (GI) tract. Metformin and thiazolidinediones agents (e.g., pioglitazone, and rosiglitazone) also increase insulin sensitivity in skeletal muscle and adipose tissue. Because metformin does not increase the release of pancreatic insulin, the risk of hypoglycemia is lower than with other sulphonylurea agents. α-Glycosidase inhibitors (e.g., acarbose and miglitol) are competitive, reversible inhibitors of intestinal α-glycosidase, which causes delayed absorption of carbohydrates, thereby blunting postprandial hyperglycemia.

Muscle

— Glucagon has no influence on muscle tissue.

Adipose tissue

— Glucagon increases lipolysis. The fatty acids produced are used for gluconeogenesis.

Pancreas

— Glucagon decreases insulin secretion.

Miscellaneous

— Glucagon decreases cholesterol synthesis.

Pathophysiology

Glucagon excess.   **Table 28.8** summarizes glucagon excess.

Table 28.8 ▶ Glucagon Excess			
Causes	**Effects**	**Symptoms**	**Treatment**
α-cell tumors ("glucagonomas")	↑ serum [glucagon] ↑ serum [glucose] ↑ serum [ketoacids] ↓ serum [amino acids] ↑ urinary nitrogen	Weight loss Skin lesions Mild diabetes Stomatitis (mouth sores) Anemia	Surgical resection Octreotide therapy to reduce glucagon secretion and to control the hyperglycemia and skin lesions

Table 28.9 summarizes the effects of insulin and glucagon on glucose homeostasis and energy metabolism.

Table 28.9 ▶ Summary of Insulin and Glucagon		
	Insulin	**Glucagon**
Stimulus for secretion	↑ serum [glucose] ↑ serum [amino acids] ↑ serum free fatty acids Glucagon GH Cortisol GIP	↓ serum [glucose] ↓ serum [amino acids] Epinephrine and norepinephrine
Actions:		
Liver	↑ glycogenesis ↓ gluconeogenesis	↑ glycogenolysis ↑ gluconeogenesis
Muscle	↑ protein synthesis ↑ glycogenesis	—
Adipose tissue	↑ lipogenesis	↑ lipolysis
Pancreas	↓ glucagon secretion	↓ insulin secretion
Miscellaneous	↑ K⁺ uptake into cells ↑ cholesterol synthesis	↓ K⁺ uptake into cells ↓ cholesterol synthesis
Net effect	↓ serum [glucose] ↑ storage of energy by increasing protein synthesis, fat synthesis, and glycogen synthesis	↑ serum [glucose] ↑ energy release by increasing protein and fat catabolism and by gluconeogenesis
Abbreviations: GH, growth hormone; GIP, gastrointestinal polypeptide.		

28.2 Effects of Other Hormones on Energy Metabolism

Several hormones besides insulin and glucagon interact to regulate blood glucose concentration under different physiological conditions. The effects of cortisol and epinephrine are listed in **Table 28.10**. Growth hormone acts similarly to insulin in the short term. Somatostain inhibits the secretion of insulin, glucagon, and gastrin.

Table 28.10 ▸ Summary of the Role of Cortisol and Epinephrine in Energy Metabolism		
	Cortisol	**Epinephrine**
Liver	↑ glycogenesis ↑ gluconeogenesis	↑ glycogenolysis
Muscle	↓ glucose uptake into cells	↑ glycogenolysis ↑ glucose uptake into cells
Adipose tissue	↑ lipolysis	↑ lipolysis
Pancreas	↑ insulin ↓ glucagon	↑ glucagon ↓ insulin
Net effect	↑ serum [glucose] ↑ energy by the catabolism of amino acids	↑ serum [glucose] ↑ energy available during stress or exercise

Review Questions

1. The major mechanism for endocrine homeostasis is
   A. neurotransmitters blocking hormone receptors.
   B. one hormone antagonizing the action of another hormone.
   C. positive feedback to accelerate a response.
   D. negative feedback of output onto input.
   E. by release of hypothalamic hormones.

2. Growth hormone (GH)
   A. decreases amino acid transport into cells.
   B. increases free fatty acids in plasma.
   C. release is decreased by glucagon.
   D. release is decreased by exercise.
   E. directly stimulates bone growth.

3. Prolactin secretion
   A. comes from the posterior pituitary.
   B. suppresses gonadotropin-releasing hormone (GnRH).
   C. stimulates glycogenolysis.
   D. stimulates gluconeogenesis.
   E. peaks early in pregnancy.

4. Oxytocin
   A. stimulates mammary myoepithelial contraction.
   B. stimulates myometrial contraction when plasma progesterone levels are high.
   C. stimulates the production of neurophysin by the liver.
   D. is a prolactin release–inhibiting factor.
   E. initiates labor.

5. Which is the most important factor in controlling vasopressin (antidiuretic hormone, ADH) secretion?
   A. Blood pressure
   B. Plasma osmolality
   C. Blood volume
   D. Angiotensin II
   E. Stress

6. One-half hour after the cessation of a release episode producing normal plasma levels of the following hormones, which one will have the highest level?
   A. Cortisol
   B. Adrenocorticotropin hormone (ACTH)
   C. Epinephrine
   D. Thyroxine (T_4)
   E. Aldosterone

7. An inhibitor of thyroid peroxidase may
   A. inhibit intracellular binding of triiodothyronine (T_3).
   B. inhibit intracellular binding of T_4.
   C. result in hypothyroidism.
   D. increase iodide uptake into the thyroid gland.
   E. diminish the sensitivity of pituitary cells to thyrotropin-releasing hormone.

8. Which of the following are characteristic of thyroxine (T_4)?
   A. Less than 50% of the hormone in circulation is bound to plasma protein.
   B. Its half-life in blood is < 1 hour.
   C. Large amounts are stored intracellularly.
   D. Conversion to triiodothyronine (T_3) takes place within target tissue cells.
   E. Its release is directly controlled by thyrotropin-releasing hormone (TRH).

Questions **9** to **11** refer to the clinical scenario that follows.
A 43-year-old man presents with lethargy and sleepiness; dry, scaly skin; sparse, dry hair; dull expression with droopy eyelids; bradycardia (60 beats/min); blood pressure of 90/70; hematocrit of 27; enlarged heart; constipation; and hypothermia. Plasma concentrations of thyroxine (T_4) and triiodothyronine (T_3) reveal a total T_4 of 2.5 µg/dL, whereas normal is 5 to 14 µg/dL. Peripheral blood has elevated levels of thyroid-stimulating hormone (TSH) levels.

9. This patient has which of the following?
   A. Addison disease
   B. Cushing disease
   C. Hyperthyroidism
   D. Hypothyroidism
   E. Type II diabetes

10. What defect might be responsible for these symptoms?
    A. Reduced synthesis of TSH
    B. Reduced synthesis of TRH
    C. Reduced synthesis of thyroid hormone
    D. Increased release of thyroid hormone
    E. Lack of nuclear thyroid hormone receptors

11. What is the best treatment for this patient?
    A. Iodinated salt
    B. Thyroid hormone (T_4)
    C. Surgery to remove a thyroid tumor, a common cause of hyperthyroidism, or drug treatment to suppress excess production of T_4
    D. Cortisol
    E. Insulin

12. Hyperparathyroidism is characterized by
    A. Ca^{2+} kidney stones and gallstones.
    B. convulsions.
    C. parethesias.
    D. numbness.
    E. tetany.

13. Calcitonin
    A. secretion is stimulated by low plasma $[Ca^{2+}]$.
    B. promotes bone formation.
    C. has membrane receptors in the intestine.
    D. decreases urinary excretion of Ca^{2+} and phosphate.
    E. is an effective treatment for osteoporosis in postmenopausal women.

For Questions **14–16**

A 55-year-old woman had a total thyroidectomy followed by thyroid hormone replacement. Forty-eight hours later, she developed laryngeal spasms, tetany, and cramps in her hands. Laboratory tests showed reduced plasma Ca^{2+} and elevated plasma PO_4^{3-}
Daily oral calcium gluconate and vitamin D were given to alleviate her symptoms.

14. What is the patient's endocrinopathy?
    A. Increased renal clearance of Ca^{2+}
    B. Excess cortisol
    C. Decreased stimulation of the anterior pituitary
    D. Lack of adequate thyroid hormone
    E. Lack of parathyroid hormone

15. Vitamin D is given with the calcium (Ca^{2+}) gluconate because vitamin D
    A. increases absorption of Ca^{2+} by the gut.
    B. binds to the Ca^{2+} in the gut.
    C. prevents destruction of Ca^{2+} in the stomach.
    D. makes the Ca^{2+} gluconate more stable in solution.
    E. decreases intestinal absorption of PO_4^{3-}.

16. What caused the tetany and muscle spasms?
    A. Hypercalcemia
    B. Hypocalcemia
    C. Hypophosphatemia
    D. Hyperphosphatemia
    E. Lack of vitamin D

17. In humans, total adrenalectomy is fatal without replacement therapy. Hypophysectomy (surgical removal of the pituitary gland) is not fatal, because in hypophysectomy
    A. the adrenal cortex undergoes compensatory hypertrophy.
    B. the secretion of aldosterone is normal.
    C. adrenal catecholamines compensate for the metabolic actions of cortisol.
    D. tissue requirements for corticosteroids fall to low levels.
    E. plasma concentration of angiotensin II increases.

18. Giving a patient a drug that inhibits angiotensin-converting enzyme (ACE) will increase which of the following?
    A. Plasma concentration of angiotensin II
    B. Na^+ clearance
    C. K^+ clearance
    D. Peripheral resistance
    E. Secretion of renin

19. A drug that prevents the binding of dexamethasone to glucocorticoid receptors would cause which outcome?
    A. A negative nitrogen balance
    B. Decreased concentration of adrenocorticotropic hormone (ACTH) in the blood
    C. Decreased concentration of cortisol in the blood
    D. Increased concentration of insulin in the blood
    E. Decreased concentration of cortisol in cell nuclei

20. Treatment of a normal person with large amounts of cortisol causes
    A. decreased hepatic glycogen content.
    B. skeletal muscle anabolism.
    C. increased adrenal size.
    D. central deposition of fat.
    E. resistance to peptic ulcers.

21. The synthesis of epinephrine depends on
    A. angiotensin II secretion.
    B. local cortisol concentration.
    C. dehydroepiandrosterone (DHEA).
    D. an intact zona glomerulosa.
    E. 11β–hydroxylase.

22. Secretion of catecholamines by cells of the adrenal medulla depends on
    A. the increased concentration of Ca^{2+} in the cytoplasm.
    B. exocytosis of calmodulin.
    C. inactivation of the release of somatostatin.
    D. hyperpolarization of the cell membrane.
    E. the release of norepinephrine from preganglionic sympathetic neurons.

23. Stimulation of cyclic adenosine monophosphate (cAMP) by epinephrine in target cells
    A. requires translocation of epinephrine into the nucleus.
    B. is mediated by a protein that binds guanosine triphosphate (GTP).
    C. is amplified by prior long-term exposure of cells to norepinephrine.
    D. is inversely related to the number of adrenergic receptors present on the cell.
    E. depends on the internalization of epinephrine-receptor complexes.

24. Ca^{2+} influx into pancreatic β cells
    A. is the result of stopping the Na^+–K^+ ATPase pump.
    B. causes glucose to enter the cells.
    C. stimulates release of glucagon.
    D. activates calmodulin.
    E. slows insulin synthesis.

25. Glucagon
    A. increases insulin secretion.
    B. operates through G proteins to increase cAMP.
    C. decreases gluconeogenesis.
    D. decreases lipolysis.
    E. decreases glycogenolysis.

26. If somatostatin were given with a β-blocker, what would you expect the plasma glucose response to be over the next 2 hours?
    A. Plasma glucose concentration would decrease.
    B. Plasma glucose concentration would increase.
    C. Plasma glucose concentration would decrease, then increase.
    D. Plasma glucose concentration would increase, then decrease.
    E. Plasma glucose concentration would not change.

For Questions **27–30**
A patient with type I diabetes is found in a coma. Blood glucose, urine glucose, blood ketones, and urine ketones are all elevated; serum HCO_3^- is < 12 mEq/L. Respirations are quick and deep with acetone breath. Blood pressure is 95/61 mm Hg, and the pulse is weak and rapid (119 beats/min).

27. What is the condition of this person?
    A. Ketoacidosis
    B. Insulin shock
    C. Heat stroke
    D. Heat exhaustion
    E. Adrenocortical insufficiency

28. What factor in this patient's condition is the major cause of their low serum HCO_3^-?
    A. The patient has been hyperventilating.
    B. The patient has been excreting acidic urine.
    C. It has been depleted to buffer ketoacids.
    D. The patient has compensated for the low CO_2.
    E. The patient tries to achieve a normal $[HCO_3^-]/P_{CO_2}$ ratio.

29. What is the cause of the patient's dyspnea, tachycardia, and hypotension?
    A. Plasma hyperglycemia
    B. High urine glucose
    C. Blood ketosis
    D. Urine ketosis
    E. Blood acidosis

30. What would you expect the plasma glucose response to be over the next 2 hours after giving insulin?
    A. Plasma glucose concentration would decrease.
    B. Plasma glucose concentration would increase.
    C. Plasma glucose concentration would decrease, then increase.
    D. Plasma glucose concentration would increase, then decrease.
    E. Plasma glucose would remain high.

31. What is the major energy source for the body after 48 hours of fasting?
    A. Muscle protein
    B. Liver glycogen
    C. Muscle glycogen
    D. Triglycerides
    E. Glucose

Answers and Explanations

1. **D**. Negative feedback control produces a relatively stable level of hormone in the bloodstream (p. 243).
   A Neurotransmitters rarely block hormone receptors.
   B Hormones do antagonize one another and minimize actions, but this does not produce stability.
   C Positive feedback produces explosive responses, not stability.
   E Most hypothalamic hormones cause release of other hormones, but their control depends on negative feedback.

2. **B**. Growth hormone (GH) increases lipolysis in the presence of cortisol (p. 248). This increases free fatty acids in plasma.
   A GH increases amino acid transport into cells.
   C,D GH release is stimulated by both glucagon and exercise.
   E Bone growth is stimulated indirectly by insulinlike growth factor (IGF).

3. **B**. Prolactin secretion suppresses GnRH. This causes lactation amenorrhea, so nursing mothers are unlikely to become pregnant (p. 249).
   A Prolactin is secreted from the anterior pituitary.
   C,D Prolactin secretion stimulates galactopoiesis (the secretion and continued production of milk by the mammary glands) rather than production of glucose or breakdown of glycogen in the liver.
   E Prolactin levels peak 2 weeks before term and then decrease, but they are increased by each session of suckling.

4. **A**. Oxytocin stimulates the contraction of the myoepithelial cells of the mammary glands (p. 250).
   B Oxytocin stimulates the contraction of an "estrogenized," not "progesteronized," uterus.
   C Neurophysin is produced in hypothalamic nuclei, not in the liver.
   D Oxytocin is a prolactin-releasing factor.
   E Oxytocin participates in but does not initiate labor.

5. **B**. Plasma osmolality is the most potent factor in controlling vasopressin secretion (p. 250).
   A,C Decrease of blood pressure or blood volume will also cause the release of greater qualities of ADH.
   D,E Angiotensin II and stress will also affect ADH release indirectly.

6. **D**. Thyroxine has the longest half-life (5–7 days) and will therefore have the highest level (p. 243).
   A,E Steroid hormones (e.g., ACTH and aldosterone) are largely bound to plasma proteins, so they have a half-life of 1 or 2 hours.
   B Protein hormones have a half-life of several minutes.
   C Catecholamines have a half-life < 1 minute.

7. **C**. Inhibition of thyroid peroxidase depresses thyroxine synthesis, producing hypothyroidism (p. 252).
   A, B Inhibition of thyroid peroxidase does not inhibit intracellular binding of T_3 or T_4.
   D,E Inhibition of thyroid peroxidase does not affect iodide uptake or sensitivity of pituitary cells.

8. **D**. Thyroxine (T_4) is converted to triiodothyronine (T_3) within target tissue cells. T_3 is the more active form of thyroid hormone (p. 254).
   A,B T_4 is > 99% bound to transport proteins in plasma and has a half-life of 7 days.
   C T_4 is stored as thyroglobulin extracellularly in the lumen of the follicle.
   E TRH acts in the anterior pituitary to control release of thyroid-stimulating hormone, which then acts on the thyroid gland to release T_3 and T_4.

9. **D**. He has the symptoms of hypothyroidism (myxedema).
   A Addison disease has a few of the same symptoms. It is caused by a lack of cortisol.
   B Cushing disease has different symptoms. It is caused by excess cortisol.
   C In hyperthyroidism, metabolism and heat production are raised.
   E Diabetic patients show drowsiness but none of the other symptoms.

10. **C**. Low levels of T_4 suggest impaired synthesis (p. 255).
    A TSH production is elevated due to disinhibition.
    B TRH release must be normal and above normal to elevate levels of TSH.
    D TRH release is a consequence of low T_4, not the primary pathology.
    E The low levels of T_4 are sufficient to explain the symptoms without a lack of nuclear thyroid hormone receptors.

11. **B**. Replacement with thyroxine (T_4) will treat the patient's symptoms (p. 255).
    A Thyroxine synthesis is operating, so it is unlikely that iodine is limiting.
    C A hypothalamic or pituitary tumor may be the primary cause, but it is not the only cause of reduced TSH.
    D Cortisol is required for treatment of Addison disease, not secondary hypothyroidism.
    E Insulin is the treatment for diabetes.

12. A. Hyperparathyroidism causes high plasma [Ca^{2+}], which leads to kidney stones and gallstones (p. 263).
B–E All the other choices occur with hypoparathyroidism and low plasma [Ca^{2+}], which destabilizes membranes of nerve and muscle cells.

13. E. Calcitonin inhibits bone reabsorption and is thus an effective treatment for osteoporosis in postmenopausal women (p. 266).
A Calcitonin secretion is stimulated by high plasma Ca^{2+}.
B Calcitonin has no effect on bone formation.
C Calcitonin has membrane receptors on osteoclasts and renal tubule cells.
D Calcitonin increases urinary excretion of Ca^{2+} and phosphate.

14. E. Lack of parathyroid hormone causes reduced plasma Ca^{2+} and elevated plasma PO_4^{3-} (p. 262).
A Increased renal clearance of Ca^{2+} cannot reduce plasma Ca^{2+} enough to produce these symptoms.
B Excess cortisol (Cushing disease) produces obesity and muscle wasting.
C Parathyroid hormone is not regulated by the pituitary.
D There are no symptoms of a lack of thyroid hormone with its replacement.

15. A. Calcitriol (vitamin D) increases the active transporters of Ca^{2+} and increases Ca^{2+} permeability of the small intestine (p. 265).
B Calcitriol does not bind to Ca^{2+} in the gut.
C Ca^{2+} is not destroyed in the stomach.
D Calcitriol does not make Ca^{2+} gluconate more stable; Ca^{2+} is more easily absorbed from the gluconate salt than from Ca^{2+} carbonate.
E Calcitriol increases absorption of PO_4^{3-}.

16. B. Hypocalcemia (low plasma Ca^{2+}) reduces the stability of skeletal muscle membranes, so they fire repetitive action potentials spontaneously (p. 260).
A Hypercalcemia (high plasma Ca^{2+}) makes excitable membranes more stable.
C–E Hypophosphatemia (low plasma PO_4^{3-}), hyperphosphatemia (high plasma PO_4^{3-}), and vitamin D have little effect on excitable membranes.

17. B. Aldosterone secretion is normal because it is regulated by plasma [K^+] and angiotensin, not the hypothalamus–pituitary–adrenal axis (pp. 270, 271).
A The adrenal cortex atrophies after hypophysectomy.
C Catecholamines and cortisol have different spectra of actions.
D Requirement for corticosteroids rises when their release is suppressed.
E Angiotensin II levels are unchanged.

18. B. ACE inhibitors slow formation of angiotensin II (A) and reduce aldosterone secretion. Less aldosterone produces less Na^+ reabsorption, so Na^+ clearance increases (pp. 270, 271).
C K^+ secretion is also reduced, so K^+ clearance decreases.
D Less vasoconstrictor action of angiotensin II produces decreased peripheral resistance.
E Secretion of renin is unchanged.

19. E. Steroids act upon target cells by binding to a cytoplasmic receptor, which is transported into the nucleus, where it induces or represses gene transcription. A glucocorticoid receptor blocker prevents cortisol from having its normal mechanism of action.
A Decreased cortisol activity produces a positive nitrogen balance as tissue is metabolized.
B,C Symptoms of decreased ACTH or cortisol (Addison disease) are similar to those with blocked receptors, but they are not caused by receptor block.
D. Insulin secretion is not increased by cortisol activity.

20. D. Large amounts of cortisol causes central deposits of fat. This leads to truncal obesity, a hallmark of Cushing syndrome (p. 276).
A, B Cortisol promotes liver glycogen deposition and muscle proteolysis.
C Cortisol causes inhibition of ACTH secretion and thus decreased adrenal size.
E Cortisol promotes peptic ulcers.

21. B. Epinephrine is formed from norepinephrine by the action of the enzyme phenylethanolamine N-methyltransferase (PNMT), which is induced by cortisol (p. 276).
A Angiotensin II facilitates synthesis of aldosterone from corticosteron E.
C DHEA is a precursor for androgens.
D Epinephrine is synthesized in the adrenal medulla, not in the zona glomerulosa of the adrenal cortex.
E 11β–hydroxylase is the final enzyme in the synthesis of cortisol.

22. A. Stimulus-secretion coupling involves depolarization of the membrane (D), followed by elevated cytoplasmic Ca^{2+}, which initiates the secretory process (p. 277).
B Calmodulin is located in muscle cells, not the adrenal medulla.
C Somatostatin has no role in secretion.
E Acetylcholine is released from preganglionic sympathetic neurons.

23. B. The binding of epinephrine to the β-adrenergic receptors results in the activation of G_s proteins, which bind GTP and subsequently activate the conversion of adenosine triphosphate (ATP) to cAMP via adenylate cyclase (p. 244).

A,C It does not require translocation of epinephrine or long-term exposure to norepinephrine.

D Stimulation is directly related to the number of adrenergic receptors.

E The Epinephrine-receptor complex remains on the cell membrane. The activated G proteins act within the cell.

24. D. Ca^{2+} influx into pancreatic β cells activates calmodulin and Ca^{2+}/calmodulin-dependent protein kinase (p. 279).

A Stopping the Na^+–K^+ ATPase pump (due to low ATP levels for example), causes Ca^{2+} influx to decrease. A rise of ATP levels causes K^+ channels to close, resulting in depolarization that opens voltage-gated Ca^{2+} channels.

B Glucose binds to a GLUT-2 transporter and enters β cells before Ca^{2+} does.

C,E The Ca^{2+} influx releases insulin (not glucagon) and increases insulin synthesis.

25. B. Glucagon activates G proteins, which stimulate an adenylate cyclase pathway to increase cAMP (pp. 283, 284).

A Glucagon decreases insulin secretion, as insulin antagonizes most of glucagon's actions.

C–E Glucagon promotes utilization of energy stores, so it increases gluconeogenesis, lipolysis, and glycogenolysis.

26. A. Plasma glucose levels would decline due to insulin action but not return to normal due to the inhibition of glucagon secretion by somatostatin and the blockade of epinephrine action by the β-blocker.

27. A. The acetone breath is typical of ketoacidosis (p. 282).

B Insulin shock results from hypoglycemia due to excessive insulin.

C,D Heat stroke or heat exhaustion are associated with normal glucose and no elevated ketones.

E Hypoglycemia is typical of adrenocortical insufficiency (lack of cortisol).

28. C. Some of the HCO_3^- release buffers the ketoacids (p. 282).

A The patient's hyperventilation slightly compensates for the acidosis and slightly reduces HCO_3^-.

B The patient will excrete an alkali urine.

D CO_2 remains low.

E The patient's compensation for the acidosis does not improve the $[HCO_3]/P_{CO_2}$ ratio by reducing CO_2, but that also reduces HCO_3^- slightly.

29. E. The acidotic state causes hyperventilation, hypotension, and tachycardia (p. 282).

A–D Ketones and glucose do not cause these symptoms.

30. C. Plasma glucose levels would decline due to insulin action, then increase (and return to normal), due to glucagon action.

31. D. Triglycerides are a potent energy source after glucose is depleted.

A Muscle protein allows amino acids to be "burned" after only 1 day, but they are less important than the breakdown of fats.

B,C Glycogen provides glucose for only 24 hours.

E Glucose in the blood is used in a few hours, although glycogenolysis keeps glucose available for almost 1 day.

29 Sexual Differentiation, Puberty, and Male and Female Reproduction

Reproduction describes processes that maintain the species rather than the individual. These processes help to assure that a viable egg meets a viable sperm. The physiology of reproduction is largely about endocrine control.

29.1 Fetal Sexual Differentiation

Genetic sex is determined by the sex chromosomes: XY in males and XX in females. All spermatozoa have either a 23X or 23Y, and all ova have a 23X. Therefore, the male determines the genetic sex of the offspring.

Gonadal sex is determined by the presence of testes in males and ovaries in females.

— The presence of a single Y chromosome causes fetal undifferentiated gonads to differentiate into testes. If the Y chromosome is absent, testes fail to develop, and ovaries form.

Phenotypic sex is determined by the formation of a male or female internal genital tract and external genitalia.

Genetic sex, gonadal sex, and phenotypic sex are summarized in **Fig. 29.1**.

Differentiation of Internal Genitalia

Male Internal Genitalia

Until the seventh week of gestation, the internal genitalia have the rudiments of both male internal genitalia (the wolffian ducts) and female internal genitalia (the müllerian ducts). A genetically male fetus with functional testes secretes antimüllerian hormone (AMH) from Sertoli cells in the testes. AMH causes regression of the müllerian ducts and differentiation of Leydig

Fig. 29.1 ▶ **Fetal sexual differentiation.**

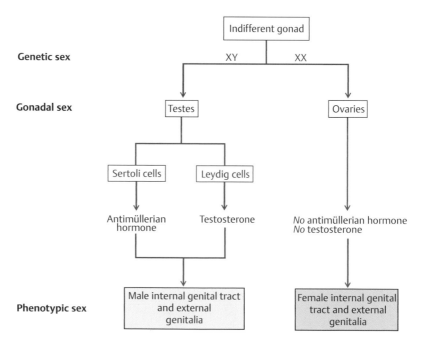

cells in the testes. Leydig cells then produce testosterone, which stimulates wolffian duct development into the male internal genitalia: the epididymis, vas deferens, and seminal vesicles.

Female Internal Genitalia

Fetal female development is genetically determined and occurs in the absence of testosterone and AMH. Without testosterone, there is regression of wolffian ducts and the formation of primordial follicles that produce estrogen. The absence of AMH allows the müllerian ducts to differentiate, forming the female internal genitalia: the fallopian tubes, uterus, cervix, and upper one-third of the vagina.

Differentiation of External Genitalia

Until the eighth week of gestation, the external genitalia of the fetus has the potential to develop along either male or female lines. This is because the external genitalia of both sexes develop from common rudiments: the urogenital sinus, the genital swellings, the genital folds, and the genital tubercle.

— In the presence of testosterone, differentiation of the external genitalia proceeds along male lines:
  — The urogenital sinus forms the prostate.
  — The genital swellings form the scrotum.
  — The genital folds form the penile urethra and shaft.
  — The genital tubercle forms the glans penis.
— In the absence of testosterone, differentiation of the external genitalia proceeds along female lines:
  — The urogenital sinus forms the lower two-thirds of the vagina.
  — The genital swellings form the labia majora.
  — The genital folds form the labia minora.
  — The genital tubercle forms the clitoris.

29.2 Puberty

Puberty is the stage of life when the reproductive system matures and becomes functional.

Up until age 9, gonadotropin-releasing hormone (GnRH) from the hypothalamus and follicle-stimulating hormone (FSH) and luteinizing hormone (LH) from the anterior pituitary are secreted at low levels and fairly evenly over a 24-hour period in both males and females. At puberty, there is a shift to pulsatile GnRH release during various stages of sleep. GnRH causes upregulation of GnRH receptors in the anterior pituitary and a pulsatile release of LH and FSH (LH > FSH). Increased secretion of LH stimulates the production of the male sex hormones testosterone and dihydrotestosterone (DHT) and the female sex hormone estrogen that are responsible for the secondary changes in males and females at puberty (**Table 29.1**).

Table 29.1 ▶ Secondary Sex Changes in Females and Males at Puberty	
Female Changes at Puberty*	**Male Changes at Puberty****
Growth of pelvis	Growth of penis
Deposition of fat over the pelvis, buttocks, and thighs	Development of body, facial, and pubic hair
Breast enlargement and development of milk ducts	Broadening of the shoulders
Increased excitability of some neurons in the brain	Thickening of the vocal cords (deepening of
Development of pubic and axillary hair**	voice)
Increased libido (sex drive)**	Increased muscle mass
	Increased libido

* Secondary sex changes in the female are mostly due to increased estrogen secretion.
** These female secondary sex changes are due to adrenal androgens.
*** Secondary sex changes in the male are due to increased secretion of testosterone and dihydrotestosterone (DHT).

Fig. 29.2 ► **Male sex organs.**
(**A**) Seminiferous structures. (**B**) Right lateral view of seminiferous structures.
From *Atlas of Anatomy*, © Thieme 2008, Illustration by Markus Voll.

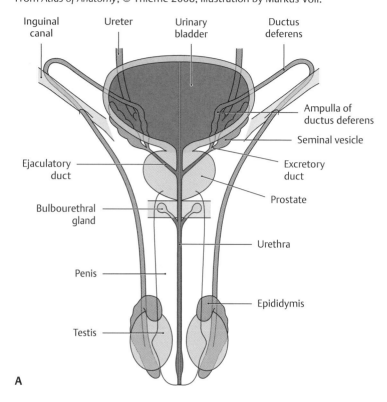

A

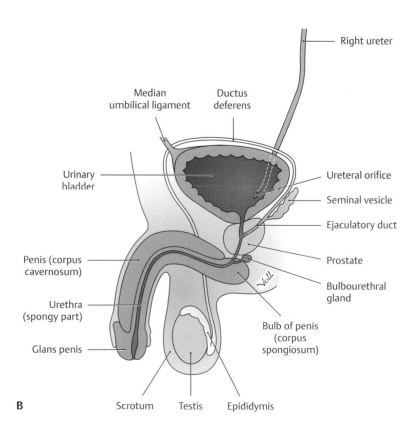

B

29.3 Male Reproduction

Male Sex Organs (Fig. 29.2)

- The penis is a copulatory and urinary organ.
- The urethra is the common pathway for expulsion of urine and sperm.
- The scrotum is involved in protection of the testes and thermoregulation.
- The testes synthesize and secrete of testosterone. They are also involved in spermiogenesis.
- The epididymis is involved in the maturation of sperm.
- The prostate gland secretes fluid that promotes sperm motility and neutralizes the acidic secretions in the vagina.
- The seminal vesicles secrete fluid that nourishes ejaculated sperm, help sperm penetrate cervical mucus, and aid in sperm propulsion (by stimulating contractions of the uterus and fallopian tubes).
- The bulbourethral glands secrete mucus.

Male Sex Hormones: Testosterone and Dihydrotestosterone

Synthesis

- Testosterone is the principal androgen synthesized and secreted by Leydig cells in the testes following stimulation by LH. LH increases the activity of cholesterol desmolase, which is needed to convert cholesterol to pregnenolone, the precursor of testosterone (**Fig. 29.3**). Other androgens synthesized by Leydig cells are androstenidone and dehydroepiandrosterone (DHEA).

Fig. 29.3 ▶ **Synthesis of testosterone and dihydrotestosterone.**
(LH, luteinizing hormone)

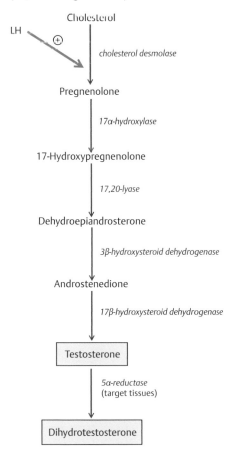

5α-reductase inhibitors

Finasteride, among other 5α-reductase inhibitors, block the conversion of testosterone to DHT by inhibiting the enzyme 5α-reductase. This results in the inhibition of androgenic activity in tissues where DHT is the active form (e.g., the prostate); however, it has little or no effect on testosterone-dependent tissues (e.g., skeletal muscle). These drugs are used in the treatment of benign prostate hyperplasia and male pattern baldness.

— Some target cells express the enzyme 5α-reductase, which reduces testosterone to its more active form, dihydrotestosterone (DHT). DHT is 29- to 50-fold more potent than testosterone on many tissues.

Note: Leydig cells in the testes cannot produce cortisol or aldosterone (as occurs in the adrenal cortex), as they lack the enzymes 21β-hydroxylase and 11β-hydroxylase (see **Fig 29.3**).

Regulation of Synthesis and Secretion

In hypothalamic–anterior pituitary control:

— GnRH release from the hypothalamus is pulsatile. In men, the pulse is relatively constant in frequency (once every 90 min), amplitude, and duration. GnRH stimulates the anterior pituitary to release LH and FSH in a pulsatile manner.

— LH stimulates testosterone synthesis in Leydig cells.

— FSH acts on Sertoli cells in the testes to promote spermatogenesis. The Sertoli cells also produce inhibin, a polypeptide hormone that is a potent and selective inhibitor of FSH release from the anterior pituitary.

In negative feedback control:

— Testosterone causes feedback inhibition of GnRH release (↓ frequency and amplitude) with subsequent inhibition of LH and FSH release.

— Testosterone directly inhibits the release of LH by the anterior pituitary.

— Inhibin acts at the anterior pituitary to inhibit the secretion of FSH (**Fig. 29.4**).

Fig. 29.4 ▶ **Control of testosterone and follicle-stimulating hormone (FSH) secretion in the testes.** (GnRH, gonadotropin-releasing hormone; LH, luteinizing hormone)

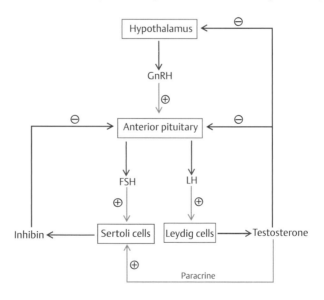

Actions

Table 29.2 lists the actions of testosterone and DHT.

Table 29.2 ▶ Actions of Testosterone and Dihydrotestosterone (DHT)	
Testosterone	**DHT**
Differentiation of the epididymis, vas deferens, seminal vesicles, and urethra	Differentiation of the penis, scrotum, and prostate
Initiation and promotion of the pubertal growth spurt (growth hormone is also involved)	Growth of the prostate gland
Secondary sex changes of male puberty (see Table 29.1)	Increased sebaceous gland activity
Closure of the epiphyseal growth plate, which marks the end of the pubertal growth spurt	Male pattern baldness
Promotion of spermiogenesis in Sertoli cells (paracrine effect)	
Regulation of testicular function (via feedback inhibition on the hypothalamus and anterior pituitary)	

Spermatogenesis

Spermatogenesis is the production of mature sperm from spermatogonia. These spermatogonia remain quiescent until they are stimulated to begin spermatogenesis by the increase in FSH and LH that occurs at puberty.

Spermatogenesis occurs in the seminiferous tubules. The process begins when spermatogonia divide to form primary spermatocytes, which migrate to Sertoli cells and become embedded within the cytoplasm. Within Sertoli cells, the following developmental changes occur:

— Primary spermatocytes undergo meiosis, forming two secondary spermatocytes, each of which has 23 chromosomes.
— The two secondary spermatocytes then undergo a second meiotic division, producing four spermatids, each of which has 23 chromosomes.
— Spermatids then form an acrosome, which is the specialized compartment at the head of the sperm containing hydrolytic enzymes and proteolytic enzymes. The nucleus and cytoplasm condense, and the flagellum is formed. The fully formed sperm are then released into the lumen of the seminiferous tubule.
— Sperm are protected from attack by the immune system by a blood–testes barrier, a physical barrier between the blood vessels and the seminiferous tubules of the testes. The barrier is formed by tight connections between Sertoli cells in the testes.

Hormonal Regulation of Spermatogenesis

FSH and testosterone are needed for spermatogenesis (**Table 29.3**).

Table 29.3 ▶ Role of Follicle-stimulating Hormone (FSH) and Testosterone in Spermatogenesis	
Hormone	**Effects on Spermatogenesis**
FSH	Stimulates spermatogonia and Sertoli cells to begin spermatogenesis
	Maintains glucose uptake and utilization by the Sertoli cell as energy sources for the developing germ cells
	Regulates ion and mineral uptake and transport to developing germ cells
	Regulates the production of seminiferous tubular fluid in Sertoli cells, which provides the "drive" for sperm to move from the testes to the epididymis
Testosterone	Stimulates spermatogonia and Sertoli cells to initiate and complete spermiogenesis

Thermoregulation of the scrotum

The temperature of the testes needs to be lower than body temperature (37°C, 98.6°F) for optimal spermiogenesis to occur. This is the reason that the testes descend to lie outside the body. There are several features of the scrotum that allow for thermoregulation: the skin of the scrotum is thin and has many sweat glands, and there is an absence of fat; the scrotum has a pampiniform plexus in which a testicular artery is surrounded by a mesh of veins, allowing for countercurrent heat exchange; contraction of the tunica dartos muscle of the scrotum decreases the surface area of scrotal skin available for heat loss and so warms the testes, whereas relaxation has the opposite effect; and contraction of the cremaster muscle causes the scrotum to rise, bringing it closer to the body, resulting in warming of the testes, whereas relaxation of this muscle allows the scrotum to hang farther from the body, resulting in cooling of the testes.

Decline of male reproductive capacity

Men experience a gradual decline in their reproductive capacity with aging. They have ~29% less sperm at age 80 years than at age 50 years, but viable sperm are still produced. Plasma levels of testosterone decrease, plasma levels of DHT remain unchanged, and plasma levels of estradiol increase. There is hyperplasia of the prostate gland, partly due to greater conversion of testosterone to DHT. This hyperplasia constricts the ureter and slows urination.

Sperm Transport and Maturation

Spermatozoa are transferred to the epididymis by Sertoli cell–derived fluid following their release from Sertoli cells. The epididymis lining has specialized ciliated epithelium and smooth muscle cells that are largely responsible for sperm movement. This lining is maintained by androgens.

Emission, Ejaculation, and Capacitation

Emission is the deposition of semen into the posterior urethra. It depends on sympathetic activation of smooth musculature of the epididymis, vas deferens, prostate, and seminal vesicles. There is a coordinated contraction of the smooth muscle of the internal bladder sphincter to prevent urine leakage from the urethra and semen entering the bladder.

Ejaculation of sperm from the urethra occurs due to a series of rapid skeletal muscle contractions. It is controlled by the somatic motor system. Ejaculated sperm are not immediately viable and cannot fertilize the ovum. To become viable, they must undergo capacitation.

Capacitation is a biochemical process that occurs within the female reproductive tract to produce sperm that are capable of fertilizing an ovum. It occurs 4 to 6 hours after ejaculation, producing "whiplike" movement of sperm. Capacitation also causes cholesterol and acrosome-stabilizing proteins to be removed from sperm membranes.

— Capacitation is a necessary precondition for the acrosome reaction, which is the final development in sperm. During the acrosomal reaction, the acrosomal membrane fuses with the outer sperm membrane, creating "pores" that allow hydrolytic and proteolytic enzymes to exit. These enzymes allow the sperm to penetrate the zona pellucida of the oocyte (**Fig. 29.5**).

Fig. 29.5 ▶ **Acrosome reaction.**
The acrosome of the sperm head contains hydrolyzing enzymes that are necessary for the penetration of the oocyte. These enzymes are firmly linked to the inner membrane of the acrosome. Just prior to encountering the oocyte, the acrosome reaction occurs, causing the cell membrane and the external membrane of the acrosome to fuse at many points (**A**). Pores develop at these points of fusion through which the enzymes are released (**B**). The perforated remains of the membrane are cast off as the sperm penetrates the corona radiata of the oocyte. The inner acrosomal membrane now covers the sperm head (**C**).

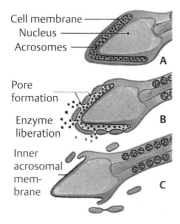

Cell membrane
Nucleus
Acrosomes
A

Pore formation
Enzyme liberation
B

Inner acrosomal membrane
C

29.4 Female Reproduction

The female reproductive system exhibits regular cyclical changes that affect the chance of fertilization and pregnancy. Its function is dependent upon the hypothalamic–anterior pituitary–ovarian axis.

Female Sex Organs (Fig. 29.6)

— The labia majora and minora are organs of copulation.
— The clitoris is an organ of copulation.
— The greater and lesser vestibular glands produce secretions.
— The mons pubis protects the pubic bone.

Fig. 29.6 ▶ **Female sex organs.**
(**A**) Internal and external genitalia. (**B**) Right lateral view of internal and external genitalia.
From *Atlas of Anatomy*, © Thieme 2008, Illustration by Markus Voll.

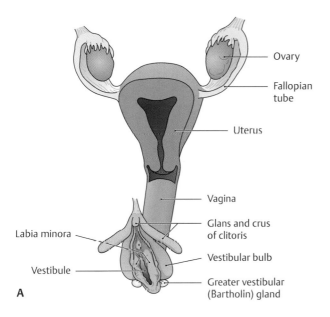

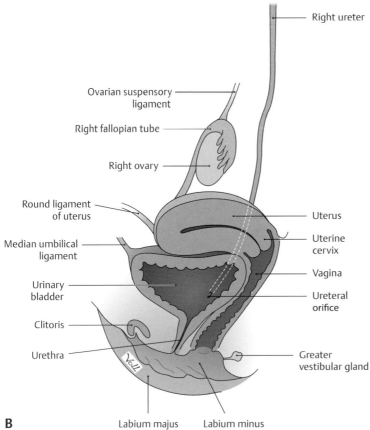

- The ovaries are involved in oocyte production and the secretion of estrogen and progesterone.
- The fallopian tubes are the site of fertilization and are involved in transportation of zygotes.
- The uterus is the site of implantation and incubation and is an organ of parturition (birth).
- The vagina is an organ of copulation and parturition.

Female Sex Hormones: Estrogens and Progesterone

The ovaries secrete estrogens (estrone [E_1], estradiol [E_2], and estriol [E_3]), and progesterone.

Synthesis

- LH stimulates the synthesis of pregnenolone, the precursor of testosterone, and progesterone by activating cholesterol desmolase (**Fig. 29.7**).
- Pregnenolone is converted to androstenedione and testosterone in thecal (ovarian endocrine) cells. Androstenedione and testosterone then diffuse to nearby granulosa (follicular) cells, where FSH stimulates secretion of aromatase that catalyzes the conversion of andro-

Fig. 29.7 ▶ **Synthesis of estradiol and progesterone.**
(FSH, follicle-stimulating hormone; LH, luteinizing hormone)

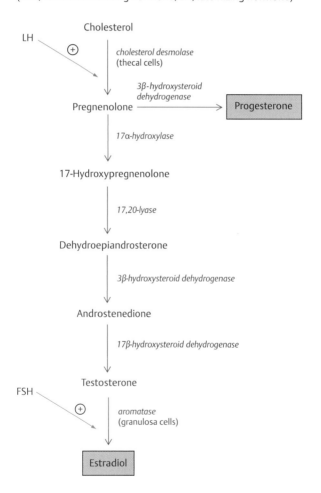

stenedione to estrone and the conversion of testosterone to estradiol (the principal estrogen secreted by the ovaries).

— Pregnenolone is converted to progesterone in thecal cells and in the corpus luteum.

Note:

— Estriol (E$_3$) is only secreted in significant amounts during pregnancy. It is made by the placenta from dehydroepiandrosterone-sulfate (DHEA-S), an androgen made in the fetal liver and adrenal glands.

— Thecal and granulosa cells in the ovaries cannot produce cortisol or aldosterone (as occurs in the adrenal cortex), as they lack the enzymes 21β-hydroxylase and 11β-hydroxylase (see **Fig 27.2**).

Regulation of Synthesis and Secretion

Hypothalamic–anterior pituitary control: GnRH release from the hypothalamus is pulsatile. In females, GnRH pulses vary in accordance with the stage of the menstrual cycle, and the ovarian production of estrogen and progesterone. GnRH stimulates the anterior pituitary to produce FSH and LH in a corresponding pulsatile manner. FSH and LH act on the ovaries to cause the following:

— FSH stimulates estradiol synthesis and the development of multiple follicles.

— LH stimulates the synthesis of pregnenolone, and the LH surge causes ovulation.

Negative feedback control:

— Low levels of estrogen inhibit GnRH release from the hypothalamus and LH release from the anterior pituitary in the early and midfollicular phase of the menstrual cycle (the phase in which follicles in the ovary mature) and in the luteal phase of the menstrual cycle (the latter phase that starts with the formation of the corpus luteum and ends in pregnancy or the onset of menses).

— Progesterone inhibits GnRH release from the hypothalamus and FSH and LH release from the anterior pituitary during the luteal phase of the menstrual cycle.

— Inhibin is a protein complex that is released from follicular granulosa cells in response to stimulation by FSH. It, in turn, is a potent inhibitor of FSH release.

Positive feedback control:

— Rising levels of estrogen stimulate GnRH release during the late follicular phase. This change in response to estrogen is critical for triggering ovulation (LH surge). There is a corresponding increase in the sensitivity of the anterior pituitary to GnRH at this time.

Actions

Table 29.4 lists the actions of estrogen and progesterone.

Table 29.4 ▶ Effects of Estrogen and Progesterone	
Estrogen Effects	**Progesterone Effects**
Negative feedback control of FSH and LH	Negative feedback control of FSH and LH secretion during the luteal phase
Positive feedback: induces LH surge, which then triggers ovulation	↑ uterine secretions during the luteal phase
Upregulation of estrogen, progesterone, and LH receptors	↑ cervical mucus production
Endometrial proliferation throughout the menstrual cycle (except during menses)	↓ uterine muscle contractility by increasing the threshold for myometrial contraction
Secondary sex changes of female puberty	Sustains pregnancy, including the ability to inhibit T-lymphocyte cell-mediated responses, which allow the body to tolerate the "foreign antigen" status of the fetus
Initiation and promotion of the pubertal growth spurt (growth hormone also involved)	
Closure of the epiphyseal growth plate, which marks the end of the pubertal growth spurt	Breast development
Breast development	↑ temperature
↑ uterine muscle contractility by lowering the threshold for myometrial contraction	
Sustains pregnancy	
↑ prolactin secretion	
Abbreviations: FSH, follicle-stimulating hormone; LH, luteinizing hormone.	

Polycystic ovarian syndrome

Polycystic ovarian syndrome (PCOS) is the most common hormonal disorder of women of reproductive age. It may occur around menarche (first menstrual cycle) or later in life following weight gain. In PCOS, the pituitary gland may secrete high levels of LH, and the ovaries may produce excess androgens (androstenedione and testosterone). The ovaries appear enlarged and polycystic (due to failure of the follicles to rupture) on ultrasound examination. Symptoms of PCOS include menstrual abnormalities, for example, prolonged cycles (> 35 days), prolonged periods, failure to menstruate for > 4 months, or fewer than eight cycles per year; excess facial and body hair (hirsutism); and adult acne. Complications of PCOS include infertility due to infrequent ovulation or lack of ovulation, type 2 diabetes, hypertension, cholesterol abnormalities, abnormal uterine bleeding, and endometrial cancer (due to continuous exposure to high levels of estrogen). Birth control pills (BCPs) may be used to regulate ovulation and reduce excess hair growth. Metformin may also be given to manage type 2 diabetes (if present) and to augment ovulation regulation with BCPs. If ovulation induction is necessary to become pregnant, then clomiphene (an antiestrogen drug) is given, followed by injectable gonadotropins (FSH and LH) if this fails.

Female hormonal contraceptives

The hormones used for contraception are usually a combination of a synthetic estrogen and progesterone (e.g., ethinyl estradiol or mestranol, combined with norethindrone, ethynodiol, or norgestrel). These agents primarily act by negative feedback inhibition of gonadotropins (FSH and LH) to inhibit ovulation, but they also inhibit the proliferation of the endometrium (to prevent successful implantation) and increase the viscosity of cervical mucus (to prevent sperm transport to the ova). Combination oral contraceptives are usually taken for 21 days, then stopped for 7 days, which induces withdrawal bleeding. If one or two contraceptive pills are missed, steroid levels decline, and gonadotropins are rapidly released. This stimulates follicular development, and ovulation may occur. Common side effects of these drugs are headache, yeast infections, nausea, depression, weight gain, and breast discomfort. Severe side effects include hypertension, thromboembolic disorders, and an increased risk of cervical cancer and breast cancer.

Pathophysiology

Figures 29.8 and **29.9** summarize the effects of estrogen and progesterone deficiency and excess, respectively.

Oogenesis

Oogenesis is the production of mature oocytes from oogonia. Oogonia within follicles in the ovary enter the prophase of meiosis and become primary oocytes approximately between 8 weeks gestation and 6 months after birth. They then remain quiescent until they complete the first meiotic division following recruitment into the menstrual cycle and ovulation many years later.

—*Ovulation* is the release of a primary oocyte from an ovary into the peritoneal cavity. The first meiotic division is completed, and the resulting secondary oocyte enters the fallopian tube. Within the fallopian tube, the second meiotic division commences but is only completed if the oocyte is fertilized by a sperm. The result is a haploid ovum with 23 chromosomes.

Fig. 29.8 ▶ **Deficiency of female sex hormones.**
Deficiency of female sex hormones leads to amenorrhea, as well as a number of other systemic complications. (25-OH-D3, calcitriol; HDL, high-density lipoprotein; LDL, low-density lipoprotein; VLDL, very-low-density lipoprotein.)

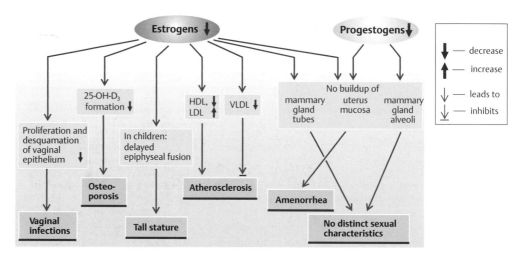

Fig. 29.9 ▶ **Excess of female sex hormones.**
An excess of estrogens and progesterone causes infertility, as well as a number of systemic complications.

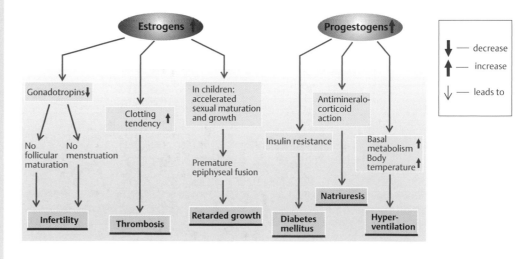

FSH, LH, and estrogen are needed for the development of a secondary oocyte from a primary oocyte. The role of these hormones in relation to the menstrual cycle is discussed in the next section.

Menstrual Cycle

The menstrual cycle occurs every 24 to 35 days, with an average of 28 days. It has two phases: the follicular and the luteal (**Fig. 29.10**).

Fig. 29.10 ▶ **Menstrual cycle.**
Over the course of an average 28-day menstrual cycle, LH (as well as FSH) peaks around cycle day 14, resulting in ovulation. Estrogen increases just prior to ovulation, and it is this positive feedback that stimulates the LH surge. Basal body temperature increases by ~0.5°C (32.9°F) probably just after ovulation and remains elevated until the end of the cycle. The progression of follicular development and endometrial changes over time is also shown.

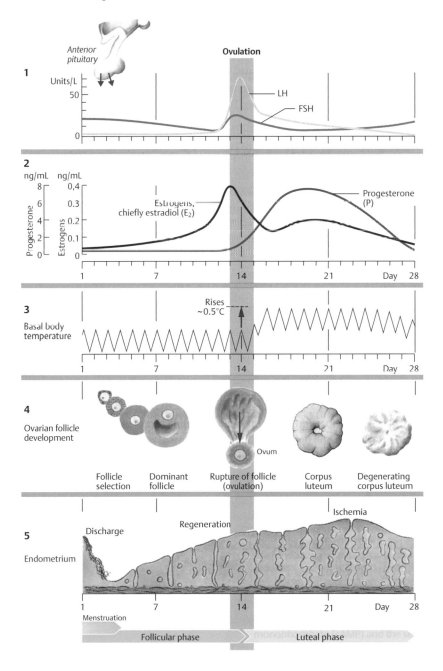

Premenstrual syndrome

Premenstrual syndrome (PMS) typically occurs 7 to 14 days before menstruation (after ovulation). It may be caused by fluctuations in progesterone and estrogen, which cause fluid retention. Symptoms of PMS include nervousness, irritability, emotional upset, depression, headaches, tissue swelling, and breast tenderness. PMS is treated symptomatically. Combination oral contraceptives reduce the variations in estrogen and progesterone levels. Fluid retention can be treated by reducing salt intake and using a mild diuretic, such as spironolactone. Nervous system symptoms are treated with relaxation techniques, antidepressants (e.g., fluoxetine), or antianxiety drugs (e.g., buspirone or alprazolam).

Mechanism of oocyte release at ovulation

The LH surge at ovulation induces prostaglandin synthesis in the granulosa cell layer. Prostaglandin E2 (PGE_2) is produced first, causing vasodilation and infiltration of the follicular wall with leukocytes and macrophages. As ovulation nears, prostaglandin $F_{2\alpha}$ ($PGF_{2\alpha}$) synthesis is preferentially induced by progesterone. $PGF_{2\alpha}$ is a potent vasoconstrictor and inflammatory mediator. Histamine and kinin release contributes to inflammation. Prostaglandins cause ischemia of the thecal layer, and tight junctions and gap junctions between granulosa cells are disrupted. These changes allow the oocyte to be released from an ovary.

Follicular Phase (Days 0–15)

Menses

— *Menses* is the vaginal discharge consisting of blood and cellular debris from the lining of the uterus (endometrium) that occurs at the end of an infertile cycle. The first day of menses marks day 1 of the menstrual cycle.

Early Follicular Phase

— FSH stimulates the growth of ~20 primordial follicles, each containing a primary oocyte.
— LH is also secreted in smaller amounts.
— FSH and LH stimulate the enzymes needed for androgen synthesis in ovarian cells, leading to estrogen production.
— FSH and LH cause feedback inhibition at the anterior pituitary.
— Estrogen causes endometrial thickening throughout the follicular phase. A thickened endometrium is necessary for successful implantation of an embryo if pregnancy is achieved later in the menstrual cycle.

Midfollicular Phase

— The follicle with the highest estrogen content is the most sensitive to FSH and therefore becomes the dominant follicle (graafian follicle). The remaining nondominant follicles undergo atresia (hormonally controlled apoptosis) and are resorbed by the ovary.
— Inhibin is released by the dominant follicle, causing FSH levels to fall.

Late Follicular Phase

— Estrogen causes LH receptors to be upregulated in the dominant follicle. The dominant follicle loses most of its responsiveness to FSH.
— Increased quantities of FSH and LH are secreted, leading to maximal estrogen production. There is positive feedback from estrogen on the anterior pituitary, which increases LH secretion (both the amount and frequency). This rapid rise in LH (LH surge) induces ovulation from the dominant follicle.
— The FSH surge does not affect the dominant follicle and is not involved in ovulation. It is necessary to stimulate the growth of a new batch of primordial follicles in the next cycle.

Ovulation

— *Ovulation* is the release of a secondary oocyte from an ovary. It marks the end of the follicular phase.
— At ovulation, basal body temperature rises (0.5°C, 32.9°F) and stays elevated until the end of the cycle, and cervical mucus changes consistency (becomes watery, clear, and elastic) to facilitate sperm transport.
— Estrogen primes the endometrium by inducing the expression of progesterone receptors and by sensitizing the endometrium to the effects of progesterone (**Fig. 29.11**), which is mainly secreted during the luteal phase.

Luteal Phase (Days 15–28)

— LH, FSH, and estrogen transform the remnants of the graafian follicle into a corpus luteum. The corpus luteum is maintained by LH.
— The corpus luteum secretes large quantities of progesterone that cause the endometrium to thicken and become more vascular and secretory as it prepares the uterus for embryo implantation and pregnancy.
— The cervical mucus becomes opaque, thickened, and dehydrated under the influence of progesterone. This type of mucus acts as a barrier to sperm.

Fig. 29.11 ▶ **Hormonal control of the menstrual cycle.**
(FSH, follicle-stimulating hormone; GnRH, gonadotropin-releasing hormone; LH, luteinizing hormone)

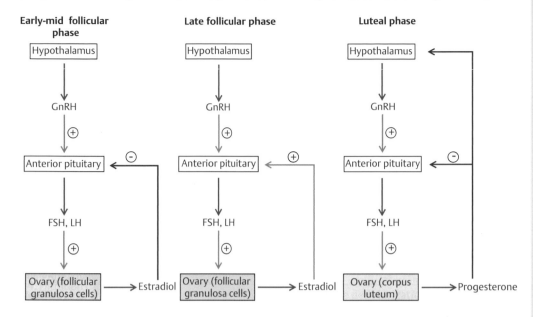

— If fertilization of the ovum does not occur, estrogen and progesterone inhibit FSH and LH both directly and by feedback inhibition on the hypothalamus and anterior pituitary. This causes a marked reduction in plasma estrogen and progesterone, which, in turn, causes the endometrial blood vessels to constrict (via PGF_α), endometrial ischemia, and discharge of the endometrium (menses). Uterine myometrial contractions, stimulated by oxytocin and prostaglandins, help slough the endometrium. The corpus luteum also regresses at this time.
— If fertilization does occur, human chorionic gonadotropin (hCG) from the developing embryo maintains the corpus luteum, allowing it to produce the progesterone needed to sustain pregnancy. The rapid rise in concentration of hCG in the urine is used to test for pregnancy even before the missed period.

Pregnancy

Fertilization

— The secondary oocyte is directed into the lumen of the oviduct by fimbria of the fallopian tube. Fertilization normally occurs in the ampulla (upper one-third of the oviduct).
— Capacitated sperm contact the surrounding corona radiata cells of the oocyte. The acrosome reaction then occurs, causing proteolytic enzymes to be released from the head of the sperm. This allows the sperm to penetrate the oocyte (corona radiata and zona pellucida).
— Oocyte activation occurs in response to the first sperm that makes contact with the oocyte's plasma membrane. This activation causes biophysical changes in the oocyte that prevent fertilization by more than one sperm.
— The sperm and oocyte membranes fuse, and the sperm is engulfed into the oocyte.
— Male and female DNA fuse within the oocyte to complete the fertilization process.
— Mitosis of the one cell zygote into a morula preembryo (16 cells) occurs within the oviduct. At the late morula stage (32 cells), the preembryo reaches the uterine lumen, where

■ **Infertility**

Infertility is the inability of a couple to conceive a baby after repeated intercourse for 1 year. It affects approximately one in five couples in the United States. More than half of couples who have not conceived after 1 year will eventually conceive. In about one-third of cases of infertility, there are problems with sperm; in about one-third, there are problems with the fallopian tubes; and in about one-sixth, there are ovulation problems. Rarely are there problems with cervical mucus. The cause of infertility is unidentified in the remainder of cases. Many causes of infertility can be reversed with hormone therapy or surgery.

blastocyst development occurs. A blastocyst consists of an outer layer of trophoectoderm (trophoblast), which will become the fetal placenta, an inner cell mass (embryoblast), which will become the fetus, and a blastocele (fluid-filled cavity) (**Figs. 29.12** and **29.13**).

Fig. 29.12 ▶ **Fertilization.**
(**1**) The acrosome reaction occurs as the sperm approaches the oocyte (see Fig. 29.5). (**2**) The corona radiata of the oocyte plays a role in chemotaxis of sperm and induction of the acrosome reaction. The sperm penetrates the epithelium of the corona radiata within a few seconds with powerful tail movements, then adheres to the zona pellucida for several minutes. (**3**) The zona pellucida is penetrated after a few minutes. Sperm pass through this layer at an angle and meet the cell membrane of the oocyte tangentially. (**4**) Contact of the sperm with the oocyte cell membrane releases cortical granules that induce an excitatory potential that is responsible for initiating the zona pellucida reaction (blocking polyspermy), removing the block on metaphase II, and activating oocyte metabolism. Embryonic development begins. (**4a**) The sperm head dips into the microvilli on the surface of the oocyte membrane. (**4b**) Incorporation of sperm into the membrane. (**4c**) Sperm head, neck, and tail sink into the yolk sac. (**5**) Fertilization causes completion of the second meiotic division, and the second polar body is expelled. (**6**) The chromosomes of the sperm and oocyte (haploid sets) decondense and form the female and male pronuclei. The flagellum disintegrates in the oocyte.

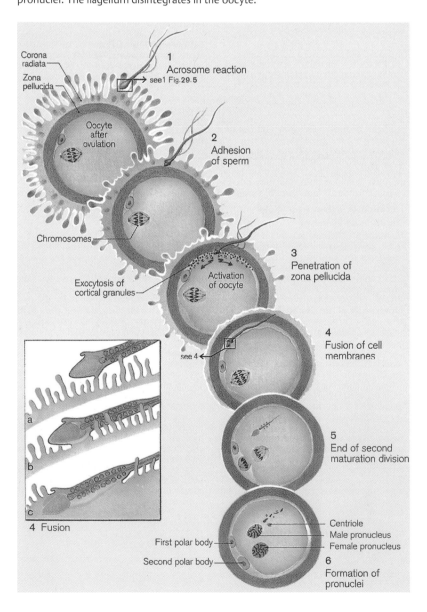

Fig. 29.13 ▶ **Development of oocyte to blastocyst.**
Stage1: Fertilization activates the oocyte. The haploid egg nucleus and the haploid sperm nucleus are transformed into female and male pronuclei. Both pronuclei go through a phase of DNA synthesis, and their replicated chromosomes are arranged on a common spindle. *Stage 2:* As the oocyte travels along the fallopian tube toward the uterus, its cells divide within the zona pellucida. *Stage 3:* A blastocyst forms consisting of an inner cell mass (embryoblast) and an outer cell mass (trophoblast). The blastocyst hatches out of the zona pellucida, allowing it to attach to the uterine endothelium.

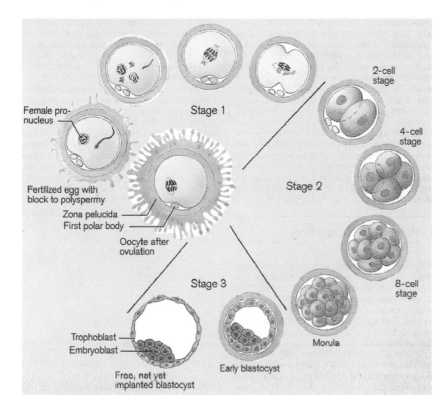

Implantation

— The blastocyst must "hatch" out of the zona pellucida before implantation into the endometrium can occur. Trophoblast cells in the attachment zone differentiate into cytotrophoblast cells. These cells fuse together to form the syncytiotrophoblast, which is able to penetrate into the endometrium. Implantation is complete by the second week of pregnancy, marking the end of the preembryonic stage.
— Stromal cells in the endometrium surround the endometrial spiral arteries and "cuff" them. This protects maternal tissues from the invading trophoblast and helps protect the fetoplacental unit from rejection by the maternal immune system.

First Trimester

— With the formation of the primary germ layers (ectoderm, mesoderm, and endoderm) and extraembryonic membranes (amnion, yolk sac, allantois, and chorion), the embryonic stage begins (weeks 3–8). During the first trimester, the placenta becomes firmly established, and embryonic/fetal organ development occurs.
— The corpus luteum is the major source of progesterone (and estrogen) during the first 6 to 8 weeks of gestation. The function of the corpus luteum is stimulated by hCG (from the syncytiotrophoblast).

Ectopic pregnancy

Ectopic pregnancy is when the fertilized ovum implants outside the uterine cavity, usually in the fallopian tubes. So-called tubal pregnancies are not viable. This is a rare condition, but it is more likely in salpingitis (inflammation of the fallopian tubes), following tubal infection, tubal damage (e.g., from previous ectopic pregnancies or endometriosis), when a woman is taking fertility medication to stimulate ovulation, and when contraceptive failure occurs. The woman may not be aware that she is pregnant, or there may be signs of early pregnancy, such as a missed period, nausea, breast tenderness, and fatigue. Other signs of ectopic pregnancy are abdominal pain, vaginal bleeding, cramping, shoulder tip pain (blood in peritoneum irritates the phrenic nerve of the diaphragm), and faintness. Treatment depends on how early the ectopic pregnancy is detected. If detected early, methotrexate may be given to arrest the development of the fertilized ovum, which is then resorbed. Laparoscopy may be needed to stop any bleeding into the peritoneum. If severe tubal damage has occurred, laparotomy is usually needed to remove the damaged tube.

Fig. 29.14 ▶ **Hormonal concentrations in plasma during pregnancy.**
Human chorionic gonadotropin (hCG), secreted by the placenta during pregnancy, is the predominant hormone during the first trimester. It stimulates the synthesis of dehydroepiandrosterone sulfate (DHEA-S) from the fetal adrenal cortex, suppresses follicle maturation in the maternal ovaries, and maintains the production of estrogen and progesterone in the corpus luteum. Maternal concentrations of human placental lactogen (hPL), corticotropin-releasing hormone (CRH), and estrogen rise sharply during the third trimester. hPL stimulates lactogenesis after parturition. CRH concentration plays a role in the timing of parturition by increasing adrenocorticotropic hormone (ACTH) production by the fetal pituitary, which increases cortisol. It also stimulates fetal lung development. Estrogen also plays a critical role in parturition by counteracting the pregnancy-sustaining effects of progesterone and helping to propagate uterine contractions.

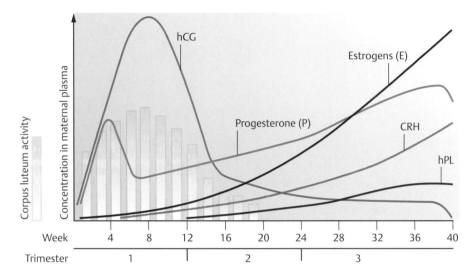

— At around the eighth week of gestation, the trophoblast takes over progesterone (and estrogen) secretion, making the placenta the main source of progesterone during pregnancy (**Fig. 29.14**).

Second and Third Trimesters

— The placenta takes over progesterone synthesis, and there is degeneration of the corpus luteum.
— The placenta is unable to convert progesterone to estrogens because of a deficiency of 17α-hydroxylase, so it must rely on the conversion of DHEA-S from both the fetal and maternal adrenal glands to synthesize estriol, estradiol, and estrone.
— Human placental lactogen (hPL) is produced by the placenta, with its peak blood concentrations occurring in the third trimester. hPL is very similar in structure and function to growth hormone and prolactin (**Fig. 29.15**). Its secretion causes an increase in lipid metabolism, enhanced carbohydrate-stimulated insulin secretion, and increased insulin resistance in some maternal tissues. Collectively, these maternal alterations in metabolism are thought to enhance maternal free fatty acid utilization while sparing glucose for use by the growing fetus. hPL may also play an important role in mammary gland development.

Fig. 29.15 ► **Hormone synthesis in the placenta, mother, and fetus.**
(**1**) The placenta produces hCG, which stimulates the synthesis of steroids, for example, dehydroepiandrosterone (DHEA) and DHEA-S, by the fetal adrenal cortex. hCG also maintains the production of estrogen and progesterone in the corpus luteum until the placenta is able to produce sufficient quantities of these hormones. (**2**) The placenta has to receive cholesterol or androgens from the maternal or fetal adrenal cortex, respectively, before it can synthesize progesterone and estrogen. Progesterone is then transported to the fetal adrenal cortex, where it is converted to DHEA and DHEA-S. DHEA and DHEA-S pass to the placenta, where they are used for estrogen synthesis. Progesterone is converted to testosterone in the testes of the male fetus.

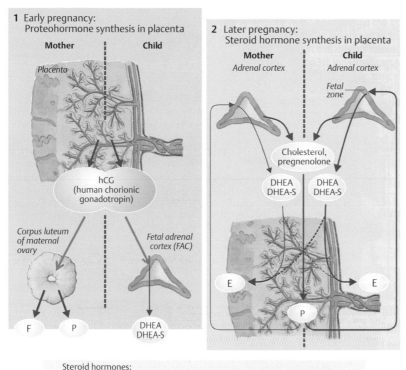

Steroid hormones:
P = progesterone; DHEA-S = dehydroepiandrosterone (sulfate); E = estrogens

Maternal Changes during Pregnancy

Pregnancy involves severe but tolerable changes from normal in the mother's physiology and anatomy. These changes are listed in **Table 29.5.**

Table 29.5 ► **Maternal Changes Associated with Pregnancy**	
System/Tissue	**Changes**
General	↑ body mass (25–29 lb)
	Changed posture (lordosis [inward curvature of a portion of the spine] and kyphosis [curvature of the upper spine])
	Nausea
	Fatigue
Hematological	↑ Blood volume (~50%)
	↑ Red blood cells (20–29% in mass)
	↑ White blood cells (neutrophils)
	↑ Need for iron
	↓ Hemoglobin (15%) (physiological anemia of pregnancy)

Continued on p. 310

Placental abruption

Placental abruption is the separation of the placenta from the uterus. The consequences of this depend on the extent of the placental separation and the amount of blood loss. In severe abruptions, the fetus may not receive an adequate supply of oxygen, causing neurologic defects or death. The mother's life may also be at risk from shock or disseminated intravascular coagulation (DIC). DIC is a pathological activation of clotting that ultimately consumes the body's supply of clotting factors and platelets, causing bleeding from the skin, mucous membranes, and viscera. Signs of placental abruption include shock that is out of keeping with visible vaginal blood loss, backache (if the abruption is posterior), abdominal pain, uterine tenderness, fetal distress, lack of fetal heartbeat, and DIC. Treatment also depends on the extent of the abruption. If it is a small abruption, then the mother is monitored frequently, but the pregnancy is allowed to progress. If severe, urgent delivery of the baby is necessary, along with supportive treatment such as blood transfusion.

Placenta previa

Placenta previa is a condition in which the placenta is situated low in the uterus, partially or completely covering the cervix. As the cervix begins to dilate later in pregnancy, the placenta stretches and tears, leading to painless vaginal blood loss. Shock may occur if the blood loss is severe. In contrast to placental abruption, there are usually no coagulation problems or uterine tenderness. Fetal distress is also less common, as the frank vaginal bleeding alerts the mother and health care providers to the problem before the fetus becomes distressed. Placenta previa is more common in women who have uterine damage (e.g., from previous births, cesarean sections, or fibroids) or when the placenta is larger than usual (e.g., with twins). Treatment depends on the severity of the placenta previa and blood loss. Minor cases are often monitored and the mother put on bed rest for the remainder of the pregnancy; more severe cases may warrant immediate delivery of the baby by cesarean section and blood transfusions.

Preeclampsia

Preeclampsia is hypertension, proteinura (excess protein in urine), and edema in pregnancy. It is seen after 20 weeks gestation. The patient presents with severe headaches, changes in vision, upper right quadrant abdominal pain, nausea and vomiting, dizziness, sudden weight gain, and swelling, particularly of the hands and face. Signs include blood pressure >140/90 mm Hg on two occasions at least 6 hours apart but no more than 7 hours apart. There is also proteinura. Preeclampsia can cause fetal distress, low birth weight, and preterm birth due to lack of blood flow to the placenta. It also increases the occurrence of placental abruption. Other complications include HELLP syndrome (hemolysis, elevated liver enzymes, and low platelet count) and eclampsia (preeclampsia symptoms plus generalized seizures). Both of these conditions are life-threatening to both mother and baby. Delivery of the baby is curative for preeclampsia, but if the baby is not mature enough for delivery, it may be appropriate in some cases to manage hypertension with antihypertensive drugs plus bed rest.

Table 29.5 ► Continued

System/Tissue	Changes
Cardiovascular	Heart moves upward and forward from growth of the uterus ↑ Heart rate (10–15%) ↑ Stroke volume (29%) ↑ Cardiac output (up to 60%) ↑ Cardiac volume (10%) ↑ Venous pressures in legs and edema Blood pressure is unchanged
Respiratory	↑ O_2 consumption (29–60 mL/min) ↑ Tidal volume (200 mL) ↑ Alveolar ventilation ↓ Residual volume (20%) ↓ Expiratory reserve volume Dyspnea is common Respiratory rate is unchanged
Renal	↑ Glomerular filtration rate (50%) ↑ Renal plasma flow (50 to 70%) ↑ Glucosuria (filtered load of glucose exceeds the resorptive capacity of the kidney)
Urinary tract	↓ Bladder capacity ↑ Residual volume ↑ Frequency of micturition ↑ Risk of urinary tract infection because of urinary stasis ↓ Bladder and urethral tone
Gastrointestinal (due to increased progesterone)	↑ Mucus production ↓ Esophageal tone ↓ Gastric acid ↓ Colon motility Rectal stasis and hiatus hernia (herniation of the upper part of the stomach through the diaphragm into the thorax) Hemorrhoids are common
Musculoskeletal (due to relaxin and progesterone)	Softening of ligaments (pubic and sacroiliac joints) Low back pain in half of pregnant women Bone density is unchanged
Skin (due to estrogen, progesterone, and melanocyte-stimulating hormone)	Midline hyperpigmentation of the abdomen and dry skin Chloasma (hyperpigmentation of the skin of the face) Striae gravidarum (mechanical stretch marks) Spider angiomata (an abnormal collection of blood vessels) Palmar erythema (red palms of hands)
Reproductive system	↑ Uterine mass (16-fold) ↑ Uterine blood flow (5- to 10-fold) ↑ Vulvar swelling and varicosities ↑ Thickness of vaginal secretions Eversion of cervix and mucus plug
Metabolic	↑ Ingestion of calories by 290 kcal/day (for fetal and placental growth) ↑ Insulin secretion (hyperplasia of pancreatic α cell) ↑ Resistance to insulin activity from increased prolactin ↑ Glucose utilization ↑ Production of ketones and free fatty acids ↑ Sex steroids ↑ Tissue glycogen storage ↑ Hepatic glucose production ↑ Corticoids ↓ Liver glycogen storage ↓ Glucose tolerance from increased production of human placental lactogen

Parturition (Birth)

— During a normal pregnancy consisting of 270 predetermined days, the secretion of progesterone prevents uterine contractions by increasing the threshold for myometrial contractility. This is referred to as the *progesterone block*.

— Just prior to parturition, placental estrogen production is increased relative to progesterone, thereby increasing the estrogen-to-progesterone ratio. This removes the progesterone block and allows estrogen to increase the synthesis of receptors for estrogen, prostaglandins, and oxytocin on myometrial cells. This upregulation of receptors is necessary for the increase in myometrial contractility at parturition.

— Myometrial stretching and pressure exerted on the cervix by the fetus cause a reflexive release of oxytocin from the posterior pituitary (Ferguson reflex). Oxytocin binds to myometrial receptors, where it stimulates the production of uterine and placental prostaglandins, which, in turn, increase intracellular Ca^{2+} and promote myometrial contractility.

— Estrogen also affects the cervix by increasing its responsiveness to relaxin (secreted by the corpus luteum and the placenta) and prostaglandins (secreted by the uterus and placenta). These hormones cause the cervix to become more vascular and change its structure. This results in cervical dilation and effacement (cervix becomes softer and shorter) during labor (**Fig. 29.16**).

Lactation

— During pregnancy, estrogen, growth hormone, hPL, and cortisol continue to stimulate the development of the mammary glands, which started at puberty. Progesterone converts duct epithelium to a secretory epithelium.

Fig. 29.16 ▶ **Endocrine control of parturition.**
(**1**) Relaxation of the cervix: the cervix remains tightly closed during pregnancy but is stimulated to relax around the time of parturition by relaxin, secreted by the corpus luteum and placenta. (**2**) Onset of labor: locally, prostaglandins cause contractions of the uterine muscles. Systemically, oxytocin, from the posterior pituitary gland, is released in response to cervical irritation caused by pressure from the fetal head (Ferguson reflex). Oxytocin causes further prostaglandin secretion.

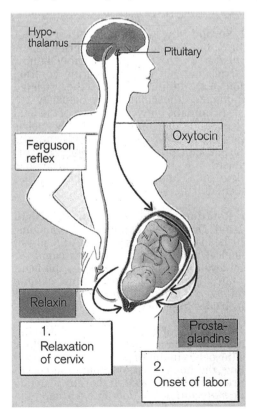

Stages of labor

Stage 1: This is the period from the onset of regular contractions until the cervix is fully dilated. Contractions originate in the fundus and progress toward the cervix, forcing the head of the fetus against the cervix. The cervix starts to dilate from the effects of estrogen and relaxin and the mechanical force from fetal pressure. During this time the cervix becomes softer and shorter (effaces). Changes in the cervix result from physical breakdown of connective tissue of the cervix with increased water content, vascularization, and mass. The fetal membranes of the amniotic sac are lost. This enhances the effects of contraction for applying fetal pressure on the cervix.
Stage 2: This is the period from full dilation of the cervix until parturition. Uterine muscle contractions are of high frequency and high amplitude. This stage typically lasts < 1 hour but can be longer.
Stage 3: The placenta separates and is delivered. This occurs within ~10 minutes after birth and is associated with weak muscle contractions.

— During the latter stages of pregnancy, estrogen acts on the anterior pituitary, causing levels of prolactin to rise. This is accompanied by a fall in prolactin-inhibiting hormone (PIH). Prolactin is the hormone after delivery that initiates lactogenesis (milk production).

— Lactation does not occur during pregnancy because placental estrogen and progesterone prevent prolactin from acting on the mammary glands. However, when estrogen and progesterone are withdrawn at birth, lactation is able to occur.

— Suckling is a mechanical stimulus for the continuation of lactation, as it stimulates increased levels of prolactin (by inhibiting PIH) and oxytocin.

— Nursing an infant is moderately effective in preventing pregnancy because prolactin inhibits ovarian function by the following mechanisms:

    — Inhibition of hypothalamic GnRH secretion

    — Indirect inhibition of FSH and LH secretion by the anterior pituitary (by decreasing GnRH)

    — Inhibition of FSH and LH actions on the ovaries

Menopause

— Female fertility declines rapidly after age 35, culminating in menopause. Menopause is a natural process in which there is a gradual reduction in the response of the ovaries to gonadotropins (FSH and LH) and a reduced number of responsive ovarian follicles. Ultimately, the ovaries become functionally inactive (ovulation ceases), and they are therefore unable to synthesize estrogen. This usually occurs between 45 and 55 years of age (mean age 51 years).

Table 29.6 lists the signs associated with premenopause and menopause.

Table 29.6 ▶ **Signs of Premenopause and Menopause**	
Signs associated with the premenopausal period	Irregular menstrual cycle lengths Lower circulating ovarian steroid concentrations Elevated gonadotropin concentrations (especially FSH) due to loss of negative feedback from estrogen and inhibin
Typical signs of menopause	Atrophy of the vagina Vaginal bleeding Osteoporosis Hot flashes Anxiety and depression Irritability and fatigue due to elevated GnRH, LH, and catecholamines in the central nervous system Increased risk of cardiovascular disease (low estrogen decreases HDLs and increases LDLs)

Abbreviations: FSH, follicle-stimulating hormone; HDL, high-density lipoprotein; GnRH, gonadotropin-releasing hormone; LDL, low-density lipoprotein; LH, luteinizing hormone.

Letdown reflex

Mechanical sensory input to the spinal cord from the breast nipple ascends to the hypothalamus and posterior pituitary, causing the release of oxytocin from the posterior pituitary. Oxytocin stimulates smooth muscle contractions. This helps shrink the uterus to pre-pregnancy size and creates high pressure in the milk ducts, which can squirt milk into the infant's mouth. This can also contribute to leakage of milk. Mechanical stimulation of the cervix can also release oxytocin.

Review Questions

1. Klinefelter's syndrome is a genetic disorder in which the patient has three sex chromosomes. Which of the following statements is true about patients with Klinefelter's syndrome?
   A. The karyotype is XYY
   B. The patient is at decreased risk of developing osteoporosis.
   C. The patient has normal muscle mass and body hair.
   D. Puberty will occur at approximately the normal age.
   E. Testosterone should be prescribed at puberty.

2. Turner's syndrome is caused by the absence of one of the sex hormones. Which of the following statements is true about patients with Turner's syndrome?
   A. They are fertile
   B. They have normal stature
   C. They are sometimes female
   D. They are hermaphrodities (have both male and female reproductive organs)
   E. They are at risk for serious vascular complications

3. Antimüllerian hormone (AMH) is
   A. secreted in response to human chorionic gonadotropin (hCG).
   B. synthesized and secreted by fetal Sertoli cells.
   C. required for differentiation of the wolffian ducts.
   D. synthesized and secreted by interstitial Leydig cells.
   E. important for differentiation of female internal genitalia.

4. Testosterone is
   A. present in the plasma as a free hormone not bound to plasma proteins.
   B. not required for the initiation and maintenance of spermatogenesis.
   C. synthesized from estradiol-17β in the testes.
   D. converted to dihydrotestosterone (DHT) by 5α-reductase in the cells of the prostate gland and other target cells.
   E. produced primarily in the Sertoli cells of the testes in response to follicle-stimulating hormone (FSH).

5. Onset of menses is associated with
   A. decreased progesterone.
   B. increased estrogen.
   C. increased LH.
   D. increased FSH.
   E. dilation of endometrial blood vessels.

6. The ovarian follicle
   A. can grow to maturation without FSH.
   B. cannot grow without locally produced androgens.
   C. secretes estradiol in response to FSH and LH.
   D. when dominant, releases inhibin, which decreases LH levels.
   E. down-regulates LH receptors before ovulation.

7. Fertilization usually occurs in what part of the female reproductive tract?
   A. Vagina
   B. Ovary
   C. Fallopian tube
   D. Uterus
   E. Corpus luteum

8. Following fertilization, the blastocyst secretes which of the following hormones?
   A. Human chorionic gonadotropin (hCG)
   B. Human placental lactogen (hPL)
   C. Oxytocin
   D. Follicle-stimulating hormone (FSH)
   E. Luteinizing hormone (LH)

9. A healthy 56-year-old man attends his urologist's office complaining of the inability to consistently develop an erection and have intercourse with his wife. Physical examination of the external genitalia and prostate are normal, and laboratory tests reveal no hormonal abnormalities or glucose intolerance. He is currently not taking any medication. He is prescribed sildenafil to take one hour before commencing sexual activity. What is the mechanism of action of this drug?
   A. Increases the conversion of testosterone from androstenidone in the testes
   B. Inhibits 5α-reductase
   C. Inhibits cGMP phosphodiesterase
   D. Stimulates adenylate cyclase
   E. Inhibits cAMP

Questions **10** and **11** refer to the clinical scenario that follows.
A 26-year-old woman complains of severe, dull pain and cramping in the lower abdomen. There are no other physical findings. A laparoscopy reveals the presence of ectopic endometrial tissue on the uterine wall and ovaries. Danazol (a synthetic androgen and inhibitor of gonadotropins), 600 mg/day, is prescribed for up to 9 months, with a 50% possibility of conception after withdrawal of the therapy.

10. Her condition is
    A. ectopic pregnancy.
    B. endometriosis.
    C. placental abruption.
    D. placenta plevia.
    E. pelvic inflammatory disease.

11. The action of danazol
    A. stimulates ovulation.
    B. stimulates before it inhibits GnRH.
    C. produces results that are different from those of combination oral contraceptives.
    D. is free of side effects.
    E. suppresses growth of abnormal endometrial tissue.

12. A 32 year old woman comes to the emergency department with severe abdominal pain, cramping, and mild, bright red, vaginal bleeding. Her history reveals that she has very irregular periods, the last one being approximately 12 weeks ago, and she is currently undergoing a course of injectable gonadotropins to treat infertility. On examination, her blood pressure is 115/75 mmHg, pulse 76/min, respirations 12/min and her temperature is 37.1°C. She has no rebound abdominal tenderness or guarding. Urinalysis shows that the patient is pregnant, but there are no proteins present in the urine. What is the condition that is most likely to be causing these symptoms?
    A. Placental abruption
    B. Pre-eclampsia
    C. Ectopic pregnancy
    D. Placenta previa
    E. Acute appendicitis

13. A woman is admitted to the OB-GYN ward for induction of labor. She is three weeks overdue. Twenty-four hours before she is to be induced, she is given an injection of dexamethasone (a glucocorticoid). The purpose of administering this drug is to
    A. cause uterine contraction
    B. prevent swelling and inflammation of the birth canal
    C. induce maturation of the fetal lungs, gut, kidney, and other organs
    D. sedate the expectant mother
    E. reduce bloating from the retention of water

Questions **14** and **15** refer to the following clinical scenario.
By the sixth month of pregnancy, a woman in her 20s feels irregular contractions of the uterus, but no complications are present. After 9 months, a healthy 7 lb, 3 oz girl is delivered with no complications. Breast feeding is planned.

14. Which hormone prevented spontaneous abortion of the implanted embryo?
    A. Estrogen
    B. Progesterone
    C. Oxytocin
    D. Prostaglandins
    E. Prolactin

15. Which hormone is the most important for milk production after birth?
    A. Human chorionic gonadotropin (hCG)
    B. Human placental lactogen (hPL)
    C. Human chorionic thyrotopin (hCT)
    D. Prolactin
    E. Oxytocin

16. Polycystic ovarian syndrome (PCOS) can cause infertility in women. Which is a symptom of this disorder?
    A. Sparse facial hair
    B. Small ovaries
    C. Regular periods
    D. Obesity
    E. Normal glucose tolerance

Answers and Explanations

1. **E** Testosterone replacement therapy may prevent some of the symptoms of the disorder.
   A The karyotype is XXY, and the patient appears to be male.
   B Osteoporosis is a complication of this disorder.
   C The patient lacks muscle mass and body hair.
   D Puberty is delayed due to a lack of testosterone.

2. **E** A serious complication of Turner's syndrome is aortic dissection, where there is tearing and bleeding between the inner wall and the outer wall of the aorta. Rupture of the outer wall of the aorta is usually fatal
   A-C Patients with Turner's syndrome are often infertile, have short stature, and are always female.
   D Patients with Turner's syndrome have the karyotype XO. Since they lack a Y chromosome, they cannot develop male internal and external genitalia and are therefore not hermaphrodites.

3. **B** AMH is secreted in response to Sertoli cells and causes regression of the müllerian ducts, an active process in male development (pp. 292–293).
   A AMH is not secreted in response to hCG.
   C Testosterone stimulates development of wolffian ducts.
   D Leydig cells secrete testosterone.
   E AMH must be absent for female genitalia to develop.

4. **D** Testosterone promotes the differentiation of the wolffian ducts and is metabolized to DHT, which causes differentiation of the external genitalia (p. 295).
   A Testosterone is bound to plasma proteins, like all steroid hormones are.
   B It is required for spermatogenesis.
   C Estradiol can be synthesized from testosterone by an aromatase reaction.
   E Testosterone is produced in the Leydig cells in response to luteinizing hormone (LH).

5. **A** If fertilization does not occur, estrogen and progesterone inhibit FSH and LH (C,D), which causes reduced plasma progesterone and estrogen (B) (pp. 304–305).

6. **C.** Estrogen secretion increases before ovulation due to FSH and especially LH (p. 304).
   A Follicles require FSH to mature.
   B FSH and LH stimulate androgen synthesis, but local androgens are not required as precursors for estrogen.
   D Inhibin negatively feeds back to FSH synthesis.
   E LH receptors are up-regulated by estrogen.

7. **C** Fertilization normally occurs in the ampulla of the fallopian tubes (p. 300).
   A The vagina is an organ of copulation and partuition (birth).
   B The ovaries secrete the female sex hormones: estrogens (estrone [E_1], estradiol [E_2], and estriol [E_3]), and progesterone.
   D The uterus is the site of implantation and incubation and is an organ of parturition (birth).
   E The corpus luteum is a temporary endocrine structure that forms from the ovarian follicle during the luteal phase of the menstrual cycle. It secretes large quantities of progesterone that cause the endometrium to thicken and become more vascular and secretory as it prepares the uterus for embryo implantation and pregnancy.

8. **A** Following fertilization the blastocyst secretes human chorionic gonadotropins (hCG) which maintains the corpus luteum and allows it to produce and secrete the progesterone needed to sustain the pregnancy (p. 305).
   B Human placental lactogen (hPL), a hormone produced by the placenta, is very similar in structure and function to growth hormone and prolactin.
   C Oxytocin released from the posterior pituitary binds to smooth muscle cell receptors in the uterus, where it stimulates the production of uterine and placental prostaglandins, which, in turn, increase intracellular Ca^{2+} and promotes contractility.
   D The pulsatile release of follicle-stimulating hormone (FSH) from the anterior pituitary is involved in the onset of puberty.
   E The pulsatile release of luteinizing hormone (LH) from the anterior pituitary is involved in the onset of puberty.

9. **C** Sildenafil inhibits cGMP-specific phosphodiesterase type 5 which delays the degradation of cGMP. This maintains the arteriolar dilation of the cavernous sinuses of the penis that are responsible for maintaining an erection. It does not have the actions of the other choices (p. 298).

10. **B** Endometrial tissue in the abdomen is endometriosis.
    A In ectopic pregnancy, the fertilized ovum implants outside the uterine cavity.
    C In placental abruption, the placenta is separated from the uterus.
    D In placenta plevia, the placenta covers the cervix.
    E Pelvic inflammatory disease is an inflammation of the fallopian tubes.

11. E Inhibition of FSH and LH reduces estrogen stimulation of the ovaries and growth of endometrial tissue (p. 301).
A Danazol suppresses ovulation.
B Danazol only inhibits GnRH; GnRH agonists stimulate release of LH and FSH before they suppress GnRH secretion by negative feedback.
C Oral contraceptives inhibit FSH and LH by negative feedback, so they can also be used to treat endometriosis.
D Danazol causes masculinization and other adverse effects.

12. C This patient is showing signs of ectopic pregnancy when the fertilized ovum develops outside the uterine cavity, usually in the fallopian tubes. This is more likely to occur in women taking fertility drugs for ovulation.
A Placental abruption is separation of the placenta from the uterus. It is unlikely in this case as the vaginal blood is bright red (in placental abruption it is usually darker) and there is no shock.
B Pre-eclampsia is a disorder of hypertension, proteinuria, and edema in pregnancy, usually in the third trimester. There is no evidence of this triad of symptoms in this patient.
D Placenta previa is a condition when the placenta lies low in the uterus, partially or completely covering the cervix. Dilation of the cervix later in pregnancy can cause vaginal blood loss. This diagnosis is unlikely, since placenta previa is usually painless.
E Acute appendicitis usually presents with colicky, central abdominal pain that then focuses on a point in the right lower quadrant (McBurney's point) as the inflammation of the appendix spreads to the overlying peritoneum. The patient will have a fever, and there will be rebound tenderness and guarding of the abdomen on palpation. The lack of these symptoms makes this diagnosis unlikely.

13. C Glucocorticoids facilitate organ maturation, e.g., they promote surfactant development in the lungs (pp. 130, 274, 308).
A Oxytocin and prostaglandins promote uterine contraction.
B Glucocorticoids reduce inflammation, but they are not used prophylactically.
D Glucocorticoids have no sedative action.
E A diuretic would reduce retention of water.

14. B High levels of progesterone maintain pregnancy and prevent myometrial contraction (p. 301).
A Estrogen levels are high, but it promotes uterine contractions.
C Oxytocin stimulates uterine contractions.
D Prostaglandins are important for parturition and menses.
E Prolactin suppresses GnRH secretion.

15. D Prolactin stimulates milk production and is released in response to suckling (pp. 311–312).
A hCG stimulates steroid production in the first trimester.
B Although hPL stimulates milk production, it is less important than prolactin.
C hCT resembles thyroid-stimulating hormone, but is a placental hormone.
E Oxytocin stimulates synthesis of galactose in the mammary glands but not milk volume.

16. D Obesity is a common symptom (p. 301).
A,C Excess facial hair and irregular periods are common symptoms
B Small ovaries are not a common symptom of polycystic ovarian syndrome (PCOS). As the name suggests, women with this condition have large, polycystic ovaries due to failure of ovarian follicles to rupture.
E Type 2 diabetes is a complication of PCOS which causes abnormal glucose tolerance.

Index

Note: Page numbers followed by "f" and "t" indicate figures and tables, respectively.
Italicized page numbers represent entries in marginal boxes.